Biogeochemistry

*To
Lisa*

Biogeochemistry
An Analysis of Global Change

William H. Schlesinger
Departments of Botany and Geology
Duke University
Durham, North Carolina

Academic Press
Harcourt Brace Jovanovich, Publishers

San Diego   New York   Boston
London   Sydney   Tokyo   Toronto

Academic Press, Inc.
San Diego, California 92101

United Kingdom Edition published by
Academic Press Limited
24–28 Oval Road, London NW1 7DX

Library of Congress Cataloging-in-Publication Data

Schlesinger, William H.
    Biogeochemistry : an analysis of global change / William H.
Schlesinger.
        p.   cm.
    Includes bibliographical references (p.)
    Includes index.
    ISBN   0-12-625156-8 (hardcover)
    ISBN   0-12-625157-6 (paperback)
    1. Biogeochemistry.   I. Title.
QH343.7.S35   1991
574.5'222–dc20   90-14401
                                                        CIP

Printed in the United States of America
91 92 93 94   9 8 7 6 5 4 3 2 1

Table of Contents

PART II

Global Cycles

Preface

This is a book about the chemistry of the surface of the Earth. Outside of a few meteors and spaceships, the Earth is a closed chemical system, in which various reactions including those that maintain life are fueled by sunlight. During the last 4 billion years there have been remarkable changes in this system. Perhaps the most provocative change was the origin of life and the appearance of oxygen in our atmosphere. Notwithstanding, the Earth system is large and most changes have occurred very slowly, with ample time for evolutionary change to keep up.

Times are different now. With the advent of industrialization and an exponentially increasing population, a single species—humans—is usurping an extraordinary portion of the resources that support life on Earth. Satellite views of the planet show broadscale destruction of tropical rain forests, expansion of deserts, and smoggy cities. The globe is changing.

Somehow, in the heated political arena, global change has become equated with climate change and global warming. This is unfortunate. Global warming is one of the most difficult trends to prove with solid scientific evidence. Yet, ample evidence that the globe has changed chemically is all about us. For me, some of the most convincing evidence is found in observations of methane—natural gas—in our atmosphere. As far as we know, natural gas is solely the result of life, both past and present on Earth. Ice core records show that the concentration was fairly constant at about 650 parts per billion (ppb) from about 1000 B.C. until early in this century. Now the concentration is 1750 ppb and increasing at 1% per year. At the bank, we would not be much impressed with this interest rate, but geologically, this rate of change is unprecedented. Coupled with increasing carbon dioxide, atmospheric methane is quite likely to lead to global warming.

The presence of oxygen in our atmosphere is strong evidence that the chemistry of the Earth is controlled by life. When we see the concentration of other biogenic gases, such as methane, increasing, we should be concerned that something has affected life on this planet—not just in the neighborhood swamp, but globally. The biosphere is unhealthy.

I wrote this book as a textbook for college-level and graduate students who are interested in global change. The book covers the basics about the effect of life on the chemistry of the Earth. The organization of this book follows the organization of a class in biogeochemistry that I have taught for several years at Duke University. With such an interdisciplinary subject, in which the atmosphere, oceans, and land are linked, it is difficult to know just where to begin. Following the class, I have organized the book into two sections. The first covers the microbial and chemical reactions that occur on land, in the sea, and in the atmosphere. Part II is a set of shorter chapters that link the mechanistic understanding of the early chapters to a large-scale, synthetic view of global biogeochemical cycles.

Throughout this book I have made a special effort to show the linkage between the elements that are important to life. In several locations I show how computer models are used to help understand elemental cycling and ecosystem function. Many of these models are based on biochemistry and interactions between the biochemical elements. The models are useful in scaling small-scale observations to the global level. Thus, I hope this book weds the disparate fields of geomicrobiology and global ecology, all of which call themselves the science of biogeochemistry.

With a look toward the future, I show how satellite technology is useful in understanding global biogeochemistry, and the important role that the Earth Observing System (EOS) will play in studies of global ecology in the next century. Although I discuss how the chemical system affects and is affected by climate, this is not a book about climate change. Similarly, I show the effects of humans on the global system, but there is little emphasis on the traditional, local problems of water and air pollution.

This text will provide only the framework for a class in biogeochemistry. It is meant to be supplemented by readings from the current literature, so that areas of specific interest or current progress can be understood in more detail. While not encyclopedic, it includes a large number of references to aid the student who wishes to enter the current literature. Reflecting its interdisciplinary subject, I have made a special effort to provide abundant cross-referencing of chapters, figures, and tables throughout the book. I hope that the book will stimulate a new generation of students to address the science and policy of global ecology.

WHS
July 15, 1990
Durham, North Carolina

Acknowledgments

My interest in ecology has been stimulated by a large number of teachers, who were influential at critical stages of my scientific career. Among them are Jim Eicher, Joe Chadbourne, John Baker, Russ Hansen, Bill Reiners, Noye Johnson, Bob Reynolds, and Peter Marks. Over the years, workshops and informal conversations with Dan Botkin, Jerry Melillo, Dan Livingstone and Peter Vitousek made me recognize that the globe could be considered as a single, interactive chemical system. A number of colleagues have provided helpful reviews of all or part of early drafts of this book. Among them are Bruce Corliss, Evan DeLucia, Steve Faulkner, Dan Livingstone, Cheryl Palm, Bill Peterjohn, Jim Siedow, and Mark Walbridge. Dawn Cardascia of NASA's Earth Science Support Office kindly provided the color plates, and Lisa Dellwo Schlesinger helped make the entire manuscript more understandable to the general reader. I thank them all.

PART I

Processes and Reactions

1

Introduction

Introduction

Whenever we read of a change that humans have made in their global environment, we build upon our recognition that living organisms, including humans, can affect the conditions of an entire planet, the Earth. There is little doubt that the composition of the atmosphere and the reactions among atmospheric constituents are changing as a result of human activities. For instance, the reduction of stratospheric ozone over the South Pole appears related to the release of chlorofluorocarbons, which are used in a variety of products in the industrialized world. Beyond human effects, the influence of all life on Earth is so pervasive that scientists have come to accept that there are few chemical reactions on the surface of the Earth not affected by biota. Many of the conditions on Earth that we now regard as "normal" are the product of at least 3.5 billion years of life on Earth (Reiners 1986). Even today, living systems exert major control on the composition of the oceans and atmosphere and on the rate of weathering of the Earth's crust. Thus, a study of the geochemistry of the surface of the Earth is the study of *bio*geochemistry.

Encompassing chemical reactions in the atmosphere, oceans, crustal minerals, and living organisms, biogeochemistry is a unique, interdisciplinary science. Traditional approaches of experimentation and strong inference cannot be used in global biogeochemistry; there is only one Earth! Working on different levels, teams of biogeochemists must assemble a model of the whole, from the reductionist study of the parts. Modeling is an essential tool of the biogeochemist. Models often help to extend the results of small-scale measurements or experiments to regional and global estimates. We can test the validity of models by observations at the global level, often using satellite technology. The ultimate goal, of course, is to understand the chemical processes controlling the environment in which we live.

A Model for the Earth as a Biogeochemical System

Garrels and Lerman (1981) offer an example of a simple model for the biogeochemistry of the Earth's surface, which includes interactions between atmospheric, oceanic, and crustal compartments and the biosphere (Fig. 1.1). The model assumes that the atmosphere and oceans have not shown large changes in their composition during geologic time. Of course, we know that this has not always been true, but for the last 60 million years or so, there is good geologic evidence that this assumption is reasonable (Holland et al. 1986). With these constraints, the model couples reactions in the atmosphere and oceans to seven compartments that represent major crustal minerals, such as gypsum ($CaSO_4 \cdot 2H_2O$), pyrite (FeS_2), and calcium carbonate ($CaCO_3$). For instance, if the weathering of limestone transfers 8 units of Ca^{2+} to the world's oceans and the Ca content of seawater does not change, then the same amount of Ca must be deposited as a sedimentary Ca mineral. All life comprises the

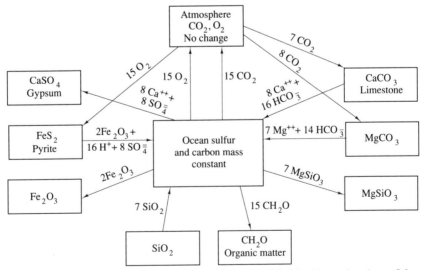

Figure 1.1   Sedimentary reservoirs and transfers in a model of the biogeochemistry of the surface of the Earth (Garrels and Lerman, 1981). Various transfers associated with an increase in the mass of the biosphere by 15 moles are indicated.

biosphere, which appears in the compartment labeled CH_2O, representing the approximate stoichiometric composition of living tissues. Changes in the mass of living material through geologic time are modeled by changes in the size of the biosphere compartment, as a result of net transfers of material in and out of that box.

Consider the increase in the total mass of organic matter that must have occurred during the Carboniferous Period when large areas of land were covered by swamps. Here, dead vegetation accumulated as peat that was later transformed into coal. Storage of carbon in dead materials, detritus, represents an increase in the mass of the biosphere. With no change in the CO_2 content of the atmosphere or CO_2 dissolved in the oceans as HCO_3^-, that carbon must have been derived from the weathering of carbonate minerals. Of course, weathering of carbonate minerals would also transfer Ca and Mg to the oceans. Ca would be deposited as $CaSO_4$, and Mg in silicate minerals through a number of reactions that occur in ocean sediments (Chapter 9). To deposit $CaSO_4$ with no change in the SO_4^{2-} content of the world's oceans, sulfur must be derived from another pool. Oxidative weathering of pyrite would supply the SO_4 to the oceans, consuming some of the oxygen that would have been added to the atmosphere by photosynthesis. The remaining oxygen would be consumed in the deposition of Fe_2O_3, so the atmospheric content of O_2 would not change. The total O_2 available for reaction is in molar stoichiometric balance with the carbon stored in organic matter by photosynthesis.

This model illustrates how minerals such as magnesium silicates, traditionally the focus of geochemical studies, are linked to the activities of the biosphere. We may ask, of course, whether this is a reasonable model for the linkage of chemical reactions on Earth. Support for the model would be found if large geologic deposits of $CaSO_4$ are associated with periods in which there were large net stores of organic carbon, since the model predicts a coupled balance:

$$\text{pyrite} + \text{carbonates} \rightleftarrows \text{gypsum} + \text{organic carbon} \qquad (1.1)$$

through geologic time. Garrels and Lerman (1981) show that the molar ratio of organic carbon and gypsum has remained fairly constant through geologic time, with large deposits of gypsum associated with the Carboniferous Period, when large amounts of organic carbon were stored in coal (Fig. 1.2).

This model also reminds us that the size of the biosphere waxes and wanes as a result of the balance between photosynthesis and respiration. The mass of the biosphere has increased at times when photosynthesis has resulted in a net storage of organic carbon and the release of free O_2 as a byproduct. Heterotrophic respiration by microbes and higher animals constitutes respiration, which converts organic carbon back to

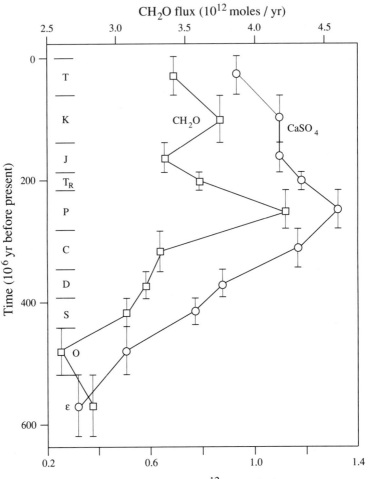

Figure 1.2 Garrels and Lerman (1981) use the isotopic ratio in the sedimentary record of organic carbon to calculate the rate of accumulation of organic carbon through geologic time. Independently, they use their model (Fig. 1.1) to calculate the sedimentary accumulation of $CaSO_4$ (gypsum) over the same interval. Deposition of organic carbon and $CaSO_4$ appears to have varied in parallel for the last 500 million years.

CO_2 and H_2O. Fires perform the same reaction abiotically and very quickly. Currently, we are burning the organic carbon stored in the biospheric compartment in the form of coal and oil deposited during earlier geologic times. As a result, concentrations of atmospheric carbon dioxide are increasing (Fig. 1.3), but presumably the linked biogeochemical system operating at the surface of the Earth will consume this CO_2 and transfer it into other compartments. When we compare the conditions on

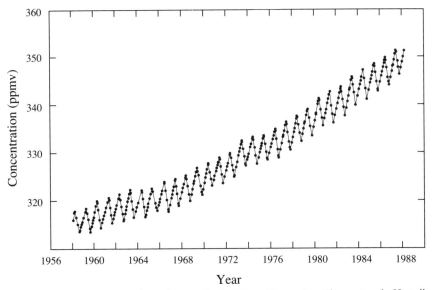

Figure 1.3 The concentration of atmospheric CO_2 at Mauna Loa Observatory in Hawaii, expressed as a mole fraction in parts per million of dry air. The annual oscillation reflects the seasonal cycles of photosynthesis and respiration by land biota in the northern hemisphere, while the overall increase is largely due to the burning of fossil fuels. From Keeling (1986).

Earth to those on other planets (Chapter 2), we will see that the storage of organic carbon and release of free O_2 are the essence of life; evidence of a significant amount of either material on another planet would be strongly suggestive of life there as well (Horowitz 1977).

Thermodynamics

Two basic laws of physical chemistry, the laws of thermodynamics, tell us that energy can be converted from one form to another and that chemical reactions should proceed spontaneously in the direction of lower free energy, G. The lowest free energy of a given reaction represents its equilibrium, and it is found in the mix of chemical species that show maximum bond strength and maximum disorder among the components. In the face of these basic laws, living systems and the conditions on the surface of the Earth exist in a nonequilibrium condition.

Even the simplest cell is an ordered system; a membrane separates an inside from an outside, and the inside contains a mix of very specialized molecules. Biological molecules are collections of compounds with relatively weak bonds. For instance, to break the covalent bonds between two carbon atoms requires 83 kcal/mole, versus 192 kcal/mole for each of the double bonds between carbon and oxygen in CO_2 (Davies 1972, Morowitz 1968).

In living tissue most of the bonds between C, H, N, O, P, and S, the major biochemical elements, are reduced, or "electron-rich" bonds (Chapter 7). It is an apparent violation of the laws of thermodynamics that the reduced bonds in the molecules of living organisms exist in the presence of a strong oxidizing agent in the form of O_2 in the atmosphere. Thermodynamics would predict a spontaneous reaction between these components to produce CO_2, H_2O, and NO_3^-. In fact, upon the death of an organism, this is exactly what happens! Living organisms must continuously process energy to counteract the basic laws of thermodynamics that would otherwise produce disordered systems with oxidized molecules.

Photosynthetic organisms capture energy in sunlight and convert the bonds between carbon and oxygen in CO_2 to the weak, reduced biochemical bonds that characterize life. Heterotrophic organisms obtain energy by capitalizing on the natural tendency for electrons to flow from reduced bonds to oxidizing substances, such as O_2. Thus, these organisms obtain energy by oxidizing the bonds in living tissues and converting the carbon back to CO_2. Various other combinations of energy transformations allow a variety of metabolic pathways in living systems (Chapter 2), but in every case metabolic energy is obtained during the flow of electrons between compounds in oxidized or reduced states. Metabolism is possible because living systems can sequester high concentrations of oxidized and reduced substances from their environment. Without membranes to compartmentalize living cells, thermodynamics would predict a uniform mix and a uniform oxidation state in an abiotic Earth. Energy transformations would be impossible.

Free oxygen appeared in the Earth's surface environments some time after the appearance of autotrophic, photosynthetic organisms. Free O_2 is one of the most oxidizing substances known, and the movement of electrons from reduced substances to O_2 releases large amounts of free energy. Thus, large releases of free energy are found in aerobic metabolism, including the efficient metabolism of eukaryotic cells. The appearance of eukaryotic cells was not immediate; the fossil record suggests that they evolved about 1.5 billion years after the appearance of the simplest living cells. Presumably the evolution of eukaryotic cells and higher organisms was possible only after the accumulation of sufficient free O_2 in the environment to allow aerobic metabolic systems with high energy yields that could, in turn, support their elaborate cellular structure.

Lovelock (1979, 1988) has reinforced our view that the chemical conditions of the present-day Earth, especially the atmosphere, are extremely unusual and in disequilibrium with respect to thermodynamics. The 20%

atmospheric content of O_2 is the most obvious result of living organisms. This O_2 level is maintained despite predicted reactions that would consume O_2 in reaction with crustal minerals and organic carbon. Lovelock suggests that the content of other gases such as N_2 and the temperature of the Earth may be strongly influenced by the biosphere as well. Reflecting the vigor and excitement of a new scientific field, other workers disagree, suggesting that abiotic factors have controlled N_2 (Walker 1984) and climatic conditions (Kasting et al. 1988) throughout much of our planetary history. The future of biogeochemistry will undoubtedly see further revisions, refinements, and new proposals for the control of chemical conditions on Earth, but the presence of O_2 speaks strongly for the role of biota.

Cycles in Biogeochemistry

Since its organization as a planet, the Earth has been exposed to cyclic phenomena (Degens et al. 1981, Harrington 1987). Some, such as the daily rotation on its axis and the annual rotation about the Sun, are now so obvious that it seems surprising that they were mysterious to philosophers and scientists throughout most of human history. Other cycles are due to processes that we still do not understand particularly well. For example, glacial–interglacial cycles seem linked to Milankovitch variations in the Earth's orbit, and the magnetic poles of the Earth have also shown cyclic reversals through geologic time.

The biosphere is always changing in response to cycles. In plants, photosynthesis dominates over respiration in the daytime; the reverse is true at night. During the summer, total photosynthesis in the northern hemisphere exceeds respiration by decomposers. This seasonal storage of carbon in plant tissues results in a seasonal decrease in atmospheric CO_2 (Fig. 1.3). The annual cycle is completed during the winter months, when atmospheric CO_2 rises as decomposition continues during the time that many plants are dormant or leafless. In the longer time frame, the biosphere has increased and decreased in size during glacial cycles and during the Carboniferous Period, when most of the major economic deposits of coal were laid down. The unique conditions of the Carboniferous are poorly understood, but it is certainly possible that such conditions are part of a long-term cycle that might return again.

All current observations of global change must be evaluated in the context of underlying cycles driven by physical processes and biotic responses. We are fairly confident that the current increase in atmospheric CO_2 is an unusual event caused by human activities. We can see that the increase is superimposed on an annual cycle of seasonal changes due to photosynthesis and respiration, and we know that CO_2 concentrations were fairly stable for about 10,000 years before the Industrial Revolution (Gammon et al. 1985). Higher atmospheric CO_2 should lead to a global warming, but any observed change in global climate must be evaluated in the context of long term cycles in climate with many possible causes (Hansen et al. 1981). Atmospheric CO_2 was lower during the last glacial period (Barnola et al. 1987), yet we do not know if that was a cause or an effect of the glacial cycle. Atmospheric CO_2 may fluctuate over the long term in response to global variations in hydrothermal activity and sea-floor spreading (Owen and Rea 1985, Berner et al., 1983; Kasting and Richardson, 1985). Concern about global change is greatest when we see changes in atmospheric constituents such as carbon dioxide (0.4%/yr), methane ($\sim$1%/yr) and nitrous oxide (N_2O; $\sim$0.3%/yr), for which we have little or no precedent in the geologic record.

Changes in atmospheric composition are perhaps our best evidence of the ability of humans to alter the environment globally. Changes in the volume and composition of river flow are also suggestive of the magnitude of human effects. As the terrestrial landscape is denuded of vegetation, greater runoff is expected due to a reduction in the transfer of water to the atmosphere by the transpiration of plants (Waring and Schlesinger 1985). Is the 3% increase in global river flux since 1900 (Probst and Tardy 1987) due to human manipulations or to a long-term underlying global cycle in precipitation and evaporation?

All human activities that increase the erosion of soil represent a change in the global rate of sediment transfer to the oceans and sedimentary deposition. Mining of fossil fuels and metal ores represents an increase in the rate at which these materials would be naturally uplifted and exposed to weathering at the surface of the Earth (Bertine and Goldberg 1971). The content of lead in coastal sediments appears directly related to fluctuations in the use of Pb by an industrialized society (Trefry et al. 1985). Recent estimates suggest that the global cycles of many metals have been

Table 1.1 Movement of Selected Elements through the Atmosphere[a]

| Element | Natural | | | Anthropogenic | | |
	Continental Dust	Volcanic Dust	Gas	Industrial Particles	Fossil Fuel	Ratio Anthropogenic:Natural
Al	356,500	132,750	8.4	40,000	32,000	0.15
Fe	190,000	87,750	3.7	75,000	32,000	0.38
Cu	100	93	0.012	2200	430	13.63
Zn	250	108	0.14	7000	1400	23.46
Pb	50	8.7	0.012	16,000	4300	345.83

[a]All data are expressed in 10^8 g/yr. From Lantzy and MacKenzie (1979).

significantly enhanced by such human activities (Table 1.1). Once again, evidence for global change induced by humans must be considered in the context of past oscillations in the rate of crustal exposure, weathering, and sedimentation due to changes in sea level (Worsley and Davies 1979).

2

Origins

Introduction

About 26 of the chemical elements comprise all of life. Six elements, C, H, O, N, P, and S, are the major constituents of living tissue and comprise 95% of the biosphere. In the periodic table all the biologically relevant elements are found at atomic numbers less than that of iodine at 53. Even though the global circulations of some of the heavier elements are affected by living organisms, we can speak of the biogeochemistry of life as the chemistry of the "light" elements (Deevey 1970a). One initial constraint on the composition of life must have been the relative abundance of the chemical elements in the galaxy; later, as the planets formed and differentiated, the composition of the crust of the Earth determined the geochemical environment in which life arose.

In this chapter we will examine models that astrophysicists suggest for the origin of the elements. Then we will examine models for the formation of the solar system and the planets. There is good evidence that conditions on the surface of the Earth changed greatly during the first billion years or so after its formation, well before life arose. These changes and changes subsequent to the origin of life strongly determine the surface conditions on Earth today. In this chapter, we consider the origin of the metabolic pathways that characterize the living systems of today. The chapter ends with a discussion of the planetary evolution that has occurred on the Earth compared to its near neighbors, Mars and Venus.

Origin of Elements

Any theory for the origin of the chemical elements must account for the variation in their abundance in the universe, shown as a function of atomic number in Fig. 2.1. Estimates of cosmic abundance are made by examining the spectral emission from stars comprising distant galaxies as well as the emission from our own Sun (Ross and Aller 1976, Wallerstein 1988; Anders and Grevesse 1989). Several points are obvious: (1) the light elements are far more abundant than the heavy elements, (2) the even-numbered elements are more abundant than odd-numbered elements with similar atomic weight, (3) the elements beyond an atomic number of 30 are more nearly similar in abundance than is the case among the light elements, and (4) three light elements, Li (lithium), Be (beryllium), and B (boron), are anomalously rare in the galaxy.

Astrophysicists seem in widespread agreement that the origin of the universe began with a "Big Bang," which initiated the fusion of hydrogen (H) to helium (He) (Fox 1988). Fusion of hydrogen is still occurring on the surface of our own Sun today. However, as the universe began to expand outward, there was a rapid decline in the temperatures and pressures that would be needed to produce elements beyond the atomic weight of helium in interstellar space. Synthesis of the elements beyond helium remained a perplexing problem until Burbidge et al. (1957) outlined a series of pathways that could occur in the interior of stars during their evolution (Penzias 1979, Fowler 1984, Woosley and Phillips 1988, Wallerstein 1988).

As a star ages the abundance of hydrogen declines, as it is converted to helium by fusion. As the heat from nuclear fusion decreases, the star begins to collapse inward under its own gravity. This collapse increases the internal temperatures until He begins to be converted, or "burn," to form carbon (C) in a two-step reaction. First,

$$^4He + {}^4He \rightleftharpoons {}^8Be \tag{2.1}$$

but ^8Be, like all the elements with potential atomic mass between 5 and 8, is unstable (Fox, 1988). Most ^8Be decays spontaneously back to helium, but the momentary existence of small amounts of ^8Be allows reaction with another helium to produce carbon:

$$^8Be + {}^4He \rightleftharpoons {}^{12}C \tag{2.2}$$

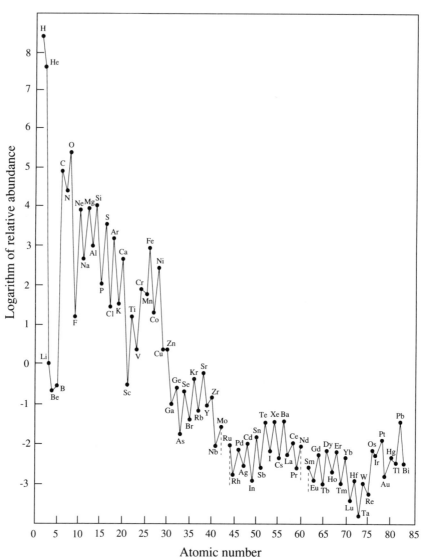

Figure 2.1 Relative abundance of elements in the Universe as a function of atomic number. Abundances are scaled so that silicon (Si) = 10000. From Mason (1966).

The main products of helium burning would be ^{12}C and ^{16}O. As the supply of helium began to decline, stellar collapse would be followed by the initiation of further "burning" reactions. First, fusion of ^{4}He with ^{12}C would produce ^{16}O (oxygen). Later, as the supply of ^{12}C dwindled, fusion of ^{4}He with ^{16}O would produce ^{20}Ne (neon). Successive fusion reactions are thought to be responsible for the synthesis of the elements up to iron (Fe), beyond which the decay of the products is in equilibrium with their formation (Fowler 1984). Fusion of elements beyond iron is impossible, since the temperatures and pressures required result in catastrophic collapse and the explosion of stars that we recognize as supernovas. Heavier elements are apparently formed by the successive capture of neutrons by light elements during Type II supernovas (Rank et al. 1988, Woosley and Phillips 1988). The explosion casts all portions of the star into space as hot gases (Chevalier and Sarazin 1987).

This model explains a number of the observations about the abundance of the light elements that are of biogeochemical interest. First, the abundance of elements declines logarithmically from hydrogen and helium, the original building blocks of the universe. However, as the universe ages, more and more of the helium is converted to heavier elements. Astrophysicists can recognize young stars that have formed from the remnants of previous supernovas because they contain a higher abundance of iron than in older stars, in which the initial helium-burning reactions are still continuing (Penzias, 1979). Second, the elements beyond helium are built by progressive fusion of He nuclei with an atomic mass of 4 (atomic number $= 2$), so the even-numbered elements are more abundant than odd-numbered elements. This continues to iron, beyond which the even/odd dicotomy is less apparent, since the elements are thought to result only from the addition of neutrons with an atomic mass of 1. Radioactive decay of heavy elements produces the odd-numbered light elements, so one would expect those to be proportionately less abundant. Note that ^{31}P, which is often limiting to plant growth on the Earth today, is much less abundant than the adjacent elements, Si and S (Fig. 2.1). Finally, the low abundance of Li, Be, and B is due to the instability of their atomic mass in the conditions of stellar interiors. These elements must have formed later, in low-density and low-temperature environments, by the decay of heavy elements.

Models for the origin and the cosmic abundance of the elements offer the initial constraints for biogeochemistry. All things being equal, we would expect that the chemical environment of life, and the composition of living tissues, would approximate the cosmic abundance of elements. It is then of no great surprise that among the light elements, no Li or Be, and only traces of B, are essential components of biochemistry.

Origin of the Solar System and the Earth

While our galaxy is probably about 10 billion years old, our own solar system appears to be only half that age, about 4.6 billion years. Current models for the origin of the solar system suggest that the Sun and the planets formed from the remnants of a stellar supernova, which left

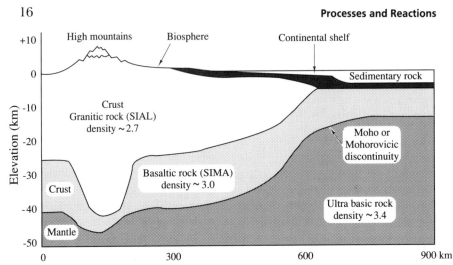

Figure 2.2 A geologic profile of the Earth's surface. On land the crust is dominated by granitic rocks, largely comprised of silicon and aluminum (SIAL). The oceanic crust is dominated by basaltic rocks, with a large proportion of silicon and magnesium (SIMA). The upper mantle has the approximate composition of olivine ($FeMgSiO_4$). From Howard and Mitchell (1985).

a collection of hot gases in space (Chevalier and Sarazin 1987). As the Sun and the planets began to condense, gravitational attraction helped add to their initial mass. The mass concentrated in the Sun apparently allowed condensation to pressures that would once again initiate the fusion of hydrogen to helium. Some astronomers suggest that the similar hydrogen-rich composition on Jupiter represents a star that was never "ignited."

The planets of our solar system appear to have formed from the coalescing of small bodies, planetesimals, that formed in interstellar space (Press and Siever 1986). The planetary compositions were affected by the position of the incipient planet relative to the Sun and the rate at which each planet grew (McSween 1989). Thus, Mercury, which is small and closest to the Sun, has an extremely dense mass, whereas the larger, outer planets are comprised of lower-density materials that could be retained under conditions of greater gravity and cooler temperature. Venus, Earth, and Mars are somewhat similar in composition and size. These planets are somewhat depleted in volatiles compared to the cosmic abundance. From an initial cosmic mix of elements, the biogeochemical environment for life on Earth represents a selective mix, peculiar to the conditions on the incipient planet.

Collision of planetesimals suggests that the Earth may have grown by homogeneous accretion; that is, the initial composition was similar throughout as a result of the collision of planetesimals of uniform composition (Stevenson 1983). Of course, we know that the Earth is not homogeneous today. After its initial condensation, several events may have acted to differentiate the core, the mantle, and the crust of the Earth. Kinetic energy generated during the collision of planetesimals (Wetherill, 1985), as well as the heat generated from radioactive decay in its interior (Hanks and Anderson 1969), are thought to have heated the primitive Earth to the melting point of iron, nickel, and other metals. These elements were "smelted" from the homogenous initial composition and sank to the interior of the Earth to form the core. As the Earth cooled, the lighter elements progressively solidified to form a mantle with the approximate composition of perovskite ($MgFeSiO_3$; Knittle and Jeanloz 1987) and olivine ($FeMgSiO_4$), and a crust dominated by aluminosilicate minerals of lower density and the approximate composition of feldspars (Chapter 4). Even today, the aluminosilicate rocks of the crust "float" on the heavier semifluid rocks of the mantle, resulting in the drift of continents on the Earth's surface (Fig. 2.2). Again, from a biogeochemical perspective, the surface of the Earth is a selective mix of the elements comprising the primordial Earth (Fig. 2.3). Despite the abundance of iron in the cosmos and in the Earth as a whole, the crust of the Earth is largely composed of Si, Al, and O (Weaver and Tarney 1984). Surprisingly, some of the radioactive elements, including uranium, are more abundant in the crust than one might expect given their atomic weight. Uranium forms relatively light crustal minerals with oxygen.

The Primitive Atmosphere and Oceans

Several lines of evidence suggest that this primitive Earth was devoid of an atmosphere, and its temperature was certainly well in excess of one that would allow a primitive ocean. For example, if the primitive Earth contained an atmosphere, we might expect that its gases would have existed in proportion to their cosmic abundance (Fig. 2.1). ^{20}Ne is of particular interest: it is not produced by any known radioactive decay, it is too heavy to have escaped from the Earth, and as an inert gas, it is not likely to have existed in any combination with crustal minerals on the primitive Earth (Walker 1977). Thus, the modern-day abundance of neon is likely to represent its primary abundance, that which remains of the initial atmosphere. Assuming that other gases were also retained, we can calculate the expected mass of the primary atmosphere by multiplying the mass of neon in today's atmosphere by the ratio of each of the other gases to neon in the cosmic abundance. For example, the cosmic ratio of N/Ne is 5.33 (Fig. 2.1). If the present-day atmospheric mass of neon, 6.5×10^{16} g, is all from primary sources, then $5.33 \times 6.5 \times 10^{16}$ g should yield the mass of nitrogen that is also of primary origin. The product, 35×10^{16} g is much less than the

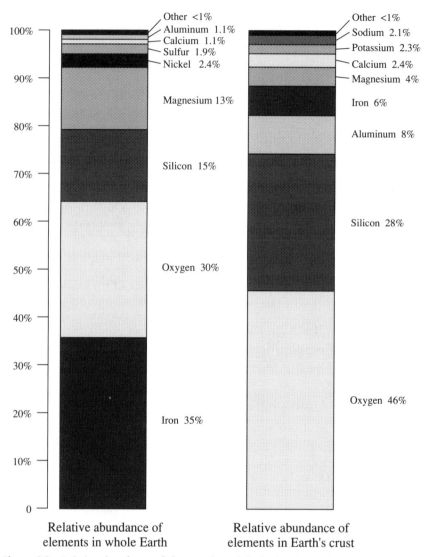

Figure 2.3 Relative abundance of elements by weight in the whole Earth and the Earth's crust. From Earth 4/E. By Frank Press and Raymond Siever. Copyright © 1986 by W. H. Freeman and Company. Reprinted with permission.

observed atmospheric mass of nitrogen, 38×10^{20} g. Thus, most of the atmosphere must have been derived later, from crustal processes in the Earth.

How were the volatiles in the atmosphere derived? Many silicate minerals carry elements such as oxygen and hydrogen as part of their crystalline

Table 2.1 Total Volatiles on the Surface of the Earth[a]

	H_2O	Carbon	CO_2	Cl	N	S	Total
Atmosphere	1.7	0.27	0.245	—	387	—	530×10^{19}
Oceans	140,000	—	13.8	2620	2.18	122	142,000
Fresh Water	4100	—	—	—	—	—	4100
Sedimentary Rocks	15,000	1000	24,200	500	100	400	41,200
Total	160,000	1000	24,000	3100	490	520	$189,000 \times 10^{19}$

[a] All data are 10^{19} g. Note that water dominates the 1.89×10^{24} g of all volatiles, while N_2 dominates the mass of the atmosphere (5.3×10^{21} g), because of the removal of H_2O, CO_2 and Cl to other spheres. From Walker (1977).

structure. Presumably planetesimals delivered these minerals during the initial accretion of the Earth. Other elements that are now found mostly in the atmosphere may have been delivered to the Earth by carbonaceous chondrites, a class of meteorite that contains simple organic molecules with C and N (Anders et al. 1973, Anders 1989). Some astrophysicists believe that these meteors may have been especially abundant among the planetesimals received during the last phases of the Earth's accretion (Anders and Owen 1977). Chyba (1987) estimates the rate of cometary bombardment from the craters on the Moon and calculates the Earth's potential receipt of water vapor from comets. His best estimate is approximately equal to seven-eighths of the mass of liquid water now present on Earth (Table 2.1), suggesting that cometary contributions to the primitive Earth must also be considered with planetesimals and meteors as a possible source of the Earth's volatiles (Chyba 1990). The receipt of volatiles occurred early in the Earth's evolution; from the age distribution of craters on the Moon, there is good evidence that the major period of planetary accretion and meteor bombardment slowed after 3.8 billion years ago (Neukum 1977).

With heating of minerals in the crust and mantle, volatile elements are released in volcanic eruptions at the surface. Table 2.2 gives the composition of volcanic gases emitted from the Nevado de Ruiz volcano in 1985 (Williams et al. 1986). Characteristically, water vapor dominates the emissions, but small quantities of C and S gases are also present. N_2 and NO_2, which were not measured, are also abundant in many volcanic gases (e.g., Inn et al. 1981). Volcanic emissions, representing degassing of the Earth's crust, are consistent with the observation that the Earth's atmosphere is of secondary origin. Today, volcanic emissions are coupled to tectonic movements of the Earth's crust. It is likely that the bulk of the atmosphere was derived from degassing early in the Earth's history, during the massive heating, fluid movement, and differentiation of the Earth's crust and mantle (Fanale 1971, Walker 1977, Stevenson 1983).

Table 2.2 Composition of Volcanic Gases (Percentages by Weight) released from the Nevado del Ruiz volcano, Columbia.[a]

Date (1985)	H_2O	CO_2	SO_2	H_2S	HCl
September 24	97.31	1.15	1.47	0.068	0.0062
October 19	95.68	1.9	2.33	0.0798	0.0036
November 12	91.62	5.68	4.43	2.25	0.0058

[a]From Williams et al. (1986). Copyright 1986 by the AAAS.

Emission of ^{40}Ar is suggestive of the extent of crustal degassing. Like neon, this noble element is too heavy for gravitational escape from the Earth's atmosphere. There is little cosmic abundance of ^{40}Ar versus ^{36}Ar; on Earth ^{40}Ar appears to be wholly the result of the decay of ^{40}K in crustal minerals. Thus, the atmospheric content of ^{36}Ar should represent the proportion that is due to the residual primary atmosphere, whereas the content of ^{40}Ar is indicative of the proportion due to crustal degassing. The ratio of $^{40}Ar/^{36}Ar$ on Earth is nearly 300, suggesting that 99.7% of our present atmosphere is derived from the interior of the Earth. The Viking spacecraft measured an $^{40}Ar/^{36}Ar$ ratio of 2750 in the atmosphere on Mars (Owen and Biemann 1976), which is consistent with the emerging view that Mars also underwent crustal degassing but has now lost its atmosphere (Carr 1987).

Once the Earth's surface temperature cooled to $100°$ C, water condensed out of the primitive atmosphere to form the oceans. This was a rainstorm of true global proportions! The earliest rock record suggests that liquid water was present on the Earth's surface 3.8 billion years ago, and has been ever since. Consistent with the belief that most of the crustal outgassing was completed early in the Earth's history, the oceans are likely to have achieved close to their present volume in a relatively short period of time (Holland 1984).

With the condensation of liquid water, various other gases followed, due to their rapid dissolution in water:

$$CO_2 + H_2O \rightleftarrows H_2CO_3 \rightleftarrows H^+ + HCO_3^- \tag{2.3}$$

$$HCl + H_2O \rightleftharpoons H_3O^+ + Cl^- \tag{2.4}$$

$$SO_2 + H_2O \rightleftharpoons H_2SO_3 \tag{2.5}$$

These reactions removed a large proportion of reactive water-soluble gases from the atmosphere, as predicted by Henry's Law for partitioning of gases between gaseous and dissolved phases:

$$S = kP \tag{2.6}$$

where S is the solubility of the gas in the liquid, k is the solubility constant, and P is the overlying pressure in the atmosphere. Under standard temperature and pressure, the solubilities of CO_2, HCl, and SO_2 are 3.2, 823, and 228 g/l, respectively. When dissolved in water, all of these gases form acids, which would be neutralized by immediate reaction with the surface minerals on Earth. The primitive atmosphere is likely to have been dominated by N_2, which has a low solubility constant in water (0.029 g/l).

An estimate of the total extent of crustal degassing must consider the content of the oceans and of some minerals such as $CaCO_3$ that have been deposited from seawater (Li 1972). Walker (1977) has compiled such an inventory (Table 2.1), which suggests that the total atmosphere of today (5.3×10^{21} g) represents less than 1% of the total degassing of the Earth's crust through geologic time. The oceans and various marine sediments represent the remainder.

It is difficult to know the exact composition of the Earth's earliest atmosphere. We might expect to deduce its composition from collections from modern terrestrial volcanoes (e.g., Table 2.2), but these may be comprised of a higher proportion of oxidized species, since they are partially derived from recent crustal sediments that have undergone subduction and heating. Collections from deep-sea hydrothermal vents at continental spreading centers contain gases such as ^3He that are certainly derived from the mantle (Lupton and Craig 1981), but it is unclear if the greater proportion of reduced species (e.g., CH_4, H_2, CO, N_2O) emitted from these vents may also be partially due to the unusual microbial communities that are found there (Lilley et al. 1982, Charlou et al. 1988, Chapter 9).

Despite uncertainty about the Earth's earliest atmosphere, several lines of evidence suggest that by the time life arose the atmosphere was only moderately reducing (Holland 1984). Despite an abundance of metallic iron in the core of the Earth, the oxidation state of iron found as olivine ($FeMgSiO_4$) in the mantle would not have allowed an abundance of highly reduced gases, such as H_2 and NH_3, in volcanic emissions. Moreover, any H_2 that was released would have been likely to escape from the Earth's gravitational field, leaving little to react with CO_2 to produce methane (Warneck 1988):

$$4H_2 + CO_2 \rightarrow CH_4 + 2H_2O \tag{2.7}$$

Ammonia (NH_3) would have been subject to photolysis in the Earth's upper atmosphere, producing N_2 with the loss of H_2 to space. Thus, it is likely that the atmosphere in which life arose was dominated by N_2, with lesser proportions of H_2O and CO_2, and trace quantities of gases from

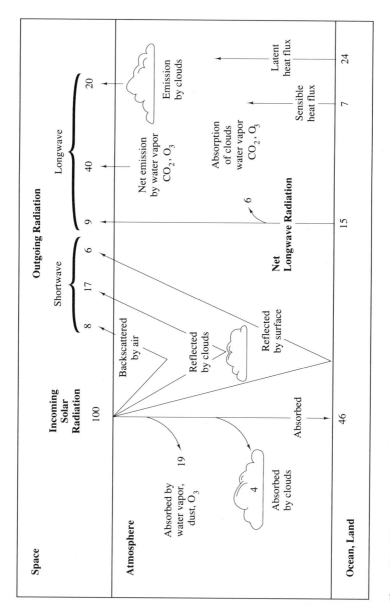

Figure 2.4 The "greenhouse effect" of CO_2 and other trace gases. The sun's radiation is dominated by short wavelengths, which are reflected or absorbed by the Earth's surface. The radiation absorbed is reradiated at longer wavelengths, which can be absorbed by atmospheric gases including CO_2. Higher concentrations of these gases in the atmosphere reduce the net emission of longwave radiation to space, warming the Earth. From MacCracken (1985).

volcanic emissions that were continuing at that time. There was certainly no O_2; the small concentrations produced by the photolysis of water in the upper atmosphere would rapidly be consumed in the oxidation of reduced gases and crustal minerals (Walker 1977, Kasting and Walker 1981).

During its early evolution as a star, the sun's luminosity was as much as 30% lower than at present. We might expect that the Earth was colder, but the fossil record indicates a continuous presence of liquid water on the Earth's surface since 3.8 billion years ago. One suggestion is that the primitive atmosphere contained somewhat higher concentrations of CO_2 than today. This CO_2 would trap outgoing infrared radiation and produce global warming through the "greenhouse" effect (Lovelock and Whitfield 1982, Kasting et al. 1988) (Fig. 2.4). In fact, even today, the presence of water vapor and CO_2 in the atmosphere create a significant greenhouse effect. Without these gases the Earth's temperature would be about 33°C cooler, and the planet would be frozen (Ramanathan 1988).

As in the present day, the composition of the earliest ocean appears to have been dominated by Na and Cl (Holland 1984), but higher concentrations of atmospheric CO_2 would have produced somewhat lower pH in seawater than that of today (Walker 1983). Deposits of calcium carbonate ($CaCO_3$) in rocks of 3.8 billion years ago suggest that the ocean must have also had higher concentrations of Ca^{2+}, since $CaCO_3$ is more soluble in acid conditions:

$$CaCO_3 + H_2CO_3 \rightleftarrows Ca^{2+} + 2HCO_3^- \tag{2.8}$$

Evaporative deposits of gypsum ($CaSO_4 \cdot 2H_2O$) of the same age suggest that the ocean contained some SO_4^{2-} as well. In the absence of free O_2, iron derived from crustal weathering would have accumulated in the ocean as Fe^{2+}, which is soluble under acid, reducing conditions and in the presence of SO_4^{2-}. In sum, with the exception of higher concentrations of reduced metals and low concentrations of sulfate, the major constituents of the earliest seawater were probably similar to those of today (Holland 1984).

Origin of Life

Working with Harold Urey in the early 1950s, Stanley Miller added probable atmospheric and ocean constituents of the primitive Earth to a laboratory flask and subjected the mix to electric discharge to represent the effect of lightning. After several days, Miller found that simple, reduced organic molecules had been produced (Miller 1957). This experiment, simulating the conditions on the early Earth, suggested that the simple constituents of living organisms could be produced abiotically. The presence of simple organic molecules, including amino acids, in carbonaceous chondrites suggests that these reactions have occurred repeatedly in the galaxy. The experiment has been repeated in many laboratories, and with many combinations of conditions (Chang et al. 1983). Ultraviolet light can substitute for electrical discharges as an energy source; a high flux of ultraviolet light would be expected on the primitive Earth in the absence of an ozone (O_3) shield in the stratosphere (Chapter 3). The mix of atmospheric constituents taken to represent the primitive atmosphere is controversial, although an acceptable yield of simple organic molecules could have been produced in the mildly reducing atmospheres that many workers now prefer (Pinto et al. 1980). These experiments are never successful when free O_2 is included as an atmospheric constituent; O_2 rapidly oxidizes the simple organic products before they can accumulate.

A wide variety of simple organic molecules has been produced in abiotic conditions in the laboratory (Dickerson, 1978). In many cases hydrogen cyanide and formaldehyde are dominant constituents. These molecules are particularly important since they polymerize to produce simple sugars such as ribose and more complex molecules such as amino acids and nucleotides. Methionine, a sulfur-containing amino acid, has also been synthesized abiotically (Van Trump and Miller 1972), and short chains of amino acids have been linked by condensation reactions involving phosphates (Rabinowitz et al. 1969, Lohrmann and Orgel 1973). The early abiotic role of organic polyphosphates speaks strongly for the origin of nucleotide triphosphates (ATP, GTP, etc.) as the energizing reactants in virtually all biochemical reactions that we know of today (Dickerson 1978).

Assembly of simple organic molecules into a metabolizing, self-replicating, and membrane-bound form that we might call life has so far eluded experimental approaches. There is strong evidence that clay minerals, with their surface charge and repeating crystalline structure, may have acted to concentrate simple, polar organic molecules from the primitive ocean, making assembly into more complicated forms, such as RNA and protein, more likely (Cairns-Smith 1985). Metal ions such as zinc and copper can enhance the binding of nucleotides to clays (Lawless and Levi 1979).

Just as droplets of cooking oil form "beads" on the surface of water, it has long been known that organic polymers will spontaneously form coacervates, which are colloidal droplets small enough to remain suspended in water. Coacervates are perhaps the simplest systems that might be said to be "bound," as if by a membrane, providing an inside and an outside. Yanagawa et al. (1988) describe several experiments in which protocellular structures with lipoprotein envelopes were constructed in the laboratory. In such structures, the concentration of substances will differ between the inside and the outside as a result of differing solubility of substances in water and in organic medium. Their resemblance to simple living organisms is striking.

For an understanding of global biogeochemistry, the exact mechanisms that organized the earliest living systems are not as important as the chemical abundances that were present at the surface of the Earth. Banin and Navrot (1975) point out the striking correlation between the abundance of elements in biota and the solubility of elements in ocean water. Fe, Al, and Si, which form insoluble hydroxides in seawater, are found in low concentration in living tissue, despite relatively high concentrations of some of these elements in the crust. Elements with low ionic potential (i.e., ionic charge/ionic radius) are found as soluble cationic species in seawater. These include Na, K, Mg, and Ca, which are common biochemical constituents. Other elements, including S, C, N, and P, that form soluble oxyanions in seawater (SO_4^{2-}, HCO_3^-, NO_3^-, and PO_4^{3-}) are also abundant biochemical constituents. Molybdenum is much more abundant in biota than one might expect based on its crustal abundance; molybdenum forms the soluble molybdate ion (MoO_4^{2-}) in ocean water.

Although phosphorus forms a soluble oxyanion, PO_4^{3-}, it may never have been particularly abundant in seawater (Griffith et al. 1977). In a delightful paper, Westheimer (1987) examines why phosphorus plays such an important role in biochemistry. With three ionized groups, phosphoric acid can link two nucleotides in DNA, with the third negative site acting to prevent hydrolysis and maintain the molecule within a cell membrane. These ionic properties also allow phosphorus to serve in intermediary metabolism and energy transfer in ATP. The unique properties of phosphorus can account for its major role in biochemistry, despite its limited solubility in seawater and its relatively low geochemical abundance on Earth.

In sum, if one begins with the cosmic abundance of elements as an initial constraint, and the partitioning of elements in the formation of the Earth as subsequent constraints, then solubility in water appears to be the final constraint in determining the relative abundance of elements in the arena in which life arose. Those elements that were abundant in seawater are also important biochemical constituents. Elements that were rare in seawater are those that are now frequently found to be poisons to living systems (e.g., Be, As, Hg, Pb, Cd).

Evolution of Metabolic Pathways

Awramik et al. (1983) report that 3.5-billion-year-old rocks collected in western Australia contain microfossils that may be our earliest direct evidence of life. Their specimens resemble some modern cyanobacteria (blue-green algae). The earliest living organisms were probably even more primitive than cyanobacteria, resembling the Archaebacteria that survive today in anaerobic and often in hydrothermal (volcanic) environments at temperatures above 90° C (Brock 1985). Archaebacteria are distinct from eubacteria by a lack of muramic acid as a component of the cell wall and distinct ribosomal ribonucleic acid (r-RNA) sequence (Fox et al. 1980). Both halophilic (salt-tolerant) and thermophilic forms of Archaebacteria are known. Huber et al. (1989) describe methanogenic Archaebacteria growing at 110° C near a deep sea hydrothermal vent in Italy. This environment may resemble some of the earliest habitats for life on Earth.

The most primitive metabolic pathway probably involved the production of methane by splitting simple organic molecules such as acetate that would have been present in the oceans from abiotic synthesis:

$$CH_3COOH \rightarrow CO_2 + CH_4 \tag{2.9}$$

These organisms were scavengers of the products of abiotic synthesis and obligate heterotrophs, sometimes classified as chemoheterotrophs. The modern fermenting bacteria in the order Methanobacteriales may be our best present-day analogs.

Longer pathways of anaerobic metabolism, such as glycolysis, probably followed with increasing elaboration and specificity of enzyme systems. The oxidation of simple organic molecules in anaerobic respiration was coupled to reduction of inorganic substrates from the environment. For example, some time after the appearance of methanogenesis from acetate splitting, methanogenesis by CO_2 reduction

$$CO_2 + 4H_2 \rightarrow CH_4 + 2H_2O \tag{2.10}$$

probably arose among early heterotrophic microorganisms. Generally this reaction occurs in two steps: eubacteria convert organic matter to acetate, H_2, and CO_2, and then primitive Archebacteria transform these to methane, following equations (2.9) and (2.10) (Wolin and Miller 1987). Note that methanogenesis by CO_2 reduction is more complicated than that from acetate splitting and would require a more complex enzymatic catalysis. Both pathways of methanogenesis are found among the fermenting bacteria that inhabit wetlands and coastal ocean sediments today (Martens et al. 1986, Crill and Martens 1986, Whiticar et al. 1986).

The sulfate-reducing pathway

$$2CH_2O + 2H^+ + SO_4^{2-} \rightarrow H_2S + 2CO_2 + 2H_2O \tag{2.11}$$

is also found in primitive Archaebacteria, and on the basis of the S-isotope ratios in preserved sediments, it appears to have arisen at least 2.4 billion years ago (Cameron 1982). Its later appearance may be related to the time needed to accumulate sufficient SO_4^{2-} in ocean waters to serve as the oxidizing substrate. This pathway has recently been found in a group of thermophilic Archaebacteria isolated from the sediments of hydrothermal vent systems in the Mediterranean Sea, where a hot, anaerobic and acidic microenvironment may resemble the conditions of the primitive Earth (Stetter et al. 1987).

Before the advent of atmospheric O_2, there is good evidence that the primitive oceans contained only low concentrations of available nitrogen in the form of nitrate (NO_3^-) (Kasting and Walker 1981, but see also Yung and McElroy 1979). Thus, even the earliest organisms may have found limited supplies of nitrogen available for protein synthesis. There is little firm evidence for the origin of nitrogen fixation, in which certain bacteria break the inert triple bond in N_2 and reduce the nitrogen to NH_4^+, but today this reaction is performed by bacteria that require strict local anaerobic conditions. The reaction involves the enzyme complex, nitrogenase, which contains a variety of metallic cofactors including iron and molybdenum. A cofactor, vitamin B_{12}, that contains cobalt is also essential. Nitrogen fixation requires the expenditure of large amounts of energy, since the bond strength in the N_2 bond is 225 kcal/mole. Modern nitrogen-fixing cyanobacteria couple nitrogen fixation to their photosynthetic reaction; other nitrogen-fixing organisms are frequently symbiotic with higher plants (Chapter 6).

Despite various pathways of anaerobic metabolism, the opportunities for heterotrophic metabolism must have been limited in a world where organic molecules were only available as a result of abiotic synthesis. Natural selection would strongly favor autotrophic systems that could supply their own reduced organic molecules for metabolism.

We might expect that the earliest photosynthetic reaction was based on sulfur, since the free energy of reaction, G, with hydrogen sulfide is less positive than with water (Schidlowski 1983). This reaction,

$$CO_2 + 2H_2S \rightarrow CH_2O + 2S + H_2O \qquad (2.12)$$

was probably performed by sulfur bacteria, not unlike the anaerobic forms of green and purple sulfur bacteria of today. These bacteria would have been particularly abundant around surface volcanic emissions of reduced gases, including H_2S.

Several lines of evidence suggest that photosynthesis by sulfur bacteria and oxygen-evolving photosynthesis by cyanobacteria were both found in the ancient seas of 3.5 billion years ago (bya). Both forms of photosynthesis produce organic carbon in which ^{13}C is depleted relative to its abundance in dissolved bicarbonate (HCO_3^-), and there are no other processes known to produce such strong fractionations between the stable isotopes of carbon. Fossil organic matter with such depletion is found in rocks

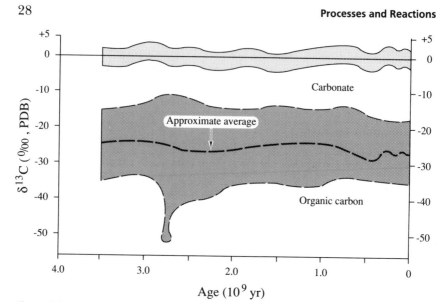

Figure 2.5 The isotopic composition of carbon in fossil organic matter and marine carbonates through geologic time, showing the mean and range (shaded) of specimens of different age. The isotopic composition is the ratio of ^{13}C to ^{12}C, relative to the ratio in an arbitrary standard (PDB belemite) that is assigned a ratio of 0. Carbon in organic matter is 2.8% less rich in ^{13}C than the standard, so this depletion is expressed as $-28‰$ ^{13}C. From Schidlowski (1983).

dating back to 3.5 bya (Fig. 2.5). This discrimination, which is about -2.8% ($-28‰$) in present-day C_3 photosynthesis, is based on the slower diffusion of $^{13}CO_2$ relative to $^{12}CO_2$ and the greater affinity of the carbon-fixation enzyme, ribulose bisphosphate carboxylase, for the more common $^{12}CO_2$ (O'Leary 1988; Chapter 5).

Evidence for oxygen-evolving photosynthesis, in particular, is also found in metamorphosed sedimentary rocks at least 3.5 billion years old that contain thick deposits of Fe_2O_3 interbedded with siliceous sediments found as chert. These deposits are known as the Banded Iron Formations (BIF), reaching a peak occurrence in rocks of 2.5–3.0 billion years ago (Walker et al. 1983). Most of the major economic deposits of iron ore in the United States (Minnesota), Australia, and South Africa date to formations of this age (Meyer 1985). Because it is difficult to envision the deposition of Banded Iron Formation by any mechanism other than the oxidation of Fe^{2+} by O_2 and the deposition of Fe_2O_3 in the sediments of the primitive ocean, these rocks are our earliest evidence for oxygen-evolving photosynthesis based on the photochemical splitting of water:

$$2H_2O + CO_2 \rightarrow CH_2O + O_2 \tag{2.13}$$

Despite the relatively large energy barrier inherent in the reaction, there must have been strong selection for photosynthesis based on the splitting of water, particularly as the limited supplies of H_2S in the primitive ocean were removed by sulfur bacteria. Water offered a nearly inexhaustable supply of substrate for oxygen-evolving photosynthesis, and the release of free oxygen as a byproduct profoundly changed the environment on the primitive Earth. The pathways of anaerobic respiration and photosynthesis by sulfur bacteria are poisoned by O_2. These organisms generally lack catalase and have low levels of superoxide dismutase, enzymes that protect cellular structures from damage by highly oxidizing compounds such as O_2 (Fridovich 1975). Today these organisms are confined to local anaerobic environments. The long co-occurrence of oxygen-evolving and obligately anaerobic organisms in the Earth's early history was due to the consumption of O_2 during the oxidation of Fe^{2+} that had accumulated in the primitive oceans as a result of millions of years of weathering in an anaerobic environment. Only when the oceans were swept clear of reduced metals could excess O_2 accumulate in ocean water and diffuse to the atmosphere.

Even then, the small initial amounts of atmospheric O_2 were further involved in oxidation reactions with reduced atmospheric gases and with exposed crustal minerals of the barren land. Oxidation of reduced minerals, such as pyrite (FeS_2), exposed on land would transfer SO_4^{2-} and Fe_2O_3 to the oceans in river flow. Deposits of Fe_2O_3 that are found in alternating layers with other sediments of terrestrial origin constitute Red Beds, which are found at 2.0 bya and indicate aerobic terrestrial weathering. It is noteworthy that the earliest occurrence of Red Beds roughly coincides with the latest deposition of Banded Iron Formation: further evidence that the oceans were cleared of reduced metals before O_2 began to diffuse to the atmosphere.

Oxygen began to accumulate to its present-day atmospheric level of 20% when the rate of O_2 production by photosynthesis exceeded its rate of consumption by the oxidation of reduced substances. Despite a few recent, highly publicized measurements of air bubbles trapped in fossil amber that suggest O_2 concentrations as high as 30% during the Cretaceous (Berner and Landis 1988), it seems likely that atmospheric oxygen has remained fairly close to the present-day level of 20% since the Silurian (Walker, 1977). What maintains the concentration at approximately 20%? Walker (1980) examined all the oxidation/reduction reactions affecting atmospheric O_2, and suggested that the balance is due to the negative feedback between O_2 and the long-term net burial of organic matter in sedimentary rocks (Chapter 3; Fig. 11.5). In the absence of photosynthesis, the continual weathering of crustal materials would consume all atmospheric O_2 in a few million years. We will examine these processes in more detail in Chapter 3, but here it is interesting to note the significance of an atmosphere with 20% O_2. Lovelock (1979) points out that with <15% O_2 fires would not burn and at >25% O_2 even wet organic matter would burn freely (Watson et al. 1978). Either scenario would result in a world with a profoundly different environment than that of today.

The release of O_2 to an anaerobic Earth is perhaps the strongest reminder we have for the influence of biota on the geochemistry of the Earth's surface. The accumulation of free O_2 in the atmosphere has established the oxidation state for most of the Earth's surface for the last 600 million years. However, of all the oxygen ever evolved from photosynthesis, only 4% remains in the atmosphere today; the remainder is bound in various oxidized sediments, including Banded Iron Formations and Red Beds (Fig. 2.6). The total inventory of free oxygen that has ever been released on the Earth's surface is, of course, balanced stoichiometrically by a storage of reduced carbon in other sediments, including coal, oil, and dispersed organic deposits known as kerogen. The sedimentary storage of carbon is estimated at 10^{22} g (Schidlowski 1983), representing the cumulative net effect of biogeochemistry since the origin of life.

Accumulations of free O_2 led to other changes in the Earth's environment and biota. Eukaryotic metabolism is possible at O_2 levels that are

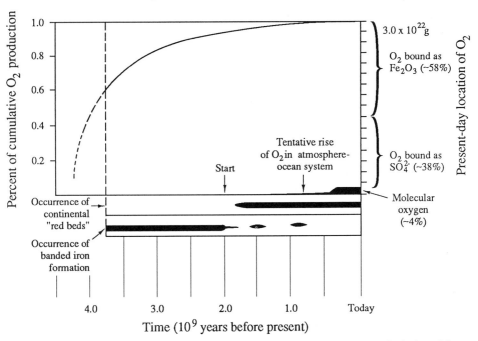

Figure 2.6 Cumulative history of O_2 released by photosynthesis through geologic time. Of more than 3.0×10^{22} g of O_2 released, 96% is contained in sedimentary rocks, beginning with the occurrence of Banded Iron formations about 3.7 billion years ago. Although O_2 was released to the atmosphere beginning about 2.0 bya, it was consumed in terrestrial weathering processes to form Red Beds, so that the accumulation of O_2 to present levels in the atmosphere was delayed, beginning about 400 million years ago. Modified from Schidlowski (1980).

about 1% of present day (Berkner and Marshall 1965, Chapman and Schopf 1983). Fossil evidence of eukaryotic organisms is found in rocks of 1.3–2.0 billion years ago (Schopf and Oehler 1976, but see Knoll and Barghoorn 1975). The more efficient photosynthesis that is possible as a result of the evolution of chloroplasts in eukaryotic cells would have further enhanced the accumulation of atmospheric oxygen, and mitochondrial respiration (Krebs cycle) is more efficient than the same reactions performed by prokaryotes.

O_2 in the stratosphere is subject to photochemical reactions leading to the formation of ozone (Chapter 3). Stratospheric ozone provides an effective shield for much of the Sun's ultraviolet radiation that would otherwise reach the Earth's surface. As the ozone shield developed, higher organisms were able to colonize land. Thus, multicellular organisms are found in ocean sediments dating to about 680 million years ago, but the colonization of land by higher plants was delayed until the Silurian at about 400 million years ago (Gensel and Andrews 1987). Some early microbial colonists may have resembled the inhabitants of desert rocks of today (e.g., Dorn and Oberlander 1981, Friedmann 1982, Bell et al. 1986, Palmer et al. 1986), but there is no fossil record of their occurrence. The colonization of land by vascular plants may have also been coincident with the development of lignin (Lowry et al. 1980) and effective symbioses with mycorrhizal fungi that obtain phosphorus from unavailable forms in the soil (Pirozynski and Malloch 1975; Chapter 6).

Oxygen also allowed several new biochemical pathways of critical significance to the global cycles of biogeochemistry. Two forms of aerobic biochemistry constitute chemoautotrophy. One based on sulfur or H_2S,

$$2S + 2H_2O + 3O_2 \rightarrow 2SO_4^{2-} + 4H^+ \qquad (2.14)$$

is performed by species of *Thiobacilli* (Ralph 1979). The hydrogen generated is coupled to energy-producing reactions, including the fixation of CO_2 to organic matter. On the primitive Earth, these organisms could capitalize on elemental sulfur deposited from anaerobic photosynthesis, but today they are confined to local environments where elemental sulfur or H_2S is present, including some deep-sea hydrothermal vents (Jannasch and Wirsen 1979, Jannasch and Mottl 1985, K. S. Johnson et al. 1986).

More important are the reactions involving nitrogen transformations by *Nitrosomonas* and *Nitrobacter* bacteria:

$$2NH_4^+ + 3O_2 \rightarrow 2NO_2^- + H_2O + 4H^+ \qquad (2.15)$$

and

$$2NO_2^- + O_2 \rightarrow 2NO_3^- \qquad (2.16)$$

Again, the energy released from these nitrification reactions is coupled to the fixation of carbon. Nitrate produced by these reactions is highly soluble in water, and it is the dominant form of inorganic nitrogen delivered in river flow to the oceans (Chapter 8).

Today, an anaerobic, heterotrophic reaction called denitrification is performed by bacteria, commonly of the genus *Pseudomonas*, found in soils and wet sediments (Knowles 1982). Although the denitrification reaction

$$5CH_2O + 4H^+ + 4NO_3^- \rightarrow 2N_2 + 5CO_2 + 7H_2O \qquad (2.17)$$

requires anaerobic environments, denitrifiers are only facultatively anaerobic. Several lines of evidence suggest that denitrification may have appeared later than the methanogenesis and sulfate-reduction pathways described earlier (Betlach l982). Most denitrifiers such as *Pseudomonas* are found among the eubacteria, which appear more advanced than Archaebacteria. Although the denitrification pathway is inactive in the presence of O_2, denitrifying organisms switch to aerobic respiration when O_2 is present, reflecting adaptation to environments in which O_2 is present during some periods. In some species the denitrification enzymes appear tolerant of low concentrations of O_2 (Bonin et al. 1989). Denitrification would have been efficient only after relatively large concentrations of NO_3 had accumulated in the primitive ocean, which is likely to have contained low NO_3 at the start (Kasting and Walker 1981). Thus, as a biochemical pathway, denitrification depends indirectly on oxygen-evolving photosynthesis, which can provide O_2 for the nitrification reactions.

The various oxidation and reduction reactions outlined in this chapter show the diversity of biochemical pathways that is possible among habitats that differ in the availability of oxygen (Fig. 2.7). Many of these reactions depend on the products of other reactions. Coupling of reducing heterotrophic pathways, in which N_2 and H_2S are produced as gases, to oxidizing pathways that yield NO_3^- and SO_4^{2-}, allows a global circulation of these elements from reduced, organic form to oxidized forms available for uptake (Chapters 12 & 13).

Comparative Planetary History: Earth, Mars, and Venus

In the release of free O_2 to the atmosphere, life has profoundly affected the conditions on the surface of the Earth. But, what might have been the conditions on Earth in the absence of life? Some indication is given by the neighboring planets, Mars, and Venus, which are our best replicates for the biogeochemical arena on Earth. We are fairly confident that there has never been life on these planets, so their surface composition represents

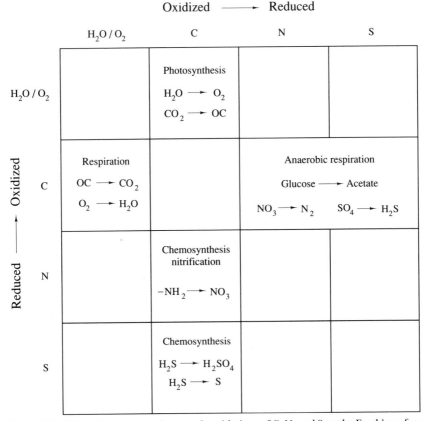

Figure 2.7   Metabolic pathways that couple oxidations of C, N, and S on the Earth's surface. For each pathway, the constituent at the top is transformed from an oxidized form obtained from the environment to a reduced form, released to the environment. At the same time, the constituent at left is transformed from a reduced form to an oxidized form. From Schlesinger (1989).

the cumulative effect of 4.6 billion years of abiotic processes (Walker 1977).

Table 2.3 compares a number of properties and conditions on Earth, Mars, and Venus. Two properties characterize the atmosphere of these planets: the total mass (or pressure) and the proportional abundance of the constituents. Overall, we might expect a less massive atmosphere on Mars than on Earth, since the gravitational field is weaker on a smaller planet. Mars probably began with a smaller allocation of primary gases during planetary formation, and we might expect that a small planet would have less internal heat to drive crustal outgassing subsequent to its origin (Anders and Owen 1977, Owen and Biemann 1976). Indeed, the present atmosphere on Mars is only about 0.76% as massive as that on

Table 2.3 Some Characteristics of the Inner Planets

	Mars[a]	Earth	Venus[b]
Distance to the sun (10^6 km)	228	150	108
Surface temperature (°c)	−53	16	474
Radius (km)	3390	6371	6049
Atmospheric pressure (bars)	0.007	1	92
Atmospheric mass (g)	2.4×10^{19}	5.3×10^{21}	5.3×10^{23}
Atmospheric composition (% wt.)			
CO_2	95	0.035	98
N_2	2.5	78	2
O_2	0.25	21	0
H_2O	0.10	1	0.05

[a]From Owen and Biemann (1976).
[b]From Nozette and Lewis (1982).

Earth (Hess et al. 1976). We might also expect that the surface temperature on Mars would be colder than on Earth, since the planet is much farther from the Sun. The average temperature on Mars, −53°C (Kieffer 1976), assures that water is frozen in the soil year-round. In the absence of liquid water, one might expect that the atmosphere on Mars would be mostly dominated by CO_2, which is mainly dissolved in seawater on Earth. Indeed, the Martian atmosphere is mostly dominated by CO_2, and the observed seasonal fluctuations in the ice cap appear wholly due to seasonal variations in the amount of CO_2 that is frozen out of its atmosphere.

Several attributes of Mars are anomalous. First, with most of the water and CO_2 trapped on the surface, why is N_2 such a minor component of the atmosphere on Mars? Second, why do the surface conditions on Mars indicate a period when liquid water was most certainly present on its surface (Carr 1987)?

While there is no evidence for tectonic activity at present, the ratio of $^{40}Ar/^{36}Ar$ of approximately 2750 on Mars suggests that there was significant crustal degassing in the past (Owen and Biemann 1976). A more massive early atmosphere may once have allowed a significant "greenhouse effect" (Fig. 2.4), and warmer surface temperatures than today. If these observations are correct, why did Mars lose most of its early complement of CO_2 and other gases?

An understanding of the long-term cycle of carbon on the Earth may help to explain the possible evolution of conditions on Mars. On Earth, CO_2 released from crustal degassing in volcanos reacts with surface minerals in carbonation weathering (Chapter 4), and rivers carry the dissolved ions to the ocean (Fig. 2.8). In the oceans, calcium carbonate is deposited in marine sediments, which in time are subducted into the upper mantle. Here, the sediments are metamorphosed; calcium is con-

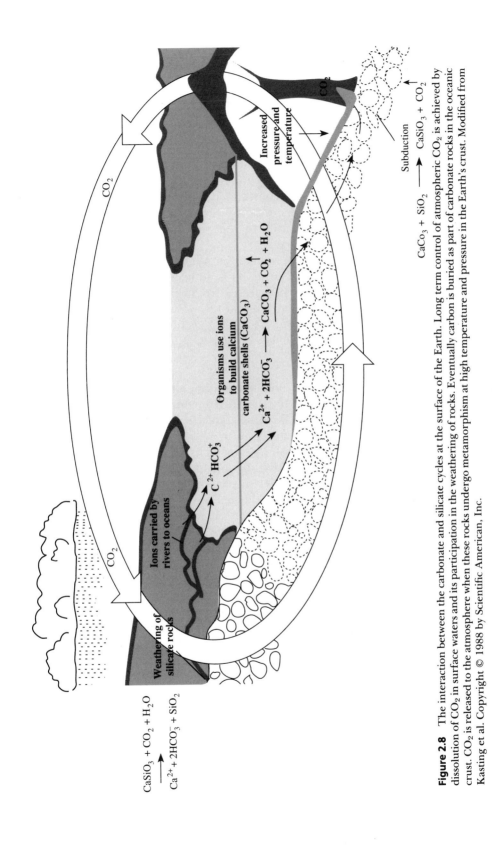

Figure 2.8 The interaction between the carbonate and silicate cycles at the surface of the Earth. Long term control of atmospheric CO_2 is achieved by dissolution of CO_2 in surface waters and its participation in the weathering of rocks. Eventually carbon is buried as part of carbonate rocks in the oceanic crust. CO_2 is released to the atmosphere when these rocks undergo metamorphism at high temperature and pressure in the Earth's crust. Modified from Kasting et al. Copyright © 1988 by Scientific American, Inc.

verted back to the primary minerals of silicate rocks, and the carbon is released in volcanic emissions at the surface. On Earth the oceanic crust appears to circulate though this pathway in about 110–170 million years (Li 1972, Howell and Murray 1986). Variations in the rate of tectonic activity may have caused portions of this cycle to operate at greater or lesser rates in the geologic past (Berner et al. 1983); indeed, CO_2 released during extensive tectonic activity may explain global warming in the Late Cretaceous and Eocene (Berner et al. 1983, Owen and Rea 1985). However, over the course of the history of the Earth, carbon has continually cycled from atmospheric CO_2, to sediments, to metamorphosed rocks, and back to the atmosphere (Kasting et al. 1988).

One possible mechanism for cooling on Mars would result from the rapid loss of internal heat and cessation tectonic activity on that planet. Carbon dioxide would cease to be released from volcanic emissions, while initial contents in the atmosphere would be consumed in reaction with crustal minerals. Mars would slowly lose its "greenhouse" warming as the absolute content of CO_2 declined, despite CO_2 remaining the major component of its atmosphere.

Losses of other atmospheric gases from Mars may have resulted from several mechanisms. A large amount of water undoubtedly resides in permafrost, but this is difficult to estimate from our space explorations to date. Some water may have been lost to space as a result of catastrophic impacts early in the evolution of Mars (Carr l987). Loss of hydrogen from Mars may have occurred as water vapor in its atmosphere underwent photolysis by ultraviolet light. Since the loss of ^1H would be more rapid than that of ^2H, we would expect this mechanism to leave a greater proportion of ^2H_2O on the present Martian surface. Recently Owen et al. (1988) found that the ratio of ^2H (deuterium) to ^1H on Mars is much higher than on Earth, suggesting that Mars may have once possessed a large inventory of water that is now lost to space.

Loss during catastrophic impacts may also explain the low abundance of nitrogen, but this gas may have also been lost as N_2 underwent photo-disassociation in the upper atmosphere, forming monomeric N. This process occurs on Earth as well, but N is too heavy to escape the Earth's gravitational field (Jeans escape), and quickly recombines to form N_2. With its smaller size, Mars allows the Jeans escape of N. Relative to the Earth, a higher proportion of $^{15}N_2$ in the Martian atmosphere is suggestive of this process, since the Jeans escape of ^{15}N would be slower than that of ^{14}N, which has a lower atomic weight (McElroy et al. 1976). In sum, various lines of evidence suggest that Mars had a higher inventory of volatiles early in its history, but most of the atmosphere is frozen or has been lost to space or to reactions with crustal minerals. Although the thin atmosphere that remains is dominated by CO_2, it offers little greenhouse warming.

Compared to Earth, Venus is a hot planet, yet its surface temperature of 474°C is much greater than one would expect based on its proximity to the Sun. In the absence of a liquid ocean, the entire inventory of volatiles on Venus should reside in its atmosphere. Indeed, the atmospheric pressure on Venus is about 100 × times that of Earth (Table 2.3). The total volatiles on Earth include those now contained in the atmosphere, the oceans, and certain crustal minerals (Table 2.1). The total mass of volatiles relative to the mass of the Earth is similar to the ratio between the mass of the atmosphere on Venus to that planet's mass (Oyama et al. 1979). This implies a similar degree of crustal degassing on these planets. The massive atmosphere on Venus is dominated by CO_2, conferring a large greenhouse warming and surface temperatures well in excess of those predicted based on the distance of the planet to the Sun (42°C). What is unusual about Venus is the low abundance of water in its atmosphere. Was Venus wet in the past?

The ratio of ^2H (deuterium) to ^1H on Venus is much higher than on Earth (Donahue et al. 1982, McElroy et al. 1982), suggesting that Venus, like Mars, may have possessed a large inventory of water in the past, but lost water through some process that acts differently on isotopes of different atomic weight (but see also Grinspoon 1987). Once again, analogous processes on Earth are instructive. In the upper atmosphere on Earth, small amounts of water vapor are subject to photodisassociation, with the loss of H_2 by Jeans escape. However, because the upper atmosphere is cold, little water vapor is present, and the process has been minor throughout the Earth's history. With the warmer initial conditions on Venus, a greater amount of the water vapor may have been subject to photodisassociation, and the planet has dried out through its history (Kasting et al. 1988). Oxygen released during the photodisassociation of water would react with crustal minerals (Donahue et al. 1982), for example,

$$2FeO + O_2 \rightarrow Fe_2O_3 \qquad (2.18)$$

As the planet has dried, continuing volcanic releases of CO_2 have accumulated in the atmosphere to produce a runaway greenhouse effect, in which increasing temperatures allow an increasing atmospheric vapor pressure of CO_2 (Fig. 2.9). At the surface temperatures found on Venus, little CO_2 can react with crustal minerals, maintaining a large CO_2 concentration in the atmosphere (Nozette and Lewis 1982). The CO_2/N_2 ratio in the atmosphere is similar to that in the total inventory of volatiles on Earth (Oyama et al. 1979). Various other gases, such as SO_2, that are

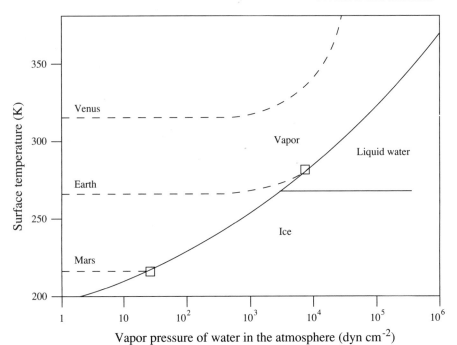

Figure 2.9 The dashed lines show how the surface temperatures on Mars, Earth, and Venus may have increased, due to the greenhouse effect, as water vapor and other gases accumulated in the atmosphere. On Mars and Earth the increase was terminated when the water vapor pressure in the atmosphere reached the saturated vapor pressure, shown as the solid line, and freezing or condensation occurred. On Venus, temperatures are higher because it is closer to the Sun and saturation was not achieved. All gases released to the atmosphere from volcanoes on Venus remain in the atmosphere, where they produce a large greenhouse effect that has increased through time. Modified from Walker (1977).

found dissolved in seawater on Earth, also reside as gases in the atmosphere on Venus.

Our look at the neighboring planets suggests that many of the conditions on Earth were set by the abiotic factors of its size and distance to the Sun. Water released by crustal degassing condensed on the surface as a liquid, and CO_2 followed by its high solubility in water. The atmosphere that remained has probably always been dominated by N_2, as it is today. However, higher early concentrations of N_2 have now been partially diluted by O_2 released from biotic activities. On a short time scale, atmospheric CO_2 is affected by biotic activities (e.g., Fig. 1.3), but on a longer time scale the dissolution of CO_2 in seawater and reaction of CO_2 with crustal materials appear to control its concentration in the atmosphere (Fig. 2.8). Variations in atmospheric CO_2 have been limited; without the present greenhouse effect the surface temperature on Earth would be 255 K, and the planet would be covered in ice (Ramanathan 1988).

Lightning bolts allow the combination of N_2 and O_2 in the atmosphere, eventually leading to the formation of NO_3^-. It is tempting to conclude that without denitrifying organisms, much of the atmospheric N_2 would have been converted to NO_3, which would be found as an extremely stable ion in seawater (Sillén 1966). However, without life there would also be no O_2, and the rate of such nitrogen fixation on an anaerobic Earth would be much lower than today (Kasting and Walker 1981). Denitrifying organisms are also important to the stability of present-day N_2, but in the absence of all life concentrations of atmospheric N_2 would probably be similar to those of today (Walker 1977, 1984).

Summary

In this chapter we have reviewed theories for the formation and differentiation of the early Earth. In the process of planetary formation, certain elements were concentrated near the surface and in forms that were readily soluble in seawater. Thus, the geochemical environment in which life arose is a special mix of the geochemical abundance of elements in the universe. Simple organic molecules can be produced by physical processes in the laboratory; presumably similar reactions occurred on the primitive Earth. Life may have arisen by the abiotic assembly of these constituents into simple forms, not unlike the most primitive bacteria that we know of today. Essential to living systems is the processing of energy, which is likely to have begun with the heterotrophic consumption of molecules found in the environment and later led to the autotrophic production of energy by various pathways, including photosynthesis. Autotrophic photosynthesis appears to be uniquely responsible for the production of O_2, which has accumulated in the Earth's atmosphere in the last 600 million years. Major biogeochemical pathways on Earth are comprised of reactions that couple oxidation and reduction.

Recommended Reading

Holland, H. D. 1984. The Chemical Evolution of the Atmosphere and Oceans. Princeton University Press, Princeton, New Jersey.

Schopf, J. W. (ed.). 1983. The Earth's Earliest Biosphere. Princeton University Press, Princeton, New Jersey.

3

The Atmosphere

Introduction

There are several reasons to begin our treatment of biogeochemistry with a consideration of the atmosphere. The atmosphere has evolved as a result of the history of life on Earth (Chapter 2), and there is good evidence that it is changing as a result of current human activities (Rasmussen and Khalil 1986). The atmosphere controls climate and ultimately determines the environment in which we live (Dickinson and Cicerone 1986, Ramanathan 1988). Further, the atmosphere is relatively well mixed, so changes in its composition can be taken as a first index of changes in biogeochemical processes at the global level. The circulation of the atmosphere transports biogeochemical constituents between land and sea, resulting in a global circulation of elements.

We will begin our discussion with a brief consideration of the structure and composition of the present-day atmosphere. We will then examine reactions that occur among atmospheric constituents, especially those of biotic origin. Many of these reactions result in the removal of constituents from the atmosphere to the land and sea, and we will treat the removal processes in some detail. In the face of constant losses, an equilibrium composition is maintained by other processes that release these constituents to the atmosphere. While we will mention the sources of atmospheric gases here briefly, they will be treated in more detail in later chapters of this book, especially as we examine the microbial reactions that occur in soils, wetlands, and ocean sediments.

Structure and Composition

The atmosphere is held on the Earth's surface by the gravitational attraction of the Earth. The mass (M) and pressure (P) of the atmosphere are thus related:

$$P = M (g) \tag{3.1}$$

where g is the acceleration due to gravity (980 cm s^{-2} at sea level). Since the acceleration due to gravity declines with increasing distance from the center of the Earth, we would expect the pressure to decline with increasing altitude. Thus, we say that the atmosphere is "thinner" at higher altitudes, and modern jet aircraft require cabin pressurization. The decline in atmospheric pressure (in atmospheres) with altitude (H in km) is approximated by the logarithmic relation

$$\log P = -0.06(H) \tag{3.2}$$

over the whole atmosphere (Garrels et al. 1975), but the decline appears nearly linear near the surface (Fig. 3.1). The lower atmosphere, the troposphere, contains about 80% of the atmospheric mass (Warneck 1988) and is the area of greatest biogeochemical interest.

Although certain atmospheric constituents, such as ozone, absorb portions of the solar radiation that the Earth receives, most radiation penetrates the atmosphere and is absorbed by the Earth's surface. The land and ocean surfaces reradiate longwave (heat) radiation to the atmosphere (Fig. 2.4), so the atmosphere is heated from the bottom, and the troposphere is warmest at the Earth's surface (Fig. 3.2). Since warm air is less dense and rises, the troposphere is well mixed. The top of the troposphere extends to 10–15 km, varying seasonally and with latitude. Above the troposphere, the stratosphere is defined by a zone in which temperatures increase with altitude, extending to about 50 km. The increase is largely due to the absorption of ultraviolet light by ozone. Vertical mixing in the stratosphere is limited, as well as exchange across the boundary

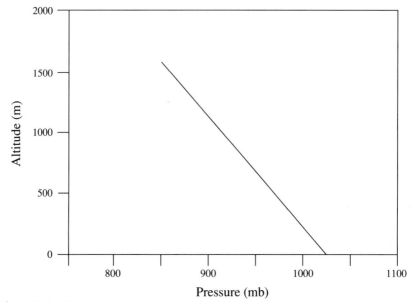

Figure 3.1 Measurements of atmospheric pressure near the surface of the Earth on November 20, 1985, using a LIDAR sounder based on NASA aircraft at 2790 m altitude.

between the troposphere and stratosphere, the tropopause. Most of our considerations of biogeochemistry will be confined to tropospheric reactions, but reactions in the stratosphere are critical to the removal of some constituents, and materials that enter the stratosphere remain for long periods, allowing transport around the globe.

The thermal instability of the troposphere is largely responsible for the global patterns of atmospheric circulation (Fig. 3.3a,b). The large annual receipt of solar energy at the equator causes warming of the atmosphere (sensible heat) and the evaporation of large amounts of water, carrying latent heat, from tropical oceans and rain forests. As this warm, moist air rises, it cools, producing a large amount of precipitation in equatorial areas. Having lost its moisture, the rising air mass moves both north and south, away from the equator. In a belt centered on approximately 30°N or S latitude, these dry air masses sink to the Earth's surface, undergoing compressional heating. It is not surprising that this latitudinal belt is the location of most of the world's major deserts. A similar, but much weaker, circulation pattern is found at the poles, where cold air sinks and moves north or south along the Earth's surface to lower latitudes. Known as direct Hadley cells, the tropical and polar circulation patterns drive an indirect circulation between 40 and 50° latitude, producing the cyclonic storm systems and the prevailing west winds that we experience in the temperate zone. As air masses move across different latitudes, they are deflected to the right by the Coriolis force, which arises because of the different speed of the Earth's rotation at different latitudes (Fig. 3.3c).

Exchange between the troposphere and the stratosphere is driven by several processes (Warneck 1988). First, the height of the tropopause varies seasonally, especially in the direct Hadley cells. When the height of the tropopause changes, tropospheric air enters the stratosphere, or

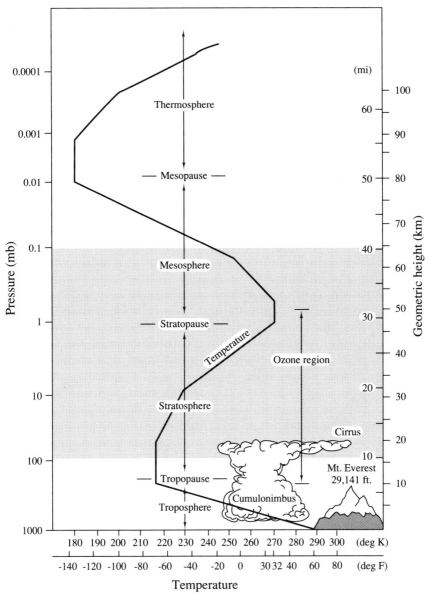

Figure 3.2 Temperature profile of the atmosphere to 100 km.

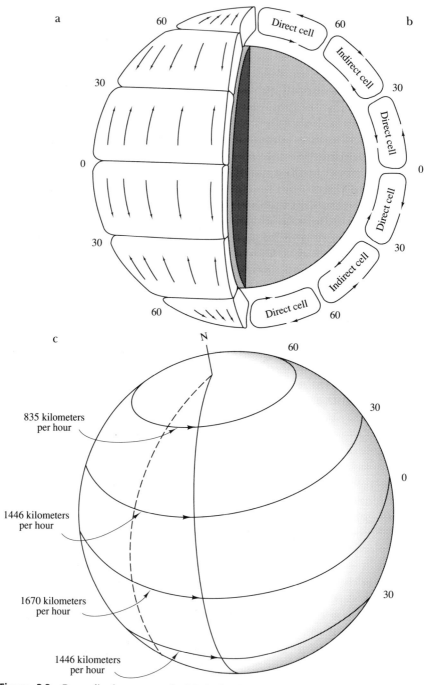

a

60

Direct cell

60

b

30

Indirect cell

30

Direct cell

0

Direct cell

0

Direct cell

30

30

Indirect cell

Direct cell

60

60

c

N

60

835 kilometers
per hour

30

1446 kilometers
per hour

0

1670 kilometers
per hour

30

1446 kilometers
per hour

Figure 3.3 Generalized pattern of global circulation showing (a) surface patterns, (b) vertical patterns, and (c) origin of the Coriolis force. Note the change in speed of the Earth's surface moving in an eastward direction going from the equator to either pole. A rocket moving from the equator to the North Pole would appear to be deflected to the right—the Coriolis effect. From Oort (copyright (c) 1970 by Scientific American, Inc.) and Gross (1982).

Table 3.1 Composition of the Atmosphere[a]

Constituent	Chemical Formula	Molecular Weight ($^{12}C=12$)	Percent by Volume in Dry Air	Total Mass (g)
Total atmosphere				5.136×10^{21}
Water vapor	H_2O	18.01534	variable	0.017×10^{21}
Dry air		28.9644	100.0	5.119×10^{21}
Nitrogen	N_2	28.0134	78.084	3.866×10^{21}
Oxygen	O_2	31.9988	20.948	1.185×10^{21}
Argon	Ar	39.948	0.934	6.59×10^{19}
Carbon dioxide	CO_2	44.00995	0.0315	2.45×10^{18}
Neon	Ne	20.183	1.818×10^{-3}	6.48×10^{16}
Helium	He	4.0026	5.24×10^{-4}	3.71×10^{15}
Methane	CH_4	16.04303	$\sim 1.5 \times 10^{-4}$	$\sim 4.3 \times 10^{15}$
Hydrogen	H_2	2.01594	$\sim 5 \times 10^{-5}$	$\sim 1.8 \times 10^{14}$
Nitrous oxide	N_2O	44.0128	$\sim 3 \times 10^{-5}$	$\sim 2.3 \times 10^{15}$
Carbon monoxide	CO	28.0106	$\sim 1.2 \times 10^{-5}$	$\sim 5.9 \times 10^{14}$
Ammonia	NH_3	17.0306	$\sim 1 \times 10^{-6}$	$\sim 3 \times 10^{13}$
Nitrogen dioxide	NO_2	46.0055	$\sim 1 \times 10^{-7}$	$\sim 8.1 \times 10^{12}$
Sulfur dioxide	SO_2	64.063	$\sim 2 \times 10^{-8}$	$\sim 2.3 \times 10^{12}$
Hydrogen sulfide	H_2S	34.080	$\sim 2 \times 10^{-8}$	$\sim 1.2 \times 10^{12}$
Ozone	O_3	47.9982	Variable	$\sim 3.3 \times 10^{15}$

[a] From Walker (1977).

vice versa. Second, rising air masses, particularly in the tropical Hadley cell, carry tropospheric air to the stratosphere. Third, there is exchange across the tropopause due to large-scale wind movements, thunderstorms (Dickerson et al. 1987), and eddy diffusion (Warneck 1988).

Table 3.1 gives the globally averaged composition of the atmosphere. While the concentration of nitrogen and oxygen are nearly invariant, the concentration of other constituents can be expected to vary in space and time. We might expect the concentration of pollutant constituents (ozone, carbon monoxide, etc.) to be especially high over cities, and the concentration of some reduced gases (methane and H_2S) to be high over swamps and other areas of anaerobic decomposition (e.g., Harriss et al. 1982, Steudler and Peterson 1985). However, most of these gases are highly reactive, so winds mix their concentrations to low mean tropospheric values within a short distance downwind of point sources. For carbon dioxide, daytime concentrations near the ground may be strongly depleted as a result of photosynthesis, while at night higher concentrations of CO_2 may accumulate under a forest canopy as a result of plant and soil respiration (Woodwell and Dykeman 1966, Reiners and Anderson 1968, Wofsy et al. 1988). This oscillation mirrors that seen on an annual time scale for concentrations in the northern hemisphere (Fig. 1.3). Nevertheless, when averaged over long periods, the concentration of CO_2 is similar throughout the atmosphere. We can best perceive changes in atmo-

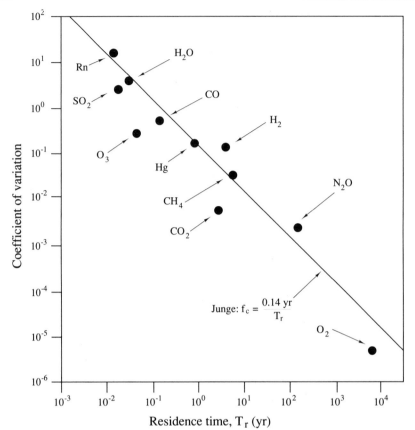

Figure 3.4 Variability in the concentration of atmospheric gases (expressed as the coefficient of variation in measurements) as a function of their estimated mean residence times in the atmosphere. From Junge (1974), as updated by Slinn (1988).

spheric composition, such as the current increase in CO_2, by averaging long-term measurements made in locations remote from known sources.

The concept of mean residence time is useful in considerations of mean atmospheric composition. For any biogeochemical reservoir, mean residence time (MRT) is defined as

$$MRT = \frac{Q}{dQ/dt} = Mass/flux, \qquad \text{Eq. 3.3}$$

where flux may be either the input or loss from the reservoir. For example, the average concentration of N_2O in the atmosphere is about 300 ppb (Warneck 1988). Multiplied by the mass of the atmosphere, we obtain 2.3×10^{15} g for the content of N_2O in the entire atmosphere. Our best estimate of the sources of N_2O suggest an annual production of at least 20×10^{12} g/yr, giving a mean residence time of over 100 yr for N_2O in the atmosphere (Cicerone, 1987, Chapter 12). With such a long residence time relative to mixing, this gas should be relatively evenly distributed within the atmosphere, showing higher concentrations only at strong point sources. In contrast, the average volume of water in the atmosphere is equivalent to ~13,000 km^3 at any time, or 25 mm above any point on the Earth's surface (Speidel and Agnew 1982). The average daily precipitation would be about 2.7 mm if it were deposited evenly around the globe. Thus, the mean residence time for water in the atmosphere is

$$\text{MRT} = 25 \text{ mm}/2.7 \text{ mm day}^{-1} = 9.3 \text{ days} \qquad (3.4)$$

This is a short time compared to the circulation of the tropospheric mass, so we would expect water vapor to show variable concentrations in space and time.

Junge (1974) related the variation in atmospheric concentration for various gases to their estimated mean residence time in the atmosphere (Fig. 3.4). Gases that have short mean residence times are highly variable from place to place, whereas those that have long mean residence times are well mixed. Mean residence time is inversely related to fractional turnover; approximately 11% of the atmospheric content of water vapor is removed each day by rainfall (viz. $k = 1/9.3$ days $= 0.11$ day^{-1}).

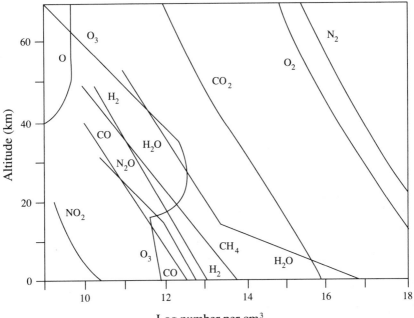

Figure 3.5 The approximate altitudinal distribution of atmospheric constituents. From Walker (1977).

Exchange between the troposphere of the northern and southern hemispheres and exchange between the troposphere and the stratosphere have been examined by following the fate of industrial pollutants released to the troposphere and radioactive contaminants released to the stratosphere during the testing of atomic weapons during the 1950s and early 1960s. These studies suggest that there is nearly complete mixing between the northern and southern tropospheric air masses each year. Exchange between the troposphere and the stratosphere accounts for 0.69–0.82 of the stratospheric mass each year, for a mean residence time of 1.2–1.4 yr for stratospheric air (Warneck 1988). These exchange times are short compared to the mean residence time for most gases, so the proportional composition of the atmosphere shows little variation in major constituents with altitude (Fig. 3.5), except that the stratosphere is too cold to contain a large amount of water vapor.

Aerosols

In addition to gaseous components, the atmosphere contains particles, known as aerosols, that arise from a variety of sources. Volcanic eruptions disperse finely divided rock materials over large areas (Table 3.2) (Fruchter et al. 1980), and soil minerals are dispersed by wind erosion (deflation weathering) from arid and semi-arid regions (Goudie 1978, Pye 1987). Particles with a diameter <1.0 μm are held aloft by Brownian motion and subject to long-range transport. Schütz (1980) notes that soil particles from the Sahara form a major component of ocean sediments in the North Atlantic. Dust from the deserts of central and eastern Asia is also observed in the Pacific ocean (Duce et al. 1980). Schütz (1980) estimates that 1×10^{15} g yr^{-1} of soil particles enter the atmosphere from arid regions; about 20% of these particles are involved in long-range transport.

An enormous quantity of particles enters the atmosphere from the ocean, as a result of tiny droplets that become airborne with the bursting of bubbles at the surface (MacIntyre 1974, Wu 1981). As the water evaporates from these bubbles, the salts crystalize to form seasalt aerosols, which carry the approximate chemical composition of seawater (Glass and Matteson 1973). As for soil dust, most of the seasalt aerosols are relatively large and settle from the atmosphere quickly, but a significant proportion remains in the atmosphere for global transport.

Smaller particles are produced by reactions between gases; for instance, particles rich in $(NH_4)_2SO_4$ are produced by the reaction of atmospheric NH_3 with derivatives of SO_2 (Quinn et al. 1987). Dimethylsulfide released from the ocean is also oxidized to form aerosols rich in SO_4 (Chapter 9). Organic particles result from the condensation of volatile hydrocarbons released in the smoke of forest fires (Hahn 1980). Hidy and Brock (1971)

Table 3.2 Composition of an Airborne Particulate Sample Collected during the Eruption of Mt. St. Helens on May 19, 1980.[a]

Constituent	Particulate Sample	Average Ash
Major elements (percent)		
SiO_2	≡65.0	65.0
Fe_2O_3	6.7	4.81
CaO	3.0	4.94
K_2O	2.0	1.47
TiO_2	0.42	0.69
MnO	0.054	0.077
Trace elements (ppm)		
S	3220	940
Cl	1190	660
Ni	<20	15
Cu	61	36
Zn	34	53
Ga	<8	18
As	22	~2.8
Se	<7	<1
Br	<8	~1
Rb	<17	32
Sr	285	460
Zr	142	170
Pb	36	8.7

[a] Average ash is shown for comparison. From Fruchter et al. (1980).

estimate that the natural sources of aerosols have been enhanced by up to 10% by the particles released from human activities.

Small particles are much more numerous in the atmosphere than large particles, but it is the large particles that contribute the most to the total airborne mass (Warneck 1988). The mass of aerosols declines with increasing altitude from values ranging from 1 to 50 $\mu g \ m^{-3}$ near unpolluted regions of the Earth's surface. While there is an inverse relation between the size of particles and their persistence in the atmosphere, the overall mean residence time for tropospheric aerosols is about 5 days (Warneck 1988). Aerosols that are transported to the stratosphere by volcanic eruptions persist for several years and undergo global transport.

The composition of the tropospheric aerosol varies greatly depending upon the proximity of continental, maritime, or anthropogenic sources. Wind-borne soil particles and particles of organic matter contribute to the global cycles of many trace elements (Nriagu 1989) and to the delivery of trace micronutrients to the sea (Chapter 9). Over the ocean, the composition of aerosols is a mixture of contributions from silicate minerals of continental origin and seasalt from the ocean (Andreae et al. 1986). Various workers have used ratios among elemental constituents to deduce the relative contribution of different sources (e.g., Moyers et al. 1977, Rahn and Lowenthal 1984, Shaw 1987).

Aerosols are important to global reflectivity (albedo), in reactions with atmospheric gases, and as nuclei for the condensation of raindrops. Raindrops are formed when water vapor begins to condense on aerosols >0.1 μm diameter. As raindrops enlarge and fall to the ground, they collide with other particles and absorb atmospheric gases. While the soil dusts comprising continental aerosols may contain a large portion of insoluble materials, seasalt aerosols and those derived from pollution sources are likely to dissolve readily in water and contribute strongly to the dissolved content of rainfall. Reactions of atmospheric gases with aerosols or raindrops are known as *heterogeneous* gas reactions. Such reactions are responsible for the ultimate removal of many reactive gases from the atmosphere.

Biogeochemical Reactions in the Atmosphere

Major Constituents

It is perhaps not surprising that the major constituents of the atmosphere, N_2, O_2, CO_2, and Ar, are relatively unreactive, all showing nearly uniform concentrations and relatively long mean residence times. From a biogeochemical perspective, N_2 is practically inert; living systems can only assimilate nitrogen that is found in other, "fixed," forms, such as NH_4 and NO_3. Despite the abundance of N_2 in the atmosphere, nitrogen is one of the primary elements that limits the growth of plants on land and in the oceans (Delwiche 1970). Natural fixation occurs by the reaction of N_2 and O_2 in lightning bolts, but the estimated rates ($10–30 \times 10^{12}$ g N/yr; Levine et al. 1984) are too low to account for a significant turnover of N_2 in the atmosphere. Certain species of bacteria and higher plants conduct biological nitrogen fixation (Chapters 2 and 6), but the total global rate is poorly known, since estimates require the extrapolation of variable small-scale measurements to large areas of the Earth's surface. Total nitrogen fixation is not likely to exceed 300×10^{12} g N/yr, and is likely to have been half that amount before the widespread production of nitrogen fertilizers and use of automotive transport (Chapter 12). Thus, the mean residence time for N_2 is over 20 million years. Only argon and the other noble gases are less dynamic.

In Chapter 2 we discussed the accumulation of O_2 in the atmosphere during the evolution of life on Earth. The atmosphere contains only a small portion of the total O_2 released by photosynthesis through geologic time. Nevertheless, the content in the atmosphere today is not determined by the current rate of photosynthesis on land and in the sea. The instantaneous combustion of all the organic matter now stored on land would reduce the atmospheric oxygen content by only 0.03% (Chapter 5). Walker (1977) suggests that atmospheric O_2 controls the storage of reduced carbon, not vice versa.

The accumulation of O_2 is the result of the long-term burial of reduced carbon in ocean sediments (Berner 1982). This organic matter is largely derived from photosynthesis in the sea, because the transport of organic carbon in the world's rivers is very small (Schlesinger and Melack 1981). The rate of burial is determined by the area of the ocean floor that is subject to anoxic conditions in organic sediments (Walker, 1977, 1980). Since that area varies inversely with the concentration of atmospheric O_2, the balance between the burial of organic matter and its oxidation maintains O_2 at a steady-state concentration of about 20% (Chapters 9 and 11).

A large amount of O_2 has been consumed in weathering of reduced crustal minerals through geologic time (Fig. 2.6); the current rate of exposure of these minerals would consume all atmosphere oxygen in about 4 million years. However, the rate of exposure is not likely to vary greatly in response to changes in atmospheric O_2, so weathering is not the major factor controlling O_2 in the atmosphere. In sum, despite the potential reactivity of O_2, the rates of reaction with reduced compounds are rather slow, and O_2 is a rather stable component of the atmosphere. The mean residence time is on the order of 10,000 years (cf. Fig. 3.4).

Carbon dioxide in the atmosphere is affected by processes that operate at different time scales, including interaction with the silicate cycle (Fig. 2.8), dissolution in the oceans, and annual cycles of photosynthesis and respiration (Fig. 1.3). The relative effect of these processes is treated in detail in Chapter 11, which examines the global carbon cycle. Here, it is important to note that CO_2 is not reactive with other atmospheric constituents, and that its mean residence time of 3 yr (Olson et al. 1985) is largely determined by exchange with seawater. However, the role of land vegetation is not insignificant, accounting for the annual oscillation in atmospheric concentrations (Fig. 1.3). The seasonal oscillation in the southern hemisphere is reversed compared to that in the northern hemisphere, and the amplitude is lower reflecting the smaller land area in the southern hemisphere (Fig. 3.6). The best estimates of net primary production on land (60×10^{15} g C/yr; Chapter 5) would suggest a mean residence time of about 13 years before a hypothetical molecule of CO_2 is stored in photosynthesis by land plants. The current increase in atmospheric CO_2 is caused by the combustion of fossil fuels and destruction of land vegetation (Houghton et al. 1983). When these processes attenuate, atmospheric CO_2 will come into equilibrium, and nearly all of the CO_2 released

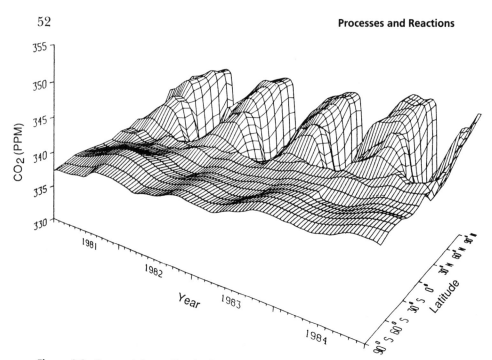

Figure 3.6 Seasonal fluctuation in the concentration of atmospheric CO_2 (1981–1984), shown as a function of 10° latitudinal belts (Conway et al. 1988). Note smaller amplitude of the fluctuations in the southern hemisphere, reaching peak concentrations during northern hemisphere minima.

will reside in the ocean. In the meantime, higher concentrations of CO_2 are likely to cause significant atmospheric warming through the "greenhouse effect" (Fig. 2.4), which is described in more detail below.

Biogenic Gases in the Troposphere

Volcanoes are the original source of volatile elements for the Earth's atmosphere (Chapter 2), and a likely source of some of the reduced gases (H_2S, H_2, NH_3) that are found in the atmosphere today. These and other trace gases [e.g., CH_4, $(CH_3)_2S$, NO_2] are found in concentrations well in excess of what is predicted under equilibrium geochemistry (Table 3.3). In most cases, the observed atmospheric concentrations are maintained by the action of living organisms, particularly microbes. Production of gases containing N and S, followed by their transport and removal from the atmosphere, contributes to the global cycling of these elements, which is certainly driven by biota (Deevey 1970b, Crutzen 1983). Unlike major atmospheric constituents, many of these gases are highly reactive, showing short mean residence times and variable concentrations in space and

Table 3.3 Actual and Equilibrium Partial Pressures and Concentrations of Nitrogen Compounds on the Earth's Surface[a]

Species	Environment	Actual Concentration	Equilibrium Concentration
N_2O	Atmosphere	0.33 ± 0.01 ppm	$10^{-12.7}$ ppm
NO	Atmosphere	0.001 ppm	$10^{-9.6}$ ppm
NO_2	Atmosphere	0.001 ppm	$10^{-3.8}$ ppm
NH_3	Atmosphere	0.006–0.020 ppm	$10^{-51.5}$ ppm
NO_3^-	Ocean water	0–30×10^{-6} mol/kg	$a_{NO3^-} = 10^{+5.7}$

[a] From Holland (1978).

time (Fig. 3.4). Losses from the troposphere are largely driven by oxidation reactions and removal of the products by rainfall. Currently the concentration of nearly all these constituents is increasing as a result of human activities, suggesting that humans are affecting biogeochemistry at the global level (Mooney et al. 1987).

Despite its abundance in the atmosphere, O_2 is too unreactive to oxidize trace reduced gases through direct reaction. However, through a variety of reactions driven by sunlight, small amounts of oxygen are converted to ozone (O_3), and further reactions yield hydroxyl radicals (OH). Ozone and OH are the primary species that oxidize many of the trace gases to CO_2, HNO_3, and H_2SO_4, which are removed from the atmosphere as dissolved constituents in rainfall. It is important to understand the natural production, occurrence and reactions of ozone in the atmosphere (Logan 1985). Nearly daily we read seemingly contradictory reports of the harmful effects of ozone depletion in the stratosphere (Bowman 1988) and harmful effects of ozone pollution in the troposphere (Environmental Protection Agency 1986). In each case, human activities are upsetting natural concentrations of ozone that are critical to atmospheric biogeochemistry and global climate.

Most ozone is produced by the reaction of sunlight with O_2 in the stratosphere, as described in the next section. Some of this ozone is transported to the Earth's surface by mixing of stratospheric and tropospheric air. However, the production of ozone in the smog of polluted cities such as Los Angeles alerted atmospheric chemists to reactions by which ozone is produced in the troposphere, by reactions involving NO_2 (Warneck, 1988). When NO_2 is present in the atmosphere, it is dissociated by sunlight,

$$NO_2 + h\nu \rightarrow NO + O \tag{3.5}$$

followed by a reaction producing ozone:

$$O + O_2 \rightarrow O_3 \tag{3.6}$$

This reaction sequence is an example of a *homogeneous* gas reaction, namely a reaction between atmospheric constituents that are all in the gaseous phase. The net reaction is:

$$NO_2 + O_2 \rightleftarrows NO + O_3 \tag{3.7}$$

which is an equilibrium reaction, so high concentrations of NO tend to drive the reaction backward. The concentration of O_3 is determined by

$$O_3 \text{ (ppm)} = 0.021 \text{ } [NO_2]/[NO] \tag{3.8}$$

Both NO_2 and NO are found in polluted air, in which they are derived from industrial and automotive emissions. The interrelation between NO and NO_2 is so strong that in many cases these gases are simply listed as NO_x. However, small concentrations of both constituents are found in the natural atmosphere, where they are derived from forest fires, lightning discharges, and microbial processes in the soil (Logan 1983; Chapter 6). Thus, the production of ozone from NO_2 is likely to occur throughout the troposphere, and the concentrations of tropospheric ozone have apparently increased as emissions have raised the tropospheric concentration of NO_2 globally (Volz and Kley 1988, Isaksen and Hov 1987).

Ozone is subject to photochemical reactions that yield hydroxyl radical,

$$O_3 + h\nu \rightarrow O_2 + O(^1D) \tag{3.9}$$

where $h\nu$ is ultraviolet light with wavelengths <310 nm and $O(^1D)$ is an excited atom of oxygen. Reaction of $O(^1D)$ with water yields hydroxyl radicals,

$$O(^1D) + H_2O \rightleftarrows 2OH \tag{3.10}$$

and OH radicals may further react to produce HO_2 and H_2O_2, which are also short-lived oxidizing radicals in the atmosphere (Walker 1977).

Hydroxyl radicals exist with a mean concentration of 7.7×10^5 cm^{-3} (Prinn et al. 1987), with higher concentrations in daylight (Platt et al. 1988) and at tropical latitudes, where the concentrations of water vapor are greatest (Hewitt and Harrison, 1985). Hydroxyl radicals are the major source of oxidizing power in the troposphere. In oxidation reactions, CH_4, CO, and other reduced gases are converted to CO_2, and ozone is produced as a byproduct. Some ozone may pass back through the reactions above to restore OH. The oxidation of carbon monoxide begins by reaction with hydroxyl radical and proceeds as follows:

$$CO + OH \rightarrow CO_2 + H \tag{3.11}$$

$$H + O_2 + M \rightarrow HO_2 + M \tag{3.12}$$

$$HO_2 + NO \rightarrow OH + NO_2 \tag{3.13}$$

$$NO_2 + h\nu \rightarrow NO + O \tag{3.14}$$

$$O + O_2 + M \rightarrow O_3 + M \tag{3.15}$$

where M represents a variety of possible catalytic molecules in the atmosphere (Warneck 1988). The net reaction is:

$$CO + 2O_2 \rightarrow CO_2 + O_3 \tag{3.16}$$

The oxidation of methane proceeds through a large number of steps, beginning by reaction with hydroxyl radicals:

$$CH_4 + OH \rightarrow CH_3 + H_2O \tag{3.17}$$

and yielding a net reaction of

$$CH_4 + 4O_2 \rightarrow HCHO + H_2O + 2O_3 \tag{3.18}$$

This reaction is the main atmospheric source of formaldehyde (HCHO), which is converted to CO and oxidized through the pathways described above. Hydrocarbons, including those released from natural vegetation (Greenberg and Zimmerman 1984, Lamb et al. 1987) are also oxidized through pathways involving CO oxidation (Zimmerman et al. 1978, Logan et al. 1981). This accounts for the high concentrations of carbon monoxide and ozone over forested regions of the Amazon Basin (Crutzen et al. 1985, Zimmerman et al. 1988) and the southeastern United States (Chameides et al. 1988).

The linkage among these biogeochemical reactions is complex (Ehhalt 1981). Higher concentrations of CO and CH_4 would be expected to lead to lower concentrations of OH in the atmosphere and higher concentrations of NO_2, which is also attacked by OH to form HNO_3. At the same time, higher concentrations of CO and CH_4 would be expected to lead to higher concentrations of O_3, from which OH is derived (Isaksen and Hov 1987, Logan 1985). Thus, Crutzen (1988) points out that depending on the concentration of NO, the oxidation of one molecule of CH_4 will consume 2–3.5 OH and 0–1.7 O_3 when NO is in low concentration and yield a net *gain* of 0.5 OH and 3.7 O_3 in polluted environments.

Currently, the tropospheric concentration of CO appears to be increasing by ~0.8–1.4%/yr (Khalil and Rasmussen 1988), presumably due to human activities, such as industrial combustion and biomass burning (Logan et al. 1981). Concentrations are higher in the northern hemisphere than the southern hemisphere (Seiler and Fishman, 1981). Greenberg et al. (1984) reported substantial releases of CO from fires used in land clearing in Brazil, with concentrations increasing by a factor of 2 during the dry season (Fishman and Browell 1988, Sachse et al. 1988). Some of the increase in CO may also come from the release of methane and hydrocarbons, from which CO is derived during their oxidation (Zimmerman et al. 1978, 1988). High concentrations of CO in the Amazon basin lead to high concentrations of O_3 by photochemical oxidation [equation (3.16)] (Crutzen et al. 1985). Ozone concentrations are highest at mid-day, and lowest at night when the forest serves as a sink for O_3 (Gregory et al. 1988, Kirchhoff 1988).

The cause of the current increase in methane (~1.0% yr^{-1}, Fig. 3.7) (Blake and Rowland 1988, Stauffer et al. 1985) is not obvious, since natural sources appear to dominate the annual production of ~500 × 10^{12} g/yr (Chapter 11). Oxidation of CH_4 and CO consumes nearly the entire annual production of OH radical in the troposphere. While the concentration of methane is much higher than that of carbon monoxide in the unpolluted atmosphere (Table 3.4), the reaction of OH

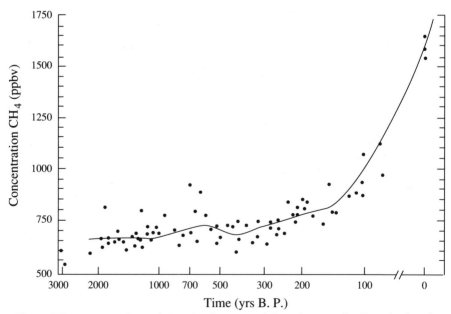

Figure 3.7 Concentrations of CH_4 in air extracted from ice cores in Greenland and Antarctica and from contemporary air samples. From Cicerone and Oremland (1988).

Table 3.4   Reaction of Atmospheric Trace Gases Containing, C, N, and S with OH[a]

Trace gas	Mixing Ratio in the Northern Hemisphere (ppb)	Mean Tropospheric Lifetime	Contribution of OH-sink Reaction (%)
CH_4	1600	10 yr	90
CO	250	60 days	100
Nonmethane hydrocarbons (NMHC), C_2–C_5	2–10	1–100 days	50–100
SO_2	0.2	14 days	50
COS	0.5	1 yr	30
H_2S		4 days	100
$(CH_3)_2S$		1 day	50
NO, NO_2	0.1	1 day	100
NH_3	~1	14 days	10
N_2O	310	150 yr	0

[a] Modified from Ehhalt (1981).

with CO is much faster. Part of the current increase in methane may be due to anthropogenic releases of CO that react with OH radicals previously available for the oxidation of methane (Khalil and Rasmussen 1985). The speed of reaction of CO with OH accounts for its short mean residence time in the atmosphere (Table 3.4). The mean residence time for methane is much longer, accounting for its more uniform distribution in the atmosphere (Fig. 3.4).

While the vast majority of the OH radical is involved in reactions with CO and CH_4, smaller quantities oxidize NO_2 and SO_2 in homogeneous gas reactions to form weak acids:

$$NO_2 + OH \rightarrow HNO_3 \tag{3.19}$$

$$SO_2 + OH \rightarrow HSO_3 \tag{3.20}$$

The reaction with NO_2 is very fast. The reaction with SO_2 is much slower, allowing long-distance transport of SO_2 as a pollutant in the atmosphere (Rodhe et al. 1981). Some SO_2 is absorbed directly by vegetation (Whelpdale and Shaw 1974, Lindberg et al. 1986) and at the surface of the ocean (Liss and Slater 1974, Beilke and Lamb 1974), but most SO_2 is eventually converted to SO_4^{2-} through a variety of heterogeneous reactions with water. For example,

$$SO_2 + H_2O \rightarrow H^+ + HSO_3^- \tag{3.21}$$

$$2HSO_3 + O_2 + M \rightarrow 2H^+ + 2SO_4^{2-} + M \tag{3.22}$$

Following these reactions HNO_3 and H_2SO_4 are removed from the atmosphere as dissolved constituents in rainfall. Hydrogen sulfide (H_2S) and dimethylsulfide ($(CH_3)_2S$), released from anaerobic soils (Chapter 7) and the ocean surface (Chapter 9), are also removed by complex reactions with OH eventually leading to the deposition of H_2SO_4 (Toon et al. 1987).

Atmospheric ammonia (NH_3) is derived from a variety of sources, of which releases from domestic animals and calcareous desert soils dominate the annual global production (Warneck 1988, Chapter 6). The solubility of NH_3 in water is rapid, and concentrations in the atmosphere fluctuate as a result of changes in relative humidity (Force et al. 1985). The major atmospheric sink is the heterogeneous reaction and removal in rainfall:

$$NH_3 + H_2O \rightleftarrows NH_4^+ + OH^- \qquad (3.23)$$

Ammonia may also react directly with sulfuric and nitric acids, forming ammonium aerosols that are deposited in rainfall and dry fallout:

$$2NH_3 + H_2SO_4 \rightarrow (NH_4)_2SO_4 \qquad (3.24)$$

Acting in this manner, NH_3 is a net source of alkalinity in the atmosphere.

Biogeochemical Reactions in the Stratosphere

Ozone is produced in the stratosphere by the disassociation of oxygen atoms exposed to short-wave solar radiation. The reaction accounts for most of the absorption of ultraviolet light ($h\nu$) at 180–240 nm wavelengths. The reaction proceeds as follows:

$$O_2 + h\nu \rightarrow O + O \qquad (3.25)$$

$$O + O_2 + M \rightarrow O_3 + M \qquad (3.26)$$

Ozone is subsequently destroyed by reaction with OH or by absorption of ultraviolet light at wavelengths between 200 and 320 nm:

$$O_3 + h\nu \rightarrow O_2 + O \qquad (3.27)$$

$$O + O_3 \rightarrow O_2 + O_2 \qquad (3.28)$$

These reactions account for the protection of the Earth's surface from the ultraviolet portion of the solar spectrum that is most damaging to living tissue. Absorption of ultraviolet radiation warms the stratosphere (Fig. 3.2), and the balance between these reactions maintains a steady-state concentration of O_3 of approximately 7×10^{18} molecules m^{-3}, peaking at 30 km altitude (Cicerone 1987). Ozone from the stratosphere accounts for about 75% of that consumed in the troposphere each year; production of O_3 in the troposphere is slower because less ultraviolet light is available (Cicerone 1987).

Although the photochemical production of O_3 is greatest at the equator, the density of the ozone layer is thickest at the poles (Cicerone 1987). Recent measurements suggest that the total density of ozone molecules in the atmospheric column has declined significantly in Antarctica (Fig. 3.8) and perhaps in other areas of the globe (Bowman 1988, Heath 1988). The decline is apparently unprecedented and represents a perturbation of global biogeochemistry. Destruction of ozone is likely to lead to an increased flux of ultraviolet radiation to the Earth's surface and to lower stratospheric temperatures that may alter global heat balance. Since previous, steady-state ozone concentrations were maintained in the face of natural photochemical reactions that produce and consume ozone, attention has focused on additional processes in which ozone is destroyed by reactions with atmospheric constituents of human origin (McElroy and Salawitch 1989).

Chlorofluorocarbons (freons) are produced as aerosol propellants, refrigerants, and solvents, and have no known natural source in the atmosphere (Prather 1985). These compounds are chemically inert in the troposphere, but they are transported to the stratosphere, where they are decomposed by photochemical reactions, producing chlorine (Cl). Chlorine acts as a catalyst to destroy ozone:

$$Cl + O_3 \rightarrow ClO + O_2 \tag{3.29}$$

$$O_3 + h\nu \rightarrow O + O_2 \tag{3.30}$$

$$ClO + O \rightarrow Cl + O_2 \tag{3.31}$$

for a net reaction of

$$2O_3 + h\nu \rightarrow 3O_2 \tag{3.32}$$

The reaction is greatly enhanced in the presence of ice particles, which may account for the early disappearance of O_3 over Antarctica (Molina et al. 1987, Tolbert et al. 1988). Eventually Cl is removed from the atmosphere as HCl. Similar reactions are possible with compounds containing bromine (Cicerone 1987, Singh et al. 1988, Barrett et al. 1988).

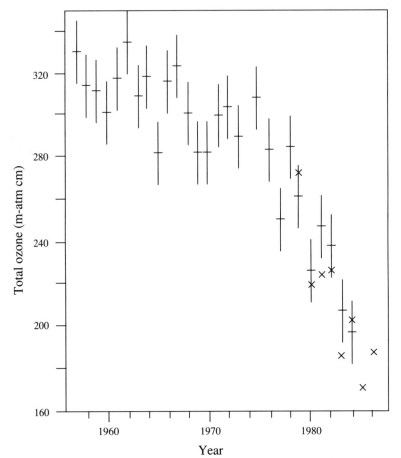

Figure 3.8 Decline in stratospheric O_3 over Antarctica during October 1957 through 1986. From Cicerone (1987). Copyright 1987 by the AAAS.

In addition to perturbation by anthropogenic chemicals, stratospheric ozone is destroyed in reaction with NO (Warneck 1988):

$$NO + O_3 \rightarrow NO_2 + O_2 \qquad (3.33)$$

$$O_3 + h\nu \rightarrow O + O_2 \qquad (3.34)$$

$$NO_2 + O \rightarrow NO + O_2 \qquad (3.35)$$

However, NO is so reactive that only a tiny portion of that produced from natural and human activities in the troposphere is transfered to the

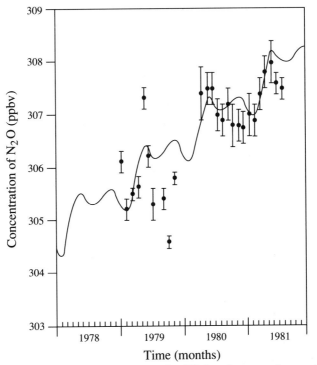

Figure 3.9 Increase and potential seasonal cycle of N_2O in the atmosphere on the coast of Oregon. From Khalil and Rasmussen (1983).

stratosphere. Stratospheric NO is produced during the degradation of nitrous oxide (N_2O), transported from the troposphere:

$$N_2O + h\nu \rightarrow N_2 + O(^1D) \tag{3.36}$$

$$N_2O + O(^1D) \rightarrow 2NO \tag{3.37}$$

Nitrous oxide is a natural constituent that is added to the atmosphere as a byproduct of nitrification and denitrification reactions in the soil (Chapters 6 and 12). Reaction with stratospheric ozone (through NO) is the only known process that destroys N_2O, so its atmospheric residence time is long, and its concentration in the atmosphere is fairly uniform (Fig. 3.4). Eventually, NO is transported to the troposphere as HNO_3, which is deposited in the heterogeneous reaction with raindrops. Large releases of N_2O are measured in tropical soils (Matson and Vitousek 1987, Keller et al. 1986), and its release is apparently stimulated by the widespread use of nitrogen fertilizers (Bremner and Blackmer 1978). Tropospheric concentrations are increasing at about 0.3%/yr (Khalil and Rasmussen 1983), and show a weak seasonal oscillation (Fig. 3.9), suggesting that biotic processes are also important sources. A change in flux of N_2O from these sources may also be responsible for the current decline in stratospheric ozone.

Most sulfur gases are so reactive that they do not reach the stratosphere, except in catastrophic events such as major volcanic eruptions (Hofmann and Rosen 1983, Legrand and Delmas 1987). An exception is carbonyl sulfide (COS), which is long-lived in the atmosphere (Table 3.4), and the most abundant sulfur gas. Carbonyl sulfide is released from microbial processes in soils and ocean waters (Adams et al. 1981, Steudler and Peterson 1985), and destroyed by vegetation uptake and photochemical reaction involving OH, leading to SO_2 in the stratosphere (Chapter 13). SO_2 is removed from the stratosphere by the reactions leading to H_2SO_4 described earlier.

Models of the Atmosphere and Global Climate

A large number of models have been developed to explain and predict physical processes and chemical reactions in the atmosphere. When these models attempt to predict the characteristics in a single column of the atmosphere, they are known as one-dimensional (1-D) and radiative–convective models. For instance, Fig. 2.4 is a simple 1-D model for the greenhouse effect, which assumes that the behavior of the Earth's atmosphere can be approximated by average values applied to the entire surface. Two-dimensional models (2-D) can be developed using the vertical dimension and a single horizontal dimension (e.g., latitude) to examine the change in atmospheric characteristics across a known distance of the Earth's surface (e.g., Brasseur and Hitchman 1988). On a regional scale these are particularly useful in following the fate of pollution emissions (e.g., Rodhe et al. 1981). Three-dimensional models (3-D) attempt to follow the fate of particular parcels of air as they move both horizontally and vertically in the atmosphere. Dynamic 3-D models are known as general circulation models (GCM) for the globe. Many models are constructed to include both chemical reactions and physical phenomena, such as the circulation of the atmosphere due to temperature differences. Chemical transformations are parameterized using the rate and equilibrium coefficients for the reactions that we have examined in this chapter. Since there is a large number of reactions, some of these models are quite complex (e.g., Logan et al. 1981, Isaksen and Hov 1987), but they offer useful predictions for future atmospheric composition when the input of several constituents is changing simultaneously.

Nearly all models suggest that substantial warming of the atmosphere (1.5–5.5°C; Bolin et al. 1986) will accompany increasing concentrations of CO_2, N_2O, CH_4, tropospheric O_3, and chloroflurocarbons (Dickinson and Cicerone 1986, Ramanathan 1988, Lashof and Ahuja 1990). The warming results from the absorption of infrared (heat) radiation, emitted from the surface of the Earth when it absorbs incident light from the Sun (Fig. 2.4). Warming will be greatest near the poles, where there is normally the greatest net loss of infrared radiation versus incident sunlight (Manabe and Wetherald 1980). Warming of the atmosphere will increase the rate of evaporation from the oceans, and will generally increase the global circulation of water in the hydrologic cycle (Chapter 10). Since water vapor also absorbs infrared radiation, it may accelerate the potential greenhouse effect (Raval and Ramanathan 1989). Thus, most models predict that with higher atmospheric concentrations of CO_2 and trace gases, the Earth will become a warmer and more humid planet (Chapter 10).

Differential warming of the atmosphere will change the global pattern of precipitation and evapotranspiration (Manabe and Wetherald 1986, Schlesinger and Mitchell 1985), causing substantial changes in the climate of most areas outside the tropics. How rapidly these changes in climate occur will be moderated by the thermal buffer capacity of the world's oceans, which can absorb enormous quantities of heat. However, the magnitude of these potential changes in climate is much larger than most changes in global climate during the last 2 million years. Currently it is difficult to demonstrate any trend toward global warming from weather records (Spencer and Christy 1990). However, many climatologists believe that we should be able to perceive a validation of this global experiment, in excess of normal climatic oscillations due to other factors, before the end of this century (Hansen et al. 1981, Ramanathan 1988).

Of course, such predictions are not made without disagreement (Luther and Cess 1985). One of the largest unknowns in climate models is the effect of changes in the global distribution of clouds (Cess et al. 1989). Increasing cloudiness may lead to cooling, due to the reflectivity of low clouds (Ramanathan et al. 1989, Slingo 1990). Based on possible increases in the atmospheric burden of aerosols (Toon and Pollack 1980) and in the reflectivity (albedo) of the Earth due to land clearing (Sagan et al. 1979), some workers predict a cool, drier climate in the future (Potter et al. 1975). It is interesting to note that both aerosol concentrations were higher (De Angelis et al. 1987) and CO_2 concentrations lower (Neftel et al. 1982, Barnola et al. 1987) during the last glacial period, but the cause and effect relation of these observations is unclear.

Future models of global climate change due to CO_2 must include interactions with other atmospheric constituents. Reductions in stratospheric ozone will cool the stratosphere and allow more solar radiation to reach the Earth's surface. However, Brasseur and Hitchman (1988) use a 2-D model to suggest that increases in CO_2 might reduce the rate of destruction of stratospheric ozone. Higher concentrations of CH_4 in the atmosphere are likely to lead to higher concentrations of H_2O in the stratosphere [equations (3.17) and (3.18)], conferring a substantial additional greenhouse effect (Ramanathan 1988, Thomas et al. 1989).

Atmospheric Deposition

Processes

Elements of biogeochemical interest are deposited on land as a result of rainfall, dry deposition (sedimentation), and direct absorption of gases. The importance of each of these processes differs for different regions and for different elements (Gorham 1961). These processes account for a large fraction of the nitrogen and sulfur that is contained in terrestrial ecosystems (Chapter 6).

The nutrient content of rainfall has received great attention, as a result of widespread concern about dissolved constituents that lead to "acid rain." The dissolved constituents in rainfall are often separated into two fractions. The rainout component consists of constituents derived from cloud processes, such as the nucleation of raindrops. The washout component is derived from below cloud level, such as the scavenging of aerosol particles and the dissolution of gases in raindrops as they fall. The relative contribution of these fractions varies, depending upon the length of the rainstorm. The concentration of dissolved constituents in precipitation is inversely related to the rate of precipitation (Gatz and Dingle 1971) and to the total volume collected (Likens et al. 1984). These relations reflect the tendency for the washout component to decline in importance as the atmosphere is cleansed. The concentration of dissolved constituents also varies inversely as a function of mean raindrop size (Georgii and Wötzel 1970), since evaporation of water from raindrops as they fall tends to concentrate the remaining salts. This inverse relation explains why extremely high concentrations of dissolved constituents are found in fog waters (Weathers et al. 1986, Waldman et al. 1982). Capture of fog and cloud water by vegetation dominates the deposition of nutrient elements from the atmosphere in some high-elevation and coastal ecosystems (Lovett et al. 1982, Azevedo and Morgan 1974, Waldman et al. 1985).

The relative efficiency of scavenging by rainwater is often expressed as the washout ratio:

$$\text{Washout} = \frac{\text{Ionic Concentration in rain (mg/liter)}}{\text{Ionic concentration in air (mg/m}^3)} . \qquad \text{Eq. 3.38}$$

With units of m^3/l, this ratio gives an indication of the volume of atmosphere cleansed by each liter of rainfall. Large ratios are generally found for ions that are derived from relatively large aerosols or from highly water-soluble gases in the atmosphere. However, some pollutant elements (e.g., Pb) have relatively small washout ratios (Peirson et al. 1973). Snowfall is generally less efficient at scavenging than rainfall.

The deposition of nutrients by precipitation is often called wetfall; dryfall is the result of gravitational sedimentation of particles during periods without rain (Hidy 1970). Dryfall of dusts in areas downwind of arid lands is often spectacular; Liu et al. (1981) reported 1 g m^{-2} h^{-1} of dustfall in Beijing, China, as a result of a single dust storm on April 18, 1980. Enormous deposits of wind-deposited soil, known as loess, were laid down during glacial periods, when large areas of semi-arid land were subject to wind erosion (Pye 1987). Today, elements necessary for plant growth are released by chemical weathering in these soils (Chapter 4).

The dryfall received in many areas contains a significant fraction that is easily dissolved by soil waters and immediately available for plant uptake. Despite the high rainfall found in the southeastern United States, Swank and Henderson (1976) reported that 19–64% of the total annual atmospheric deposition of ions such as Ca, Na, K, and Mg and up to 89% of the deposition of P were derived from dryfall. Dry deposition contributes about 30% to the total input of acidic substances in southern Canada (Sirois and Barrie 1988). Dryfall is often measured in collectors that are designed to close during rainstorms. When open to the atmosphere, these instruments capture particles that are deposited vertically, known as sedimentation. In natural ecosystems, dryfall is also derived by the capture of particles on vegetation surfaces. When vegetation captures particles that are moving horizontally in the airstream, the process is known as impaction (Hidy 1970). Impaction is a particularly important process in the capture of seasalt aerosols near the ocean (Art et al. 1974, Potts 1978).

In addition to the uptake of CO_2 in photosynthesis, vegetation also absorbs N- and S-containing gases directly from the atmosphere (Hosker and Lindberg 1982, Lindberg et al. 1986). Recapture of NH_3 by vegetation accounts for a large proportion of that released by agricultural soils (Hutchinson et al. 1972, Denmead et al. 1976). Uptake of pollutant SO_2 by vegetation is particularly important in humid regions (McLaughlin and Taylor 1981), where plant stomata remain open for long periods. Lovett and Lindberg (1986) found that HNO_3 vapor accounted for 75% of the annual dry deposition of nitrogen (4.8 kg/ha) in a deciduous forest in Tennessee, and dry deposition was nearly half of the total annual deposition of nitrogen from the atmosphere. Since vegetation is also a source of some reduced gases such as NH_3 (Farquhar et al. 1979) and H_2S (Wilson et al. 1978, Winner et al. 1981), the net biogeochemical input from the atmosphere is poorly known in most cases.

Total capture of dry particles and gases by land plants is difficult to measure. Rainfall collections under the canopy contain materials washed from plant surfaces, but also large quantities of elements that are derived from plant root uptake and leached from the leaf cells (Parker 1983, Chapter 6). Artificial collectors (surrogate surfaces) are often used to approximate the capture by vegetation (White and Turner 1970, Vandenburg and Knoerr 1985, Lindberg and Lovett 1985). The capture on known surfaces can be compared to the airborne concentrations to calculate a deposition velocity (Sehmel 1980):

$$\text{Deposition Velocity} = \frac{\text{Rate of dryfall (mg/cm}^2\text{/s)}}{\text{Concentration in air (mg/cm}^3\text{)}} . \qquad (3.39)$$

In units of cm/s, these velocities can be multiplied by the estimated area of vegetation surface (cm^2) and the concentration in the air to calculate total deposition for an ecosystem. For example, Lovett and Lindberg (1986) used a deposition velocity of 2.0 cm/s to calculate a deposition of 3.0 kg NO_3-N/ha in a forest with a leaf area of 5.8 m^2/m^2 and an ambient concentration of 0.82 μg N m^{-3}. It is often unclear if deposition velocities on artificial surfaces are realistic under other conditions, and accurate estimates of the surface area of vegetation are difficult (Whittaker and Woodwell 1968). Clearly, biogeochemistry needs further work on these processes.

Regional Patterns

Regional patterns in rainfall chemistry in the United States reflect the relative importance of different constituent sources and deposition processes in different areas (Munger and Eisenreich 1983) (Fig. 3.10). Coastal areas are dominated by atmospheric inputs from the sea, with large inputs of Na, Mg, Cl, and SO_4, which are the major constituents in the seasalt aerosol (Junge and Werby, 1958). Areas of arid and semi-arid land show high concentrations of soil-derived constituents, such as Ca (Young et al. 1988, Ganor and Mamane 1982, Löye-Pilot et al. 1986). Areas downwind of regional pollution show exceedingly low pH and high concentrations of SO_4^{2-} and NO_3^- (Cogbill and Likens 1974, Gorham et al. 1984, Schwartz, 1989).

The ratio among ionic constituents in rainfall can be used to trace their origin. Except in unusual circumstances nearly all the sodium (Na) in rainfall is derived from the ocean. When magnesium is found in a ratio of 0.12 with respect to Na, one may presume that it is also of marine origin (Fig. 3.11). In the southeastern United States, however, Mg/Na ratios in wetfall range from 0.29 to 0.76 (Swank and Henderson 1976). Here the Mg content has been increased relative to Na, presumably because the airflow that brings precipitation to this region has crossed the United States and contains Mg from soil dust and other sources. Schlesinger et al.

(1982) used this approach to deduce nonmarine sources of Ca and SO_4 in the rainfall in coastal California (Fig. 3.11). Similarly, Fe and Al are largely derived from the soil, and ratios of various ions to these elements in soil can be used to predict their expected concentrations in rainfall when soil dust is a major source (Lawson and Winchester 1979, Warneck 1988). For instance, high concentrations of Al in dryfall on Hawaii were used to trace its origin to springtime dust storms in the central plains of China (Parrington et al. 1983). Windborne particles of soil and vegetation contribute significantly to the global transport of trace metals in the atmosphere (Nriagu 1989).

In many areas downwind of pollution, a strong correlation between H^+ and SO_4^{2-} is the result of the production of H_2SO_4 during the oxidation of SO_2 and its dissolution in rainfall (Cogbill and Likens 1974, Gorham et al. 1984, Irwin and Williams 1988). Nitrate (NO_3^-) also contributes to the strong acid content in rainfall (HNO_3). These constituents depress the pH of rainfall below 5.6, which would be expected for water in equilibrium with atmospheric CO_2 (Galloway et al. 1976). The pH of such rainfall is determined by the concentration of strong acid anions that are not balanced by NH_4^+ and Ca^{2+} (or other cations), namely (from Gorham et al. 1984),

$$H^+ = [NO_3^- + 2SO_4^{2-}] - [NH_4^+ + 2Ca^{2+}] \qquad (3.40)$$

In the western United States, high concentrations of SO_4^{2-} are apparently due to industrial pollution (Epstein and Oppenheimer 1986, Oppenheimer et al. 1985), but the rainfall is less acidic because the acid-forming anions have reacted with soil aerosols containing $CaCO_3$ (Young et al. 1988, Schlesinger and Peterjohn 1988).

The concentration of constituents in the Greenland snowpack reflects the changes in the atmospheric burden of anthropogenic pollutants during industrialization (Herron et al. 1977, Mayewski et al. 1986). Similarly, recent lake sediments contain higher concentrations of many trace metals (Galloway and Likens 1979). Long-term records of precipitation chemistry are rare, but the collections at the Hubbard Brook Ecosystem in central New Hampshire suggest a recent decline in the concentrations of Pb and SO_4 that may reflect improved control of emissions. Over most of the same period, however, concentrations of NO_3 have increased, so the acidity of rainfall shows little change (Likens et al. 1984, Hedlin et al. 1987).

The long-term records suggest that many natural ecosystems, land and water, currently receive a greater input of N, S, and other elements of biogeochemical importance than before widespread emissions from human activities. Pollutant emissions have nearly doubled the annual input of S-containing gases to the atmosphere globally (Möller 1984), but the input and deposition of these compounds is more localized (Barrie and

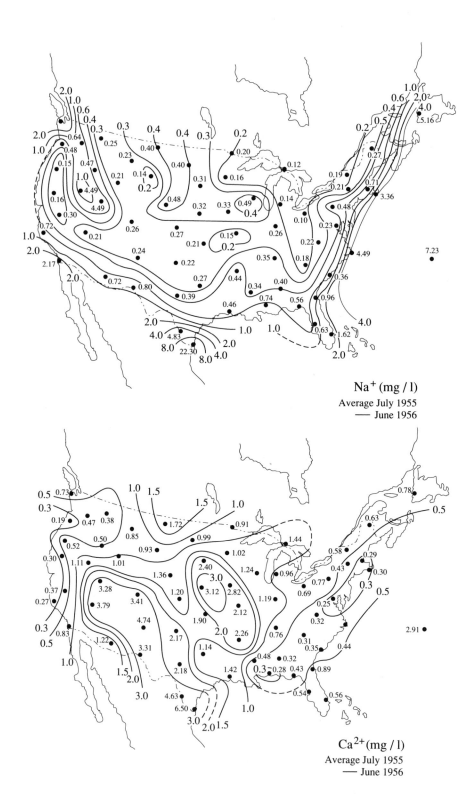

Na⁺ (mg / l)

Average July 1955
— June 1956

Ca²⁺ (mg / l)

Average July 1955
— June 1956

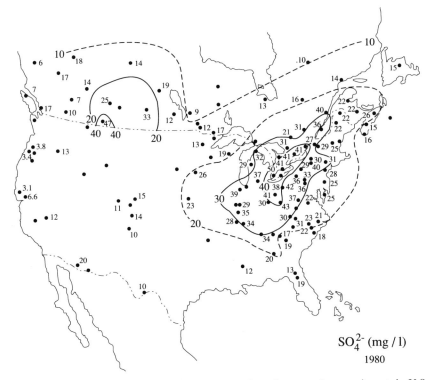

Figure 3.10 Geographic pattern in the concentration of some major constituents in U.S. precipitation. Na and Ca are from Junge and Werby (1958) and SO_4 is modified from Barrie and Hales (1984).

Hales 1984). Galloway et al. (1984) calculate that the deposition of SO_4^{2-} in the eastern United States has been enriched by 2 to 16 times over background conditions. The western North Atlantic ocean receives about 25% of the sulfur and nitrogen oxides emitted in eastern North America (Galloway and Whelpdale 1987). Deposition of nitrogen in fixed compounds might be expected to enhance the growth of forests, but current investigations suggest that in combination with acidity, this fertilization effect quickly leads to deficiencies of P, Mg, and other plant nutrients (Waring and Schlesinger 1985). These interactions are discussed in more detail in Chapters 6 and 9.

Summary

In this chapter we have examined the physical structure, circulation, and composition of the atmosphere. Major constituents, such as N_2, are rather unreactive and have long mean residence times in the atmosphere. CO_2 is largely controlled by dissolution in waters on the surface of the Earth. The atmosphere contains a

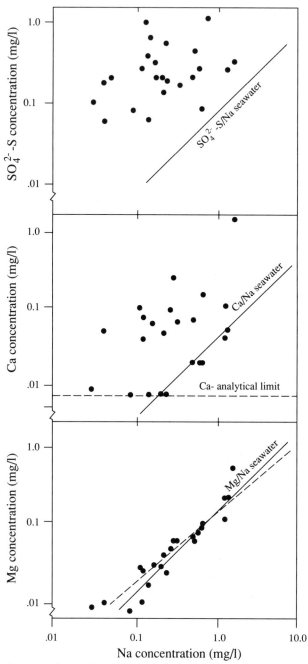

Figure 3.11 Concentrations of SO₄, Ca, and Mg in wetfall precipitation near Santa Barbara, California, plotted as a logarithmic function of Na concentrations in the same collections (Schlesinger et al. 1982). The solid line represents the ratio of these ions to Na in seawater. Ca and SO₄ are enriched in wetfall relative to seawater, whereas Mg shows a correlation (dashed) that is not significantly different from the ratio expected in seawater.

variety of minor constituents, many of which are reduced gases. These gases are highly reactive in homogeneous reactions with hydroxyl (OH) radicals and heterogeneous reactions with aerosols and cloud droplets, which scrub them from the atmosphere. Changes in the concentration of many trace gases are indicative of global change, perhaps leading to future climatic warming and higher surface flux of ultraviolet light. The oxidized products of trace gases are deposited in land and ocean ecosystems, resulting in the input of N, S, and other elements of biogeochemical significance. Pollution of the atmosphere by the release of oxidized gases containing N and S as a result of human activities results in acid deposition in downwind ecosystems. The enhanced deposition of N and S represents altered biogeochemical cycling on a regional and global basis.

Recommended Reading

Walker, J.C.G. 1977. Evolution of the Atmosphere. MacMillan, New York.
Warneck, P. 1988. Chemistry of the Natural Atmosphere. Academic Press, London.

4

The Lithosphere

Introduction

Since early geologic time, the atmosphere has interacted with the exposed crust of the Earth through the process of rock weathering. Many of the volcanic gases in the earliest atmosphere dissolved in water to form acids that could react with surface minerals (Chapter 2). Later, as oxygen accumulated in the atmosphere, rock weathering also occurred when oxygen reacted with reduced minerals, such as pyrite, that were exposed at the surface of the Earth. Since the advent of land plants, soil minerals have been exposed to high concentrations of carbon dioxide maintained in soil pores as a result of decomposition and the metabolic activities of roots. Carbonic acid (H_2CO_3), derived by reaction of this CO_2 with soil water, determines the rate of rock weathering in most ecosystems (Schwartzman and Volk 1989). Current human activities that cause the release of reactive N and S gases to the atmosphere result in the deposition of acid rain (Chapter 3) and potential increases in the rate of rock weathering in downwind areas (Cronan 1980).

Recognizing the close linkage between atmospheric constituents and rock weathering, Siever (1974) summarized a portion of the global sedimentary cycle as

$$\text{Igneous rocks + acid volatiles = sedimentary rocks + salty oceans} \qquad (4.1)$$

This formula recognizes that through geologic time the primary minerals of the Earth's crust have been exposed to reactive, acid-forming C, N, and S gases of the atmosphere. The products of the reaction are carried to the ocean, where they accumulate as dissolved salts or in deposits of sedimentary rock (Li 1972). Through geologic time, large amounts of sedimentary rock have been deposited. Geologic processes return some of these rocks to the deep Earth, where the constituents are converted back to primary minerals under great heat and pressure (Fig. 2.8; see also Siever 1974). However, about 75% of the rocks now exposed on land are sedimentary rocks that have been subjected to geologic uplift (Blatt and Jones 1975). These sedimentary rocks are subject to further weathering reactions with acid volatiles, in accord with Siever's basic equation.

In this chapter we will review the basic types of rock weathering on land and the processes that drive the weathering reactions. Rock weathering is important for the release of biochemical elements that have no gaseous forms (e.g., Ca, K, Fe, and P). Reactions between soil waters and the minerals found in soil determine the availability of these elements to biota and the losses of these elements in runoff. Conversely, land biota and soil microbes affect rock weathering and soil development. In this chapter, we will examine soil development in the major ecosystems on Earth. Finally, we will examine the rates of weathering in an attempt to determine the supply of biochemical elements on land and the global loss of weathering products to rivers and the ocean.

Rock Weathering

Upon uplift and exposure, all rocks undergo weathering, a general term that encompasses a variety of geological processes by which parent rocks are broken down. Mechanical weathering is the fragmentation of materials with no chemical change; in laboratory terminology it is equivalent to a physical change. Chemical weathering occurs when parent rock materials react with acidic and oxidizing substances. Usually chemical weathering involves water, and mineral constituents are released as dissolved ions. Chemical weathering also includes the formation of new, secondary minerals that are more stable at the physical conditions on the surface of the Earth.

Mechanical weathering includes wind abrasion and rock splitting by the freezing of water and by the growth of roots in rock crevices. Mechanical weathering is important in extreme and highly seasonal climates and in areas with much exposed rock. Fragmented rock often forms the lower horizons of soil profiles. The sand and silt fractions of soils are largely derived from the mechanical weathering of primary minerals, especially quartz. Thus, mechanical weathering is important in exposing parent rock to other weathering processes (e.g., Miller and Drever 1977). Finely divided rock and soil can also be removed by erosion, the transport of particulate solids from the ecosystem.

Catastrophic mechanical weathering events such as landslides remove large amounts of material from forest ecosystems (Swanson et al. 1982). Gregor (1970) suggests that the rate of sediment transport, as a measure of mechanical weathering, may have been about four times greater before the land surface was colonized by plants. Today, the highest concentrations of suspended sediment are seen in rivers draining arid and semi-arid regions where vegetation is sparse (Milliman and Meade 1983).

Chemical Weathering

Chemical weathering releases elements from the crust of the Earth, making them available for uptake by biota. Weaver and Tarney (1984) calculate the approximate composition of the continental crust (Table 4.1), which is comprised of various aluminosilicate minerals that contain small quantities of the important biochemical elements (see also Fig. 2.3). The crust of the Earth is composed of a variety of rock types. Igneous and metamorphic rocks contain primary minerals (e.g., olivine and plagioclase) that were formed under conditions of great temperature and pressure deep in the Earth. Most sedimentary rocks, including limestone and shales, consist largely of weathering products, known as secondary minerals, formed at the Earth's surface. However, some sedimentary rocks (e.g., sandstone) also contain some resistant primary minerals (quartz) that remain following the weathering of igneous rocks.

Rates of chemical weathering are strongly dependent on rock type. Both metamorphic and igneous rocks consist of various primary silicate minerals that are crystalline in structure. The primary silicate minerals form two classes—the ferromagnesian or mafic series, and the plagioclase or felsic series—depending on crystal structure and the presence of magnesium versus aluminum in the crystal lattice (Fig. 4.1). Among these

Table 4.1 Approximate Mean Composition of the Earth's Continental Crust[a]

Constituent	Percentage Composition
SiO_2	63.2
Al_2O_3	16.1
FeO	4.9
CaO	4.7
Na_2O	4.2
MgO	2.8
K_2O	2.1
TiO_2	0.6
P_2O_5	0.19
MnO	0.08

[a] Data from Weaver and Tarney (1984). Reprinted by permission from Nature vol. 310 p. 576, copyright (c) 1984 Macmillan Magazines Ltd.

minerals the rate of weathering tends to follow a reverse sequence of their formation during the original cooling and crystallization of rock; that is, minerals that condensed first are the most susceptible to weathering reactions (Goldich 1938). Formed through rapid, early crystallization at high temperatures, these minerals contain few bonds that link the units of their crystalline structure. Thus, olivine, which is formed under conditions of great heat and pressure deep within the Earth, is most likely to weather rapidly when exposed on the surface. Primary minerals that are most susceptible to weathering are those in which various cationic elements (e.g., Ca, Na, K, and Mg) or trace metals (Fe and Mn) are substituted in the silicate crystal structure.

In rocks or soils of mixed composition (e.g., on granite), chemical weathering is concentrated on the relatively labile minerals (April et al. 1986). Quartz is very resistant to chemical weathering and often remains when other minerals are lost (Fig. 4.1). Quartz is a relatively simple silicate mineral consisting only of silicon and oxygen in tetrahedral crystals that are linked in three dimensions. In the process of chemical weathering, primary minerals are altered to more stable forms, ions are released, and secondary minerals are formed as weathering products.

Since chemical weathering involves chemical reactions, it is not surprising that it occurs most rapidly under conditions of higher temperature and precipitation. Thus, on a worldwide basis, climate is a major determinant of weathering rates (Peltier 1950). Chemical weathering is more rapid in tropical forests than in temperate forests, and more rapid in most forests than in grasslands or deserts.

Ferromagnesian Series

Felsic Series

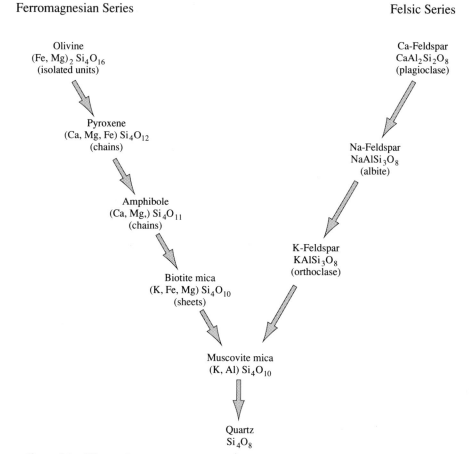

Figure 4.1 Silicate minerals are divided into two classes, the ferromagnesian series and the felsic series, based on the presence of Mg or Al in the crystal structure. Among the ferromagnesian series, minerals that exist as isolated crystal units (e.g., olivine) are most susceptible to weathering, while those showing linkage of crystal units and a lower ratio of oxygen to silicon are more resistant. Among the felsic series, Ca-feldspar (plagioclase) is more susceptible to weathering than Na-feldspar (albite) and K-feldspar (orthoclase). Quartz is the most resistant of all. This weathering series also follows the order in which these minerals are precipitated during the cooling of magma.

The dominant form of chemical weathering is the carbonation reaction, driven by the formation of carbonic acid, H_2CO_3 in the soil solution:

$$H_2O + CO_2 \rightleftarrows H^+ + HCO_3 \rightleftarrows H_2CO_3 \tag{4.2}$$

Because plant roots and decomposing soil organic matter release CO_2 to the soil air, the concentration of H_2CO_3 in soil waters is often much greater than that in equilibrium with atmospheric CO_2 at 0.035% concen-

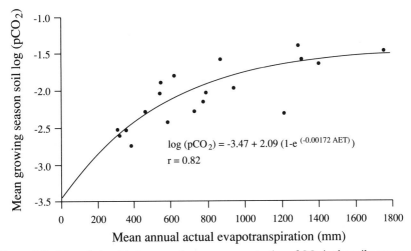

Figure 4.2   The relationship between the mean concentration of CO_2 in the soil pore space and the actual evapotranspiration of the site for various ecosystems of the world. From Brook et al. (1983).

tration (Castelle and Galloway 1990). Buyanovsky and Wagner (1983) report seasonal CO_2 concentrations of greater than 7.0% in the soil beneath wheat fields in Missouri. Such high concentrations of CO_2 can extend to considerable depths in the soil profile, affecting the weathering of underyling rock (Sears and Langmuir 1982). Wood and Petraitis (1984) found CO_2 concentrations of ~1.0% at 36 m, which they link to the downward transport of organic materials that subsequently decompose at depth. Solomon and Cerling (1987) found that high concentrations of CO_2 accumulated in the soil under a mountain snowpack, potentially leading to significant weathering during the winter. Brook et al. (1983) suggest that the average concentration of soil CO_2 varies as a direct function of actual evapotranspiration of the site (Fig. 4.2), consistent with observations of the greatest plant growth in warm and wet climates, which also have the greatest rates of carbonation weathering (Johnson et al. 1977). However, even in arid regions, rock weathering appears to be controlled by carbonation weathering under high partial pressures of CO_2 in the soil profile (Routson et al. 1977). By maintaining high concentrations of soil CO_2, living organisms exert control over the geochemical process of rock weathering on land—a good example of the importance of understanding biogeochemistry (Schwartzman and Volk 1989).

Carbonic acid attacks silicate rocks. For example, weathering of the Na-feldspar, albite, proceeds as

$$2NaAlSi_3O_8 + 2H_2CO_3 + 9H_2O \longrightarrow$$
$$2Na^+ + 2HCO_3^- + 4H_4SiO_4 + Al_2Si_2O_5(OH)_4 \qquad (4.3)$$

During this process a primary mineral is converted to a secondary mineral, kaolinite, by the removal of Na^+ and soluble silica. A sign that carbonation weathering has occurred is that HCO_3^- is the dominant anion in runoff waters. Precipitation of the secondary mineral, kaolinite, involves hydration with H^+ and water. The secondary mineral also has a lower ratio of Si to Al, as a result of the loss of some Si to stream waters. Since only some of the constituents of the primary mineral are released, this type of weathering reaction is known as an incongruent dissolution. Under conditions of high rainfall, as in the humid tropics, kaolinite may undergo a second incongruent dissolution to form another secondary mineral, gibbsite:

$$Al_2Si_2O_5(OH)_4 + 5H_2O \rightarrow 2H_4SiO_4 + Al_2O_3 \cdot 3H_2O \qquad (4.4)$$

Some weathering reactions involve congruent dissolutions. Limestone undergoes congruent dissolution during carbonation weathering:

$$CaCO_3 + H^+ + HCO_3^- \rightarrow Ca^{2+} + 2HCO_3^- \qquad (4.5)$$

Olivine ($FeMgSiO_4$) undergoes congruent dissolution in water, releasing Fe, Mg, and Si (Grandstaff 1986). Magnesium and silicon are lost in runoff waters, but the Fe^{2+} usually reacts with oxygen, resulting in the precipitation of Fe_2O_3 in the soil profile. Similarly, pyrite (FeS_2) undergoes a congruent reaction during oxidation:

$$2FeS_2 + 4H_2O + 6O_2 \rightarrow Fe_2O_3 + 8H^+ + 4SO_4^{2-} \qquad (4.6)$$

The H^+ produced in this reaction accounts for the acidity of runoff from many mining operations. As in the case of olivine, Fe_2O_3 is subsequently precipitated in the soil profile or stream bed (Garrels and MacKenzie 1971).

In addition to carbonic acid, living organisms release a variety of organic acids to the soil solution that can be involved in the weathering of silicate minerals. Many simple organic compounds, including acetic and citric acids, are released from plant roots (Smith 1976). Organic acids from plant roots and microbes can weather biotite mica, releasing K (Boyle and Voigt 1973, April and Keller 1990). Phenolic acids (i.e., tannins) are also released during the decomposition of plant remains (Schlesinger 1985a), and many fungi release oxalic acid that results in chemical weathering (Cromack et al. 1979, Lapeyrie et al. 1987). Soil microbes produce fulvic and humic acids during decomposition of plant remains (Chapter 5).

In addition to their contributions to total acidity, organic acids speed the weathering reactions by combining with some weathering products in a process called chelation. When Fe and Al combine with fulvic acid, they are mobile and move to the lower soil profile in percolating rainwater. When these elements are involved in chelation, their inorganic concentration in the soil solution remains low and an equilibrium between dissolved products and primary mineral forms is not achieved. Grandstaff (1986) found that additions of small concentrations of EDTA (an organic chelation agent) to weathering solutions increased the dissolution of olivine by 110 times over inorganic conditions. Fulvic and humic acids increase the weathering of a variety of silicate minerals, including quartz, particularly when the soil solution is acid (Baker 1973, Tan 1980, Bennett et al. 1988).

Organic acids often dominate the acidity of the upper soil profile, while carbonic acid is important below (Ugolini et al. 1977). In general, organic acids dominate the weathering processes in cool temperate forests where decomposition processes are slow and incomplete, whereas carbonic acid dominates weathering in tropical forests where lower concentrations of fulvic acids remain after surface litter decomposition (Johnson et al. 1977).

Secondary Minerals

Many types of secondary minerals can form in soils through weathering processes. Temperate forest soils are often dominated by layered silicate or "clay" minerals. These exist as small (<0.002 mm) particles that control the structural and chemical properties of soils. Although weathering removes Si as a dissolved constituent of runoff, some Si is retained through its role in the formation of secondary minerals. In general, two types of layers characterize the crystalline structure of secondary, aluminosilicate clay minerals—Si layers, and layers dominated by Al, Fe, and Mg. These layers are held together by shared oxygen atoms. Clay minerals and the size of their crystal units are recognized by the number, order, and ratio of these layers (Birkeland 1984). Moderately weathered soils are often dominated by secondary minerals such as montmorillonite and illite, which have a 2:1 ratio of Si- to Al-dominated layers. More strongly weathered soils, such as in the southeastern United States, are dominated by kaolinite clays with a 1:1 ratio of layers, reflecting a greater loss of Si.

When secondary minerals incorporate elements of biochemical interest, one cannot assume that the release of those elements from primary minerals is immediately reflected by an increase in the pool of ions available for uptake by plants (Olsson and Melkerud 1989). Potassium is fixed in the crystal lattice of illite, whereas montmorillonite contains Mg. These minerals are common in temperate soils. Similarly, while little nitrogen is contained in primary minerals, some 2:1 clay minerals incorporate N as fixed ammonium (NH_4) in their crystal lattice. Fixed ammonium can represent up to 10% of the total N in some soils (Stevenson 1982, Antisari and Sequi 1988). The release of fixed ammonium from clay minerals is slow, but recognizing the widespread nitrogen limitation for the growth of land plants (Chapter 6), the dynamics of fixed ammonium may play an important role in determining the availability of N for plant growth (Mengel and Scherer 1981, Baethgen and Alley 1987, Jensen et al. 1989).

In contrast to the loss of Si and other cations (e.g., Ca and Na) to runoff waters, Al and Fe are relatively insoluble unless they are involved in chelation relations with organic matter (Huang 1988). When these elements are released during weathering, they tend to accumulate in the soil as oxides. Initially free Fe and Al accumulate in amorphous and poorly crystallized forms, known as ferrihydrite, which are often quantified by extraction in a weak oxalate solution (Birkeland 1984, but see also, Parfitt and Childs 1988). With increasing time, most Fe and Al is found in crystalline oxides and hydrous oxides, which are traditionally extracted using a reducing solution of citrate–dithionate. Some of these transformations involve bacteria, and thus are biogeochemical (Fassbinder et al. 1990). Crystalline oxides and hydrous oxides of Fe (e.g., goethite and hematite) and Al (e.g., gibbsite and boehmite) are common in many tropical soils, where high temperatures and rainfall cause relatively rapid organic decomposition and few organic acids remain to chelate Fe and Al. Under these climatic conditions, the secondary clay minerals typical of temperate zone soils are subject to weathering, with the nearly complete removal of Si, Ca, K, and other basic cations in stream water.

Phosphorus Minerals

Phosphorus deserves special attention, since it is often in limited supply for plant growth. The only primary mineral with significant phosphorus content is apatite, which can undergo carbonation weathering in a congruent reaction, releasing P:

$$Ca_5(PO_4)_3OH + H_2CO_3 \longrightarrow$$
$$5Ca^{2+} + 3HPO_4^{2-} + 4HCO_3^- + H_2O \tag{4.7}$$

While this phosphorus may be accumulated by biota, a large proportion is involved in reactions with other soil minerals, leading to precipitation in unavailable forms. Phosphorus may be bound by iron and aluminum oxides, accounting for the low availability of phosphorus in many tropical soils (Sanchez et al. 1982a, Smeck 1985). This occluded phosphorus is essentially unavailable to biota. Nonoccluded phosphorus includes forms that are held on the surface of soil minerals by a variety of reactions, including anion absorption (see below). As seen in Fig. 4.3, phosphorus

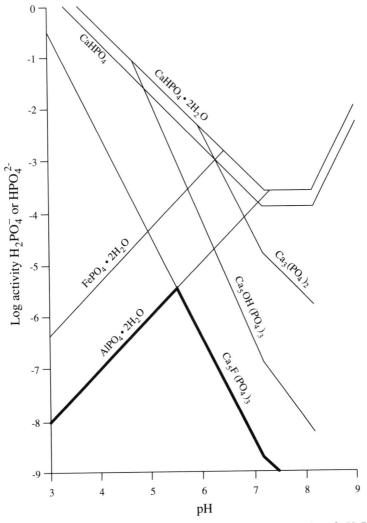

Figure 4.3 The solubility of phosphorus in the soil solution as a function of pH. Precipitation with Al sets the upper limit on dissolved phosphate at low pH (bold line); precipitation with Ca sets a limit at high pH. Phosphorus is most available at pH ~5.7. Modified from Lindsay and Vlek (1977).

availability is controlled by direct precipitation with iron and aluminum in acid soils (Lindsay and Moreno 1960), while in arid soils most phosphorus is held on the surface of $CaCO_3$ or precipitated as calcium phosphate (Lajtha and Schlesinger 1988, Lajtha and Bloomer 1988).

Walker and Syers (1976) diagram the general evolution of phosphorus availability during the weathering of rocks containing apatite (Fig. 4.4).

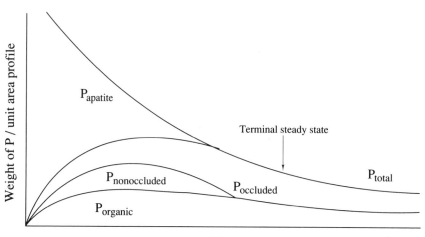

Figure 4.4 Changes in the forms of phosphorus found during soil development on sand dunes in New Zealand. Modified from Walker and Syers (1976).

Apatite weathers rapidly, giving rise to phosphorus contained in other forms and to a decline of total phosphorus in the system due to losses in runoff. Phosphorus released from apatite is initially held in nonoccluded forms or taken up by biota (organic P). With time, oxide minerals accumulate, and phosphorus is precipitated in occluded forms. At the later stages of weathering and soil development, occluded and organic P dominate the forms of P remaining in the system. At this stage almost all available phosphorus is found in a biogeochemical cycle in the upper soil profile, while phosphorus found at lower depths is primarily involved in geochemical reactions with secondary minerals (Wood et al. 1984). Plant growth depends on the rapid root uptake of phosphorus released from dead organic matter in the biogeochemical cycle (Tiessen et al., 1984). In animals phosphorus is incorporated in secondary hydroxyapatite (bones) and fluoroapatite (teeth), which are rather resistant to weathering and sometimes useful in determining past human settlement patterns (Sjöberg 1976).

As seen for the weathering of silicate minerals, organic acids can influence the availability of phosphorus in the upper soil layers. Organic acids can inhibit the crystallization of Al and Fe hydrous oxides, reducing the rate of phosphorus occlusion (Schwertmann 1966, Kodama and Schnitzer 1977, 1980). Jurinak et al. (1986) show how the production of oxalic acid can lead to the weathering of P from apatite. Following its release, P may be more available in the presence of organic acids, such as oxalate, which also remove Fe and Ca from the soil solution by chelation and precipitation (Graustein et al. 1977). The production and release of oxalic acid by mycorrhizal fungi (Chapter 6) explain their importance in maintaining and supplying phosphorus to plants (Bolan et al. 1984, Cromack et al. 1979), and explain the greater availability of phosphorus under fungal mats (R. F. Fisher 1972, 1977). Some workers believe that the biogeochemical control of phosphorus availability by symbiotic fungi was a precursor to the successful establishment of plants on land (Pirozynski and Malloch 1975; Chapter 2).

Soil Chemical Reactions

Following release by weathering, the availability of essential biochemical elements to biota is controlled by a number of reactions that maintain an equilibrium between concentrations in the soil solution and contents that are associated with the soil mineral or organic fraction. In contrast to the kinetics of weathering reactions, soil exchange reactions occur relatively rapidly. The specific soil reactions differ depending on how the soil development is affected by climate, age, biota, and the parent material of the soil.

Cation Exchange Capacity

The layered silicate clay minerals that dominate temperate zone soils possess net negative charge that attracts and holds cations dissolved in the soil solution. The negative charge has several origins. Most negative charge arises from ionic substitutions within silicate clays, especially 2:1 clays. For example, when Mg^{2+} substitutes for Al^{3+} in montmorillonite, there is an unsatisfied negative charge in the internal crystal lattice. This negative charge is permanent in the sense that it arises inside the crystal structure and cannot be neutralized by covalent bonding of cations from the soil solution. Permanent charge is expressed as a zone or "halo" of negative charge surrounding the surface of clay particles in the soil.

A second source of negative charge is found at the edges of clay particles, where hydroxide (—OH) radicals are often exposed to the soil solution. Depending on the pH of the solution, the H^+ ion may be more or less strongly bound to this radical. In most cases, a considerable number of the H^+ are dissociated, leaving negative charges (—O$^-$) that can attract and bind cations (e.g., Ca^{2+}, K^+, and NH_4^+). This cation exchange capacity is known as pH-dependent charge. The binding is reversible and exists in equilibrium with ionic concentrations in the soil solution. This form of cation exchange capacity is especially important on kaolinite and on iron and aluminum oxide minerals, which are known as variable-charge minerals.

In many temperate soils, a large amount of cation exchange capacity is also contributed by soil organic matter. These are also pH-dependent charges originating from the phenolic (—OH) and organic acid (—COOH) radicals of soil humic materials. In some sandy soils, as in central Florida, and in most soils of the humid tropics nearly all cation exchange is the result of soil organic matter. Organic matter is also the major source of cation exchange in desert soils that contain a relatively small proportion of secondary clay minerals as a result of relatively limited chemical weathering.

The total negative charge is expressed as meq/100 g or cmol(+)/kg of soil and comprises cation exchange capacity (CEC). Exchange of cations occurs as a function of chemical mass balance with the soil solution. Elaborate models of ion exchange have been developed by soil chemists (Sposito 1984). In general, cations are held and displace one another in the sequence

$$Al^{3+} > H^+ > Ca^{2+} > Mg^{2+} > K^+ > NH_4^+ > Na^+ \qquad (4.8)$$

on cation exchange sites. This sequence assumes equal molar concentrations in the initial soil solution and can be altered by the presence of large quantities of the more weakly held ions. Agricultural liming, for example, is an attempt to displace and neutralize H^+ ions from the exchange sites by "swamping" the soil solution with excess Ca^{2+}. In most cases, few cation exchange sites are actually occupied by H^+, which acts to weather soil minerals releasing Al and other cations.

Cations other than Al and H are informally known as base cations, since they tend to form bases [e.g., $Ca(OH)_2$] when they are released to the soil solution (Birkeland 1984). The percentage of the total cation exchange capacity occupied by base cations is termed base saturation. Both cation exchange capacity and base saturation increase during initial soil development on newly exposed parent materials. As the weathering of soil minerals continues, cation exchange capacity and base saturation decline (Bockheim 1980). Temperate forest soils dominated by 2:1 clay minerals have greater cation exchange capacity than those dominated by 1:1 clay minerals such as kaolinite. Tropical forest soils dominated by aluminum oxide minerals have essentially no cation exchange capacity from the mineral fraction at their natural soil pH. The cation exchange capacity of these soils is almost wholly derived from organic matter.

Soil Buffering

Cation exchange capacity acts to buffer the acidity of many temperate soils. When H^+ is added to the soil solution, it exchanges for cations, especially Ca, on clay minerals and organic matter (Bache 1984, James and Riha 1986). Over a wide range of pH, temperate soils maintain a constant value (k) for the expression

$$pH - \tfrac{1}{2}(pCa) = k \qquad (4.9)$$

which is known as the lime potential. This expression suggests that when H^+ is added to the soil solution (lower pH), the concentration of Ca^{2+} increases in the soil solution (lower pCa), so that k remains constant. The $\tfrac{1}{2}$ reflects the valence of Ca versus H. As long as there is sufficient base saturation, buffering by CEC explains why many temperate soils that are exposed to acid rain show little change in soil pH (Federer and Hornbeck 1985).

In strongly acid soils, as in the humid tropics, there is little CEC to buffer the soil solution. These soils are buffered by various geochemical reactions involving aluminum (Fig. 4.5). Aluminum is not a base cation

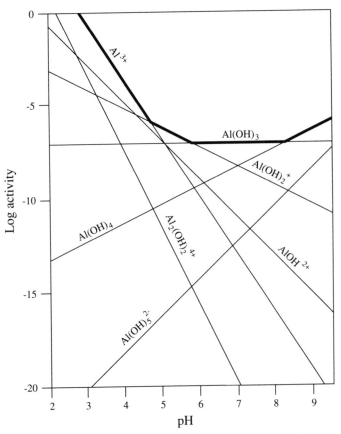

Figure 4.5 The solubility of aluminum as a function of pH. For pH in the neutral range, gibbsite $(Al(OH)_3)$ controls aluminum solubility, and there is little Al^{3+} in solution. Al^{3+} becomes more soluble at pH < 4.7. From Lindsay (1979).

inasmuch as its release to the soil solution leads to the formation of H^+ as Al is precipitated as aluminum hydroxide:

$$Al^{3+} + H_2O \rightleftarrows Al(OH)^{2+} + H^+ \qquad (4.10)$$

$$Al(OH)^{2+} + H_2O \rightleftarrows Al(OH)_2{}^+ + H^+ \qquad (4.11)$$

$$Al(OH)_2{}^+ + H_2O \rightleftarrows Al(OH)_3 + H^+ \qquad (4.12)$$

These reactions account for the acidity of many soils in the humid tropics (Sanchez et al. 1982a), but the reactions are reversible, so that the soil solution is buffered against additions of H^+ by the dissolution of aluminum hydroxide. The acid rain received by the northeastern United States appears to dissolve gibbsite (Al_2O_3) from many forest soils, leading to high concentrations of Al^{3+} that are toxic to fish in streams and lakes at high elevations. As stream waters flow to lower elevations, H^+ is consumed in weathering reactions with various silicate minerals, stream-water pH increases, and aluminum hydroxides are precipitated (N. M. Johnson et al. 1981).

Anion Absorption Capacity

In contrast to the permanent negative charge in soils of the temperate zone, tropical soils dominated by oxides and hydrous oxides of iron and aluminum show variable charge, depending on soil pH (Uehara and Gillman 1981, Sollins et al. 1988). In acid conditions these soils possess positive charge, as a result of the association of additional H^+ with the surface hydroxide radicals (Fig. 4.6). With experimental increases in pH, the soil passes through a zero point of charge (ZPC), and develops cation exchange at relatively high pH. For gibbsite, ZPC occurs around pH 9.0, so significant anion absorption capacity (AAC) is present in acid tropical soils in most field situations. It is important to recognize that these reactions occur on all soil constituents, but the ZPC of hydroxyl groups on layered silicate minerals or soil organic matter occurs at pH < 2.0, so anion absorption capacity is absent in nearly all natural situations (Sposito 1984).

   The ZPC of a bulk soil sample will depend on the relative mix of various minerals and organic matter. Tropical soils in Costa Rica show ZPC at pH ~4.0, due to a mix of soil organic matter and gibbsite (Sollins et al. 1988). Some anion absorption capacity can occur in temperate soils when iron and aluminum hydroxides are found in the soil profile (D. W. Johnson et al. 1981, 1986). Anion absorption capacity is typically greater on poorly crystalline forms of Fe and Al (oxalate-extractable), which have greater surface area than crystalline forms (dithionate-extractable) (Parfitt and

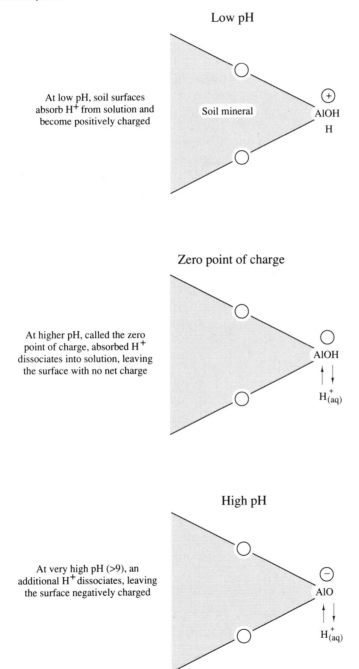

Figure 4.6 Variation in surface charge on iron and aluminum hydroxides as a function of the pH of the soil solution. From Johnson and Cole (1980).

Smart 1978, D. W. Johnson et al. 1986). Potential absorption of sulfate from acid rain is positively correlated to the oxalate-extractable Al in a variety of soils (Harrison et al. 1989, Courchesene and Hendershot 1989, MacDonald and Hart 1990).

Anion absorption follows the sequence

$$PO_4^{3-} > SO_4^{2-} > Cl^- > NO_3^- \qquad\qquad (4.13)$$

which accounts for the low availability of phosphorus in many tropical soils. Frequently anion exchange is described using the Langmuir model (Fig. 4.7), in which the content of anions held on exchange sites is expressed as a function of the content in the solution (Travis and Etnier 1981, Reuss and Johnson 1986). Phosphorus, sulfate, and selenite (SeO_4) are so strongly held that the binding is known as specific absorption or ligand exchange and is thought to replace —OH groups on the surface of the minerals (Fig. 4.8) (Hingston et al. 1967). Thus, the absorption of SO_4^{2-} from acid rain is associated with an increase in soil pH and a decline

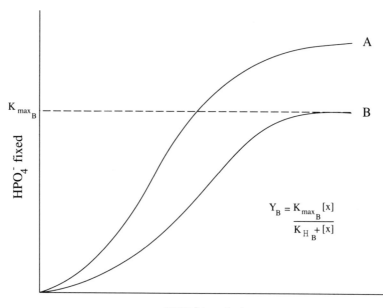

Figure 4.7 The Langmuir adsorption isotherm is used to compare the affinity of soils for anions as a function of the concentration of the anion in solution. In this diagram, soil B has a lower affinity for phosphorus than soil A; at equal concentrations of phosphorus in solution, more P will be available in soil B. Conversely, if these soils are exposed to long-term additions of solution with a given phosphorus concentration, it will take longer for soil A to equilibrate with that solution (see Johnson and Cole, 1980; Reuss and Johnson, 1986).

in apparent ZPC. All these anions are also involved in nonspecific absorption, which is more readily reversible with changes in concentration in the soil solution. Phosphorus held on anion absorption sites by either mechanism is known as nonoccluded phosphorus (see above).

Anion absorption capacity is inhibited by organic matter, which also binds to the surface of Fe and Al minerals (Johnson and Todd 1983, Singh 1984). Thus, soils rich in organic matter are less efficient in anion absorption than those dominated by Fe and Al oxide minerals. Percolating waters often carry anions from the upper organic layers of the soil to lower depths, where they are held on Fe and Al minerals.

Biogeochemical control over the exchange of soil cations and anions is most easily seen in tropical soils, where CEC is wholly the result of soil organic matter. In these soils, the AAC is also determined by the effects of soil organic matter on the ZPC of the bulk soil and the binding of organic matter to the anion exchange sites.

Soil Development

The soil in a terrestrial ecosystem usually consists of a number of layers, or horizons, that collectively comprise the complete soil profile, or pedon. Recognition of the processes that occur in these horizons is an essential part of understanding the biogeochemical cycles on land. Conversely, knowledge of such processes as rock weathering, water movement, and decomposition is essential to understanding the development of the soil profile under varying climatic conditions (Jenny 1980). In this section, we consider soil development in forests, grasslands and deserts.

Forests

In forests it is often easy to separate an organic layer, the forest floor, from the underlying layers of mineral soil, but these two major categories

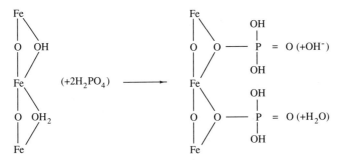

Figure 4.8 Specific adsorption of phosphate by iron sesquioxides may release OH^- or H_2O to the soil solution. From Binkley (1985).

can be further subdivided. In the forest floor, the L or O_i layer consists of fresh, undecomposed litter, easily recognized by species. The F or O_e layer lies immediately below the L layer and consists of fragmented organic matter in a stage of partial decomposition. This layer is dominated by organic materials in cellular form, and fungi and bacteria are common. The designation of F layer is derived from "fermentation," but this does not imply that the environment for microbial processes is anaerobic. Beneath the F layer lies the H, O_a, or humus layer, primarily consisting of amorphous, resistant products of decomposition, and with lower proportions of organic matter in cellular form. The lower portion of the H horizon often shows an increasing proportion of inorganic mineral soil constituents. Thus, the differentiation of the H layer from the uppermost layer of mineral soil is sometimes difficult, but a greater predominance of organic content versus mineral content is a useful criterion.

Not all forest soils show the differentiation of all layers of the forest floor. The thickness and presence of the layers also vary seasonally, especially in regions where litterfall is strongly seasonal. In some tropical forests decomposition of fresh litter is so rapid that there is little forest floor development (Olson 1963, Vogt et al. 1986). On the other hand, slow decomposition in coniferous forests, especially in the boreal zone, results in the accumulation of a thick forest floor, known as a mor, that is sharply differentiated from the underyling soil (Romell 1935). Much of the arctic zone is characterized by waterlogged soils, in which the entire rooting zone is composed of organic materials. Such peatland soils are known as Histosols. We will treat the properties of waterlogged organic soils in Chapter 7.

The upper mineral soil is designated as the A horizon, which contains a significant organic fraction. It may vary in thickness from several centimeters to 1 m. The A horizon is recognized as a zone of removal or *eluvial* processes. In most temperate regions soil water percolating through the forest floor contains organic acids derived from the microbial decomposition of litter. These organic acids dominate the weathering of soil minerals in the A horizon. Solutions collected in the A horizon contain cations and silicate, derived from weathering reactions (Table 4.2). Iron and Al may also be removed from the A horizon by chelation with fulvic acids that percolate downward from the forest floor (Ugolini et al. 1977, Antweiler and Drever 1983, Driscoll et al. 1985). Downward transport of Fe and Al in conjunction with organic matter is known as podzolization (Chesworth and Macias-Vasquez 1985).

Although it is found throughout the world, soil podzolization is particularly intense in the region extending from the arctic to cool temperate forests. (e.g., Ugolini et al. 1987, Evans 1980, De Kimpe and Martel 1976). Much of this area is characterized by coniferous forests, which

Table 4.2 Chemical Composition of Precipitation, Soil Solutions, and Groundwater in a 175-yr-old *Abies amabilis* Stand in Northern Washington[a]

Solution	pH	Total Cations (mEq/l)	Soluble Ions (mg/l)			Total (mg/l)	
			Fe	Si	Al	N	P
Precipitation							
Above canopy	5.8	0.03	<0.01	0.09	0.03	0.60	0.01
Below canopy	5.0	0.10	0.02	0.09	0.06	0.40	0.05
Forest floor	4.7	0.14	0.04	3.50	0.79	0.54	0.04
Soil							
15 cm E	4.6	0.12	0.04	3.55	0.50	0.41	0.02
30 cm B_s	5.0	0.08	0.01	3.87	0.27	0.20	0.02
60 cm B3	5.6	0.25	0.02	2.90	0.58	0.37	0.03
Groundwater	6.2	0.26	0.01	4.29	0.02	0.14	0.01

[a] Data from Ugolini et al. (1977), Soil Sci. **124**, 291–302. Copyright (1977) Williams & Wilkins.

produce litterfall that is rich in phenolic compounds and organic acids (Cronan and Aiken 1985). In these ecosystems, decomposition is slow and incomplete, and large quantities of fulvic acid percolate from the forest floor into the underlying A horizon. The pH of the soil solution is often as low as 4.0 (Dethier et al. 1988). When the removal of Fe, Al, and organic matter is very strong, a whitish layer is easily recognized at the base of the A horizon. This horizon is sometimes designated as an A_e or E (eluvial) horizon and may consist of nearly pure quartz, which is relatively insoluble compared to other soil constituents (Pedro et al. 1978).

In the deciduous forests of the warm temperate regions to the south, decomposition is more rapid. Smaller quantities of fulvic acids remain to percolate through the A horizon, podzolization is less intense, and there is no sharly defined E horizon (Pedro et al. 1978). The dominant process is the downward movement of secondary clay minerals that are formed from weathering in the A horizon.

Podzolization is found in some tropical soils, especially those that develop on sandy parent materials (Bravard and Righi 1989). It is absent in other areas of the tropics, where decomposition is so complete that there is almost no soluble organic acid percolating through the soil profile. In the absence of podzolization, iron and Al are not removed from the A horizon by chelation but are precipitated as oxides and hydrous oxides in the soil profile near the site of their release by weathering. Cations and Si are lost to runoff.

During soil development, substances leached from the A and E horizons are deposited in the underlying B horizons. This is defined as the zone of deposition or the illuvial horizon, where secondary clay minerals accumulate. Clay minerals arrest the downward movement of dissolved organic compounds that are carrying Fe and Al (Greenland 1971,

Chesworth and Macias-Vasquez 1985, Cronan and Aiken 1985). Typically Fe minerals precipitate first, and Al moves lower in the profile (Adams et al. 1980, McDowell and Wood 1984), but the mechanism underlying this difference is not entirely clear. Strongly podzolized soils, Spodosols, are characterized by a dark spodic horizon designated B_s that is rich in Fe and organic matter. Examining an age sequence of soils in Ontario, Protz et al. (1984, 1988) showed that the development of spodic horizons required about 2000–3000 yr. In New England forests, the accumulation of organic matter in the B_s horizon appears to control the loss of dissolved organic carbon in streams (McDowell and Wood 1984; Chapter 8).

Where podzolization is less intense, forest soils are often classified as Alfisols, on the basis of an accumulation of clay in a B horizon, B_t, and high base saturation in the subsoil. In the southeastern United States, soils known as Ultisols are widespread. These soils are recognized by a predominance of iron oxide in the lower profile, which gives a yellowish to deep-reddish coloration to the B horizon. Ultisols have low base saturation and are transitional intermediates to the highly weathered soils of the tropics.

Soils in tropical forests may be many meters in depth, since in many areas they have developed over millions of years, without disturbances such as glaciation (Birkeland 1984). In the absence of clear zones of eluviation and illuviation, the distinction of A and B horizons is difficult. Long periods of intense weathering have removed cations and silicon from the soil profile. Many tropical soils are known as Oxisols, on the basis of high contents of Fe and Al oxides throughout the soil profile (Richter and Babbar, in press). Over large portions of the lowland tropical rainforest region, soils are acid (from Al buffering), low in base cations, and infertile with P deficiency (Sanchez et al. 1982a). In extreme conditions, these soils are known informally as laterite.

A comparative index of soil formation and the degree of weathering is seen in the ratio of Si to sesquioxides (Fe and Al) in the soil profile (Table 4.3). In boreal forest soils, Si is relatively immobile and Fe and Al are removed, which results in high values for this ratio in the A horizon. The accumulation of secondary minerals such as montmorillonite in the moderately weathered soils of the glaciated portion of the United States yields Si/sesquioxide ratios of 2–4, as a result of the ratio of Si to Al in the crystal lattice. Silicon/sesquioxide ratios are lower in more highly weathered soils. In the southeastern United States, low ratios characterize soils in which kaolinite (a 1:1 mineral) has accumulated as a secondary mineral. Tropical Oxisols and Ultisols have very low values for this ratio in all horizons, since they are dominated by kaolinite and iron and aluminum hydroxides.

Below the B horizons, the C horizon consists of coarsely fragmented

Table 4.3 Silicon/Sesquioxide (Al_2O_3 + Fe_2O_3) Ratios for the A and B Horizons of Some Soils in Different Climatic Regions[a]

Region	Number of Sites	Mean Si/Sesquioxide Ratio		Reference
		A Horizon	B Horizon	
Boreal	1	9.3	6.7	Leahey (1947)
Cool-temperate	4	4.07	2.28	Mackney (1961)
Warm-temperate	6	3.77	3.15	Tan and Troth (1982)
Tropical	5	1.47	1.61	Tan and Troth (1982)

[a] Note that the removal of Al and Fe results in high values in boreal and cool temperate soils, especially in the A horizon. Lower values characterize tropical soils and there is little differentiation between horizons as a result of the removal of Si from the entire profile in long periods of weathering. From Waring and Schlesinger (1985).

soil material with little organic content. When the soil has developed from local materials, the C horizon shows mineralogical similarity to the underlying parent rock, but when the parent materials have been deposited by transport, there may be little resemblance between the C horizon and the underlying bedrock. In either case, carbonation weathering tends to predominate in the C horizon (Ugolini et al. 1977).

The distribution of soil groups forms a continuous gradient over broad geographic regions, in response to parent materials, topography, climate, vegetation, and time (Jenny 1980). For example, the degree of podzolization varies strongly beneath deciduous and coniferous forests in the same geographic region (Stanley and Ciolkosz 1981, De Kimpe and Martel 1976). Soil profile development on steep slopes is often incomplete as a result of landslides and other mechanical weathering events. Soils in floodplain areas and those exposed to deposits of volcanic ash may have "buried" horizons. In areas of recent disturbance, soils with little or no profile development are known as Inceptisols and Entisols, respectively. In all cases one must remember that soil profile development is slow compared to changes in vegetation.

Soil profile development is also affected by human activities. In many areas of the Piedmont region of the southeastern United States, the forest floor often resides directly on top of the B horizon. Here, the A horizon has been lost by erosion during past agricultural use. In the northeastern United States, forest Spodosols are now exposed to acid rain. At high elevations the acidity of the solution percolating through the forest floor and A horizon is not dominated by organic acids or bicarbonate, but by "strong" acids such as H_2SO_4 (see Chapter 13). At normal levels of acidity, aluminum is mobile only as an organic chelate, and it is precipitated in the B horizon. Under the present conditions of higher acidity, Al is mobile as Al^{3+}, which is carried through the lower profile to stream waters with SO_4^{2-} as a balancing anion (Fig. 4.5) (Johnson et al. 1972, Cronan 1980, Reuss et al. 1987). The total rate of rock weathering has increased (Cronan 1980, April et al. 1986).

Grasslands

In contrast to soil development in forests, where precipitation greatly exceeds evapotranspiration and excess water is available for soil leaching and runoff, soil development in grassland proceeds under variations of relative drought. Under these conditions, there is lower plant production, and smaller quantities of plant residue are transferred to the pool of soil organic matter each year. Nevertheless, grassland soils contain large stores of soil organic matter, because the limited availability of water results in slower rates of decomposition (Chapter 5). Significantly, there is limited production of mobile organic acids during the decomposition of organic matter in these soils. These conditions yield slower rates of weathering and reduce the loss of weathering products through the soil to runoff. Most grassland soils in the temperate zone are classified as Mollisols on the basis of a high content of organic carbon and base saturation in the surface layers.

These trends are seen by examining soils in a transect across the mid-portion of the United States, along which mean annual precipitation decreases from the tallgrass prairie to the shortgrass prairie at the base of the Rocky Mountains. Honeycutt et al. (1990) show that the thickness of the soil profile—measured as depth to the peak content of clay—decreases from east to west along this gradient (Fig. 4.9). Along the same gradient, base saturation, pH, and the content of calcium increase (Ruhe 1984). Similar trends are seen along a gradient of precipitation that includes the tropical grasslands of eastern Africa (Scott 1962).

In conditions of near neutral pH, most organic acids are found complexed with clay minerals, and fulvic acids are relatively immobile in the soil profile. Lower levels of free acidity and less rainfall lead to less intense weathering and increasing stability of clay minerals. High contents of Ca also tend to flocculate clay minerals in the upper soil profile, so that downward movement as colloids in the soil solution is retarded. Often there is little development of a B_t horizon until the upper profile has been leached free of Ca. In the western Great Plains, the leaching of the soil profile is so limited that Ca precipitates as $CaCO_3$ that accumulates in the lower profile in calcic horizons designated B_k and known informally as caliche.

As seen for forest ecosystems, soil properties vary as a result of underlying parent materials and hillslope positions. Schimel et al. (1985) show how the downslope movement of materials results in thin soils on hillslopes and accumulations of organic matter and nitrogen in local depres-

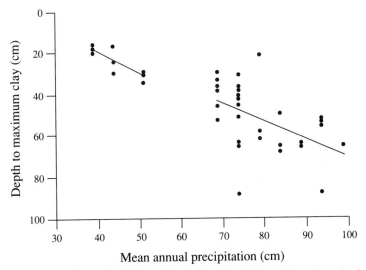

Figure 4.9 Depth to the peak content of clay in the soil profile, an index of weathering and soil development, decreases from east to west across the Great Plains of the United States as a function of the decrease in mean annual precipitation. From Honeycutt et al. (1990).

sions. Similarly, Aguilar and Heil (1988) found that soils derived from sandstone had lower contents of organic carbon, nitrogen, and total P than soils derived from fine-textured materials, such as shale, which have a higher content of clay minerals. The proportion of P contained in organic forms was greater on soils derived from sandstone. Such differences in soil properties can strongly affect the local productivity of grasslands.

Deserts

Trends in soil development that are seen with increasing aridity in grasslands reach their extreme expression in desert soils. There is little runoff from most desert regions, so the soils retain a record of the vertical leaching and horizonal redistributions of materials across the landscape, much like a chromatographic column used in a laboratory. Over much of the southwestern United States, soils are deposited by alluvial transport from adjacent mountain ranges (Cooke and Warren 1973), and soil development can be studied by examining an age sequence of soils that have been deposited from similar parent materials (Fig. 4.10). Recently deposited soils show little profile development and are classified as Entisols; older soils show several distinct horizons and are classified as Aridisols (Dregne 1976).

Chemical weathering proceeds slowly in deserts, but the small amounts of water percolating through these soils transport substances both ver-

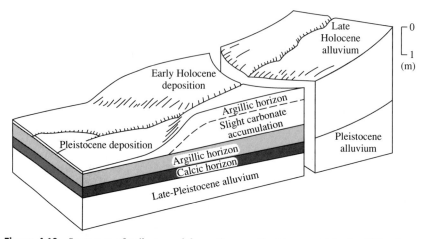

Figure 4.10 Sequence of soil age and formation on alluvial material in the Chihuahuan desert of New Mexico. From Lajtha and Schlesinger (1988).

tically in the profile and horizontally across the landscape. As water is removed by plant uptake, soluble substances precipitate. Typically, well-developed soils contain $CaCO_3$ horizons, known as caliche, that show progressive development and cementation through time (Gile et al. 1966). Depth to the $CaCO_3$ horizon shows a direct relation to mean annual rainfall and wetting of the soil profile (Arkley 1963, 1967, Schlesinger 1982). Absorption on $CaCO_3$ controls the availability of P in most desert soils (Cole and Olson 1959, Lajtha and Schlesinger 1988). Beneath the $CaCO_3$, one may find horizons in which $CaSO_4 \cdot 2H_2O$ (gypsum) or $NaCl$ is dominant, reflecting the greater solubility and downward movement of these salts (Yaalon 1965). Similar patterns are seen across the landscape, where Na, Cl, and SO_4 are carried to intermittant lakes in basin lows, while Ca remains in the upland soils of the adjacent piedmont (Drever and Smith 1978, Eghbal et al. 1989).

Despite sparse plant cover, much of the nutrient cycling in desert ecosystems is controlled by biota. With widespread root systems, desert shrubs accumulate nutrients from a large area, and concentrate dead organic matter in the local area beneath their canopy. Most of the annual turnover of N, P, and other elements is controlled by biogeochemical processes in these "islands of fertility" (Lajtha and Schlesinger 1986, Klopatek 1987, Burke 1989). In the soil solution beneath shrubs, the deposition of $CaCO_3$ is affected by the presence of dissolved organic materials in the soil solution (Inskeep and Bloom 1986, Reddy et al. 1990). Since $CaCO_3$ precipitates in equilibrium with CO_2 derived from plant root respiration;

$$2CO_2 + 2H_2O \rightleftarrows 2H^+ + 2HCO_3^- \tag{4.14}$$

$$Ca^{2+} + 2HCO_3^- \rightleftarrows CaCO_3 + H_2O + CO_2 \tag{4.15}$$

soil carbonate carries a carbon isotopic signature that can be traced to photosynthesis (Schlesinger 1985b, Quade et al. 1989).

Soil development in desert ecosystems occurs slowly, due to limited weathering and leaching of the soil profile. However, desert soils frequently contain clay minerals deposited in eolian dust, giving the appearance that substantial weathering has occurred. The horizons of many desert soils in the United States are thought to have been formed under conditions of greater rainfall during the latest Pleistocene glaciation. For instance, many desert soils contain horizons of illuvial clay, indicating greater rates of weathering and illuviation than occurs in the modern climate (Nettleton et al. 1975). Similarly, most calcic horizons are >10,000 yr old and the $CaCO_3$ has accumulated at rates of 1.0–5.0 $g/m^2/yr$ from the downward transport of Ca-rich minerals deposited from the atmosphere (Schlesinger, 1985b). McFadden and Hendricks (1985) describe a pattern of increasing accumulation and crystallization of iron oxyhydroxides in desert soils that is analogous to the trends we have discussed for soils is mesic regions.

Models of Soil Development

The processes underlying soil profile development are conducive to simulation modeling. Models of soil chemistry include the equilibrium constants for the weathering reactions described earlier in this chapter and reactions for the exchange of cations and anions between the soil solution and the mineral phases. Depending on the time scale of the simulation, the model usually routes daily or annual precipitation sequentially through the soil profile, during which the solution achieves an equilibrium with soil minerals.

Water is removed from the profile by calculations of evaporation from the surface and plant uptake, or by runoff to streams. Models have been constructed to simulate soil profile development and to calculate losses of dissolved constituents in forested regions subject to acid rain (Reuss 1980, Cosby et al. 1985, 1986, David et al. 1988) and in arid regions in which gypsum is accumulating in the soil profile (Robbins et al. 1980).

Long-term development of arid soil profiles is simulated in a model, CALDEP, developed by Marion et al. (1985), in which daily precipitation achieves equilibrium with carbonate biogeochemistry as it percolates through the soil profile. Plant root respiration is explicitly included in the calculation, and varies seasonally and with depth in the profile. Plants also control the loss of water from the soil surface in transpiration. The model suggests that the development of the $CaCO_3$ horizon will be deeper in the

profile in coarse-textured parent materials, which allow greater percolation, and when plant root respiration varies seasonally, showing high
values of soil CO_2 during the growing season. Using current climatic
conditions to parameterize precipitation and evaporation, the model was
run to simulate 500 yr of soil profile development (Fig. 4.11). It predicted
mean depths to the $CaCO_3$ that were much shallower than observed in a

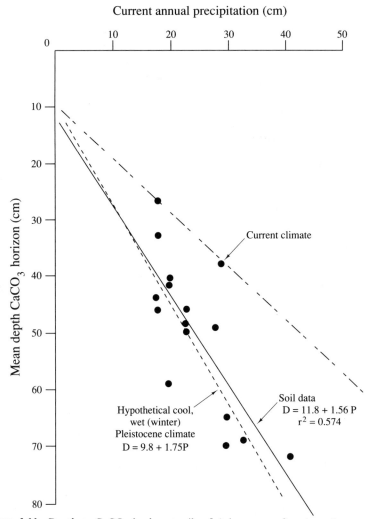

Figure 4.11 Depth to $CaCO_3$ in desert soils of Arizona as a function of mean annual
precipitation. The dashed line (-—-) shows the prediction from the CALDEP model using
current precipitation regimes. The solid line shows the best fit to actual data reported from
the field. The short-dash line (----) shows the predictions when the model is run with
hypothetical climatic data from the latest Pleistocene pluvial period. Modified from Marion
et al. (1985).

sample of 16 desert soils from Arizona. When the model was reparameterized using the cool, wet conditions that are thought to have been widespread in the southwestern United States during the latest Pleistocene glaciation, the predicted depth to $CaCO_3$ closely matched that found in the field. These conditions produced greater percolation of soil moisture and lower rates of evaporation from the soil surface. Such models are only as good as the data used in the simulations, and rarely can models establish the importance of processes unequivocally. Nevertheless, models are useful for hypothesis development and for organizing research priorities. CALDEP suggests that most $CaCO_3$ horizons were formed during conditions of greater precipitation during the Pleistocene. This suggestion is consistent with the ^{14}C age of many caliche layers (Schlesinger, 1985b).

Weathering Rates

Rock weathering and soil formation are difficult to study because the proceses occur slowly and the soil profile is impossible to sample without disturbing many of the chemical reactions of interest. Often we must interpret the probable weathering reactions and estimate weathering rates from what remains in the soil profile and what is lost to stream water. Estimates of weathering are essential to understanding the biogeochemistry of local watersheds, where essential biochemical elements are derived from the underlying rock. Estimates of the dissolved and suspended load of rivers allow us to calculate a global estimate of weathering, which supplies nutrient elements and sediments to the oceans.

Chemical Weathering Rates

One of the best-known attempts to calculate the rate of rock weathering in forests began in 1963, when Gene Likens, Herbert Bormann, and Noye Johnson quantified the chemical budgets for a forest ecosystem in New Hampshire (Likens et al. 1977). Here, a number of comparable watersheds are underlaid by an impermeable bedrock with no flow to groundwater. These workers reasoned that if the atmospheric inputs of chemical elements were subtracted from the stream water losses, the difference should reflect the annual release from rock weathering. They were able to calculate the rate of rock weathering using the equation

$$\text{Weathering} = \frac{(\text{Ca lost in streamwater}) - (\text{Ca received in precipitation})}{(\text{Ca in parent material}) - (\text{Ca in residual material in soil})}$$

The solution of this equation shows rather different amounts of bedrock weathering when the calculations are performed using different rock-forming elements (Table 4.4). Losses of calcium and sodium imply higher

Table 4.4 Calculation of the Rate of Primary Mineral Weathering, Using the Stream-Water Losses and Mineral Concentrations of Cationic Elements[a]

Element	Annual Net Loss (kg/ha/yr)	Concentration in Rock (kg/kg of rock)	Concentration in Soil (kg/kg of soil)	Calculated Rock Weathering (kg/ha/yr)
Ca	8.0	0.014	0.004	800
Na	4.6	0.016	0.010	770
K	0.1	0.029	0.024	20
Mg	1.8	0.011	0.001	180

[a] Data from Johnson et al. (1968).

rates of weathering than those found using potassium and magnesium. Johnson et al. (1968) suggest that the latter elements are accumulating in secondary minerals (illite and vermiculite) in the soil. In addition, plant nutrients may accumulate in long-term biotic storage (e.g., wood growth) in this ecosystem (Chapter 6).

A similar mass-balance approach was used by Garrels and MacKenzie (1967) to elucidate the weathering reactions responsible for the dissolved ions found in springs of the Sierra Nevada (Table 4.5). Reasoning that the weathering reactions would begin with the most labile minerals (Fig. 4.1), these workers suggested that all of the Ca in stream water was derived from the weathering of Ca-feldspar (plagioclase), which they found in the parent rock, to produce kaolinite, which they found in soils. Using a backward approach, they calculated the amount of feldspar that must have been weathered to explain the observed loss of Ca in stream waters. That amount also explained some or all of the loss of other ions as well. Similarly, they reasoned that biotite weathering was likely to begin after Ca-feldspar; backward calculations to reform biotite were used to explain Mg in spring waters. When weathering of K-feldspar was used to explain the loss of the remaining ions, the overall balance was exceptionally good. Only a small amount of silicate, perhaps from the weathering of quartz, remained unexplained.

A large number of watershed studies have been conducted, allowing compilations of weathering rates for a variety of ecosystems (Likens et al. 1977, Henderson et al. 1978, Feller and Kimmins 1979). Weathering rates and the release of biochemically important ions depend on climate and the composition of the underlying parent materials. In most areas of silicate rock, the loss of elements in streamwater relative to their concentration in bedrock follows the order

$$Ca > Na > Mg > K > Si > Fe > Al \qquad (4.17)$$

Table 4.5 Reconstruction of source minerals for ephemeral springs of the Sierra Nevada[a]

Reaction (coefficients × 10⁴)	Na$^+$	Ca^{2+}	Mg^{2+}	K$^+$	HCO$_3^-$	SO$_4^{2-}$	Cl$^-$	SiO$_2$	Notes
Initial concentrations in spring water	1.34	0.78	0.29	0.28	3.28	0.10	0.14	2.73	Derived from rock
Minus concentrations in snow water	1.10	0.68	0.22	0.20	3.10	—	—	2.70	
Change kaolinite back into plagioclase					Minus plagioclase				
Kaolinite									
$1.23\ Al_2Si_2O_5(OH)_4 + 1.10\ Na^+ + 0.68\ Ca^{2+}$ $+ 2.44\ HCO_3^- + 2.20\ SiO_2 =$	0.00	0.00	0.22	0.20	0.64	0.00	0.00	0.50	$1.77\ Na_{0.62}Ca_{0.38}$ feldspar
Plagioclase									
$1.77\ Na_{0.62}Ca_{0.38}Al_{1.38}Si_{2.62}O_8 + 2.44\ CO_2 +$ $3.67\ H_2O$									
Change kaolinite back into biotite					Minus biotite				
Kaolinite									
$0.037\ Al_2Si_2O_5(OH)_4 + 0.073\ K^+ + 0.22\ Mg^{2+}$ $+ 0.15\ SiO_2 + 0.51\ HCO_3^- = 0.073$	0.00	0.00	0.00	0.13	0.13	0.00	0.00	0.35	0.073 biotite
Biotite									
$KMg_3AlSi_3O_{10}(OH)_2 + 0.51\ CO_2 + 0.26\ H_2O$									
Change kaolinite back into K-feldspar					Minus K-feldspar				
K-feldspar									
$0.065\ Al_2Si_2O_5(OH)_4 + 0.13\ K^+ + 0.13\ HCO_3^-$ $+ 0.26\ SiO_2 =$	0.00	0.00	0.00	0.00	0.00	0.00	0.00	0.12	0.13 K-feldspar
K-feldspar									
$0.13\ KAlSi_3O_8 + 0.13\ CO_2 + 0.195\ H_2O$									

[a] From Garrels and MacKenzie in W. Stumm (ed.) Equilibrium Concepts in Natural Water Systems. Copyright 1967 American Chemical Society.

but the order is affected by the specific composition of bedrock and the secondary minerals that are formed in the soil profile (Holland 1978, Harden 1988, Olsson and Melkerud 1989). This general order reflects the tendency for Ca and Na silicates to weather easily and for little involvement of Ca and Na in the formation of secondary minerals. In most cases, Fe and Al are retained in the lower soil profile as oxides, and are essentially immobile (Chesworth et al. 1981).

Release from rock weathering is the dominant source of Ca, Mg, K, Fe, and P for the Hubbard Brook Experimental Forest, whereas deposition from the atmosphere is the dominant input for Cl, S, and N, which have a small content in rocks (Table 4.6). In forests not subject to severe input of acid rain, the proportion of sulfur that is derived from the atmosphere is lower (e.g., Mitchell et al. 1986). High stream-water contents of SO_4^{2-} in watersheds of the eastern United States (Table 4.7) probably reflect the widespread deposition of SO_4 in acid rain, despite an attempt to correct these data for atmospheric inputs (see Likens et al. 1977). A high stream-water concentration of HCO_3^- in the rain forest of Venezuela (Table 4.7) reflects the importance of carbonation weathering in tropical ecosystems. The net mobilization of cations and silicon is also high in Venezuela, consistent with our concepts of soil profile development in tropical climates.

Since temperate forest soils are dominated by clay minerals with permanent negative charge, the loss of cations to stream waters is determined by the availability of mobile anions that pass through the soil profile (Gorham et al. 1979, Johnson and Cole 1980, Terman 1977). Losses of cations due to elevated inputs of SO_4^{2-} in acid rain are lower in soils that possess anion exchange capacity due to iron and aluminum minerals

Table 4.6 Inputs and Outputs of Elements from the Hubbard Brook Experimental Forest, New Hampshire[a]

	Inputs (%)		Output as a Percent of Input
	Atmosphere	Weathering	
Ca	9	91	59
Mg	15	85	78
K	11	89	24
Fe	0	100	25
P	1	99	1
S	96	4	90
N	100	0	19
Na	22	78	98
Cl	100	0	74

[a] Data from Likens *et al.* (1981).

(Reuss and Johnson 1986, Harrison et al. 1989, Cronan et al. 1990). However, when the anion absorption capacity is saturated, increasing concentrations of SO_4^{2-} are expected in stream water (Ryan et al. 1989). In most soils, the dominant anion in soil water is bicarbonate (HCO_3^-), but elevated losses of nitrate (NO_3^-) may increase the loss of cations following forest cutting (Chapter 6). Thus, plant uptake of available nitrogen can control the apparent chemical weathering of the landscape. In the absence of nitrogen uptake, nitrification rates increase, nitrate is lost, and the soil is left with a lower base saturation and pH.

Losses of dissolved constituents from terrestrial ecosystems represent the products of chemical weathering and constitute chemical denudation of the landscape. Despite exchange reactions that may retain weathering products for short periods of time, the eventual loss of cations to riverflow explains the decline in base saturation and pH during soil development (Bockheim 1980). In comparisons of ecosystems of the world, total chemical denudation is found to increase with increasing runoff (Holland 1978). Alexander (1988) found that chemical denudation ranged from 19 to 264 kg/ha/yr in 18 undisturbed ecosystems, and used rates of chemical weathering to calculate the rate of soil formation in different regions.

Total chemical denudation transports about 4×10^{15} g of dissolved substances to the ocean each year (Table 4.8). The chemical weathering of

Table 4.7 Net Transport (Export Minus Atmospheric Deposition) of Major Ions, Soluble Silica, and Suspended Solids from various Watersheds of Forested Ecosystems

Watershed Characteristics	Caura River, Venezuela	Gambia River, W. Africa	Catoctin Mtns., Maryland	Hubbard Brook, New Hampshire
Size (km²)	47,500	42,000	5.5	2
Precipitation (cm)	450	94	112	130
Vegetation	Tropical forest	Savanna forest	Temperate forest	Temperate forest
Soluble transport (kg/ha/yr)				
Na	19.4	3.9	7.3	5.6
K	13.6	1.4	14.1	1.0
Ca	14.2	4.0	11.9	11.5
Mg	5.7	2.0	15.6	2.5
HCO_3^-	124.0	20.3	78.1	7.7
Cl^-	−1.4	0.6	16.6	−1.6
SO_4^{2-}	1.5	0.4	21.2	14.8
SiO_2	195.7	15.0	56.1	37.7
Solids transport (kg/ha/yr)	274	49	—	33

a From Lewis et al (1987).

Table 4.8 Chemical and Mechanical Denudation of the Continents

Continent	Chemical Denudation[a]		Mechanical Denudation[b]		Ratio Mechanical / Chemical
	Total (10^{14}g/yr)	Per Unit Area (Mg/km²/yr)	Total (10^{14}g/yr)	Per Unit Area (Mg/km²/yr)	
North America	7.0	33	14.6	84	2.1
South America	5.5	28	17.9	100	3.3
Asia	14.9	32	94.3	304	6.3
Africa	7.1	24	5.3	35	0.7
Europe	4.6	42	2.3	50	0.5
Australia	0.2	2	0.6	28	3.0
Total	39.3		135.0		3.4

[a] From Garrels and MacKenzie (1971).
[b] From Milliman and Meade (1983).

primary minerals in igneous rocks accounts for 27% of the dissolved constituents delivered to the ocean, while chemical weathering of sedimentary rocks accounts for the remainder (Li 1972), roughly in proportion to their exposure on land (Blatt and Jones 1975). Since chemical weathering involves the reaction between atmospheric constituents and rock minerals, weathering of 100 kg of igneous rock results in 113 kg of sediments that are deposited in the ocean and about 2.5 kg of salts that are added to seawater (Li 1972). Thus, a significant fraction of the transport of total disolved substances in rivers (Table 4.8) is derived from the atmosphere and does not represent true denudation of the continents (Berner and Berner 1987). The global rate of chemical denudation is important to biogeochemistry since it determines the supply of many nutrient elements to land biota (Chapter 6), rivers (Chapter 8), and the ocean (Chapter 9).

Total Denudation Rates

In addition to chemical denudation, a large amount of material derived from mechanical weathering is removed from land and carried in rivers as particulate or suspended load. These materials have received less attention by biogeochemists, because their elemental contents are not immediately available to biota; however, the *total* denudation of land is dominated by the products of mechanical weathering, which exceeds chemical weathering by three to four times, worldwide (Table 4.8). The importance of mechanical weathering increases with increasing elevation; differences in mean elevation among continents explain much of the variation in mechanical weathering (Table 4.8). Milliman and Meade (1983)

Table 4.9 Concentrations of Major Elements in Continental Rocks and Soils and in River Dissolved and Particulate Matter[a]

Element	Continents		Rivers				Element Weight Ratio	
	Surficial Rock Concentration (mg/g)	Soil Concentration (mg/g)	Particulate Concentration (mg/g)	Dissolved Concentration (mg/l)	Particulate Load (10^6 tons/yr)	Dissolved Load (10^6 tons/yr)	River Particulate/ Rock	Particulate/ (Particulate + Dissolved)
Al	69.3	71.0	94.0	0.05	1457	2	1.35	.999
Ca	45.0	15.0	21.5	13.40	333	501	0.48	.40
Fe	35.9	40.0	48.0	0.04	744	1.5	1.33	.998
K	24.4	14.0	20.0	1.30	310	49	0.82	.86
Mg	16.4	5.0	11.8	3.35	183	125	0.72	.59
Na	14.2	5.0	7.1	5.15	110	193	0.50	.36
Si	275.0	330.0	285.0	4.85	4418	181	1.04	.96
P	0.61	0.8	1.15	0.025	18	1.0	1.89	.82

[a] From Berner and Berner (1987).

calculate the total transport of suspended materials in all rivers of the world as 13.5×10^9 tons/yr, 70% of which is carried by the rivers of southeast Asia. Assuming that the specific gravity of suspended sediment is 2.5 g/cm^3, their estimate is four times higher than an estimate (1.27 km^3/yr) of the volume of deep ocean sediments derived from land (Howell and Murray 1986). Presumably, the remainder is deposited near the shore, in continental shelf sediments (Chapter 9).

Particulate and suspended sediments account for the bulk of the removal of Fe, Al, and Si from terrestrial ecosystems, since these elements are poorly soluble in water (Table 4.9). In addition, the loss of phosphorus and trace metals from land is largely carried in the suspended load of rivers, since these elements tend to absorb to the surface of particulates and organic matter (Martin and Meybeck 1979, Avnimelech and McHenry 1984). Other products of chemical weathering are found almost entirely in the dissolved load.

Summary

In this chapter we have seen that the rate of weathering and soil development is strongly affected by biota, particularly through carbonation weathering and the production of organic acids. It is tempting to speculate that the rate of carbonation weathering was lower before the advent of land plants, when it depended solely on the downward diffusion of atmospheric CO_2 through the soil profile. However, at periods in the Earth's history the concentration of atmospheric CO_2 was most certainly higher than today, yielding high rates of carbonation weathering. Weathering is also driven by the availability of water. The high concentration of CO_2 on Venus (Table 2.3) is ineffectual in weathering because the surface of the planet is dry (Nozette and Lewis 1982).

Human activities have increased the rate of both chemical and mechanical weathering. Fossil fuel combustion and mining have added significant quantities of dissolved materials to global riverflow (Bertine and Goldberg 1971, Martin and Meybeck 1979). Exposure and erosion of soils have increased the global denudation due to mechanical weathering by a factor of about 2 (Gregor 1970), leading to increases in the rate of sediment accumulation in estuaries and river deltas.

Chemical weathering is a source of essential elements for the biochemistry of life, but stream-water runoff removes these elements from the land surface. Chemical reactions among soil constituents and uptake by biota determine the rate of loss, but the inevitable removal of cations results in lower soil pH and base saturation through time (Bockheim 1980). Phosphorus is particularly critical as a soil mineral, since it is not abundant in crustal rocks and easily precipitated in unavailable forms in the soil. Old soils in highly weathered landscapes are formed from the accumulation of resistant, residual Fe and Al oxide minerals. In these soils, P is often deficient for plant growth.

Recommended Reading

Birkeland, P.W. 1984. Soils and Geomorphology. Oxford University Press, Oxford.

Garrels, R.M. and F.T. MacKenzie. 1971. Evolution of Sedimentary Rocks. W.W. Norton Company, New York.

Reuss, J.O. and D.W. Johnson. 1986. Acid Deposition and the Acidification of Soils and Waters. Springer-Verlag, New York.

5

The Terrestrial Biosphere

Introduction

Photosynthesis is the biogeochemical process that acts to transfer carbon from its oxidized form, CO_2, in the atmosphere to reduced forms in the tissues of plants. Directly or indirectly, photosynthesis provides the energy for all other forms of life in the biosphere, and the use of plant products for food, fuel, and shelter links biogeochemistry to our daily lives. The growth of plants also affects the composition of the atmosphere (Chapter 3) and the development of soils (Chapter 4), linking photosynthesis to other aspects of global biogeochemistry. Indeed, the presence of organic carbon is the basis for the striking contrast between the biogeochemistry on Earth and the simple geochemistry that controls processes on our neighboring planets.

In this chapter we will consider the measurement of net primary production, defined as the accumulation of reduced carbon in the tissues of land plants. Similar treatment of photosynthesis in the world's oceans is given in Chapter 9. The rate of net primary production varies widely over the land surface. Deserts and continental ice masses may have little or no net production, while tropical rainforests may show annual production of > 1000 g C m^{-2}. We will consider the factors that determine the net primary productivity of land plant communities. As any home gardner knows, light and water are important, but plant growth is also determined by the stock of available nutrients that are ultimately derived from the atmosphere and from the underlying bedrock. Finally we will attempt to estimate the global rate of net primary productivity and the total storage of reduced carbon in plant tissues (biomass), dead plant parts (detritus), and soil organic matter. The storage of carbon on land is determined by the balance between primary production and decomposition, which returns carbon to the atmosphere as CO_2 (Schlesinger 1977).

Photosynthesis

Photosynthesis occurs in chloroplasts of plant leaf cells (Fig. 5.1). A pigment, chlorophyll, absorbs sunlight energy, especially in red and blue wavelengths, and transfers that energy to chemical reactions, where it is captured in chemical bonds of adenosine triphosphate (ATP) and other reduced coenzymes. Chlorophyll contains magnesium (Mg), an essential nutrient that plants take up from the soil (Chapter 4). The fact that chlorophyll absorbs only some wavelengths of light is the basis for measuring leaf area and photosynthsis from satellites, as we will see in a later section.

During the capture of sunlight energy, O_2 is released by the splitting of water molecules. The water molecule is split by an enzyme, which contains manganese, located in the membranes of the plant chloroplast (George et al. 1989). With the aid of another enzyme, ribulose bisphosphate carboxylase, the high-energy compounds are then used to build carbohydrate molecules from CO_2. The net reaction is

$$CO_2 + H_2O \rightarrow CH_2O + O_2 \tag{5.1}$$

but we should remember that the process occurs in two stages—the capture of light energy followed by carbon reduction.

Carbon dioxide for photosynthesis diffuses into plant leaves through pores, stomates, that are generally found on the lower surface of broadleaf species. One factor that determines the rate of photosynthesis is the stomatal aperture, which plant physiologists express as stomatal conductance in units of cm/s. Stomatal conductance is controlled primarily by the availability of water and the concentration of CO_2 *inside* the leaf, where it

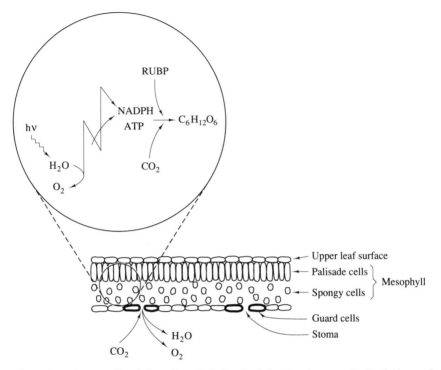

Figure 5.1 Cross-sectional view of a typical plant leaf, showing the upper (palisade) layer of cells, in which photosynthesis occurs, and guard cells, which control the diffusion of CO_2 (in) and H_2O and O_2 (out) through stomates on the lower surface. A summary of the photosynthetic reaction, occurring in the chloroplasts of the palisade cells is shown in the insert.

is consumed in photosynthesis. When well-watered plants are actively photosynthesizing, internal CO_2 is relatively low and stomates show maximum conductance. Under such conditions, the amount and activity of the primary carboxylase enzyme, ribulose bisphosphate carboxylase,[1] which adds CO_2 to small carbohydrate units, may determine the rate of photosynthesis (Sharkey 1985). In most cases, however, the rate of diffusion of CO_2 through the stomates determines the rate of photosynthesis.

Water and Nutrient Use Efficiency in Photosynthesis

When plant stomates are open, O_2 and H_2O diffuse outward to the atmosphere. The loss of water through stomates, transpiration, is a major

[1] For understanding global biogeochemistry, we will consider only photosynthesis in C3 plants, which comprise the overwhelming proportion of net primary productivity and plant biomass on Earth. While the overall reaction for photosynthesis in C4 plants is identical, the biochemical pathway is different, with different water and nutrient use efficiency and different isotopic fractionation in plant carbon.

mechanism by which soil moisture is returned to the atmosphere (Chapter 10). In the Hubbard Brook Experimental Forest in New Hampshire (see Chapter 4), about 25% of the annual precipitation is lost by plant uptake and transpiration; stream flow increased by 26–40% when the forest was clear-cut (Pierce et al. 1970). Since water is often in short supply for plant growth (Kramer 1982), the large losses of water by plants are somewhat surprising. One might expect natural selection for more efficient use of water by plants, especially in droughty environments.

Plant physiologists express the loss of water relative to photosynthesis as water use efficiency (WUE):

$$\text{WUE} = \text{mmoles of } CO_2 \text{ fixed/moles of } H_2O \text{ lost} \qquad (5.2)$$

This provides a measure for the loss of water at a given rate of photosynthesis. For C3 plants, water use efficiency typically ranges from 0.86 to 1.50 mmol/mol, depending upon environmental conditions (Osmond et al. 1982). Water use efficiency is lower at higher stomatal conductance.

Estimation of water use efficiency is difficult, because stomatal conductance changes continuously as plants respond to environmental conditions. Equation (5.2) is largely used by plant physiologists working in the laboratory. For the biogeochemist, long-term average water-use efficiency may be estimated from the carbon isotope composition of plant tissues. This method is based on the observation that the diffusion of $^{12}CO_2$ is more rapid than that of $^{13}CO_2$, which comprises about 1.1% of atmospheric CO_2. Thus, in a given period of time more $^{12}CO_2$ enters the leaf than $^{13}CO_2$. Inside the leaf, ribulose bisphosphate carboxylase also has a higher affinity for $^{12}CO_2$. As a result of these factors, plant tissue contains a lower proportion of $^{13}CO_2$ than the atmosphere by about 2% (= 20‰)(O'Leary, 1988). The discrimination (fractionation) between carbon isotopes is expressed relative to an accepted standard as

$$\delta^{13}C = \left[\frac{^{13}C/^{12}C_{\text{sample}} - {}^{13}C/^{12}C_{\text{standard}}}{{}^{13}C/^{12}C_{\text{standard}}} \right] \times 1000 \qquad \text{Eq. 5.3}$$

and expressed in parts per thousand parts (‰). Since atmospheric CO_2 shows an isotopic ratio of $-8.0‰$ versus the standard, most plant tissues show $\delta^{13}C$ of $\sim -28‰$ [i.e., $(-8‰) + (-20‰)$]. Sedimentary organic carbon with the isotopic signature of photosynthesis is useful in determining the antiquity of photosynthesis as a biochemical process (Fig. 2.5).

The discrimination between $^{12}CO_2$ and $^{13}CO_2$ during photosynthesis is greatest when stomatal conductance is high (Fig. 5.2). When stomates are partially or completely closed, nearly all of the CO_2 inside the leaf reacts with ribulose bisphosphate carboxylase and there is little fractionation of the isotopes. Thus, the isotopic ratio of plant tissue is directly related to

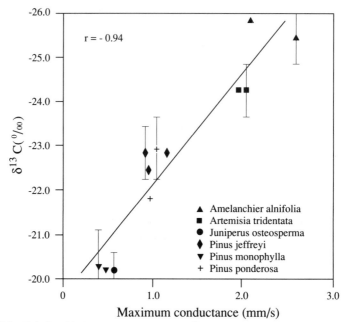

Figure 5.2 Relationship between the content of ^{13}C in plant tissues (expressed as $\delta^{13}C$) and stomatal conductance for a variety of plant species in western Nevada. From Delucia et al. (1988).

the average stomatal conductance during its growth, providing a long-term index of water use efficiency (Farquhar et al. 1989).

Over a broad range of plant species, the rate of photosynthesis is directly correlated to leaf nitrogen content (Fig. 5.3). Most leaf nitrogen is contained in enzymes; ribulose bisphosphate carboxylase alone accounts for 6–20% of leaf nitrogen. Seemann et al. (1987) found that the photosynthetic potential is directly related to the content of ribulose bisphosphate carboxylase and leaf nitrogen in several species. These data suggest that the availability of nitrogen determines the leaf content of ribulose bisphosphate carboxylase and the rate of photosynthesis in land plants. In addition to nitrogen, leaf phosphorus content may be an important determinant of photosynthetic capacity in some species (Reich and Schoettle 1988). Other essential elements, such as magnesium and manganese, are seldom in short supply for plant growth.

Since most land plants grow under conditions of nitrogen deficiency, we might expect adjustments in nutrient use to maximize photosynthesis. The rate of photosynthesis per unit of leaf nitrogen is one measure of nutrient use efficiency (NUE). Subtle variations in the slope of the relationship in Fig. 5.3 reflect differences in NUE during photosynthesis among plants grown in different environments (Evans 1989). For many

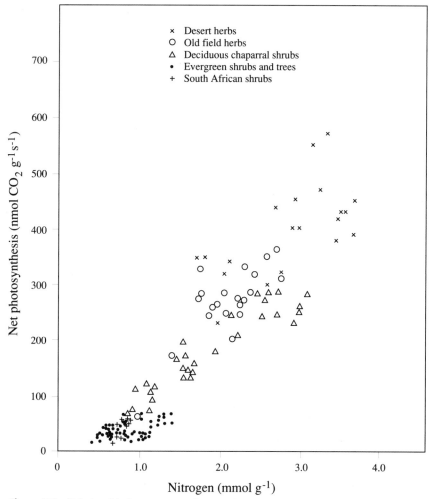

Figure 5.3 Relationship between net photosynthesis and leaf nitrogen content among 21 species from different environments. From Field and Mooney (1986).

plant species, when leaf nutrient content increases (by fertilization), NUE declines (Ingestad 1979b, Lajtha and Whitford 1989). Nutrient use efficiency also appears inversely correlated to WUE across many species (Field et al. 1983, Lajtha and Whitford 1989, Reich et al. 1989, DeLucia and Schlesinger, 1991).

Respiration

Photosynthesis is usually measured by placing leaves or whole plants in closed chambers and measuring the uptake of CO_2 or release of O_2. The rates are a measure of *net* photosynthesis by the plant, that is, the fixation of carbon in excess of the release of carbon by plant respiration. Respiration is largely the result of mitochondrial activity in plant cells, and in woody plants a large fraction of the respiration is contributed by stems and roots (Waring and Schlesinger 1985). For leaf tissues, rates of respiration are higher in the daytime than during the night as a result of the additional process of photorespiration (Sharkey 1988).

Independent measures of respiration suggest that about one-half of the gross carbon fixation by photosynthesis is used by plants, so the actual rate of photosynthesis is often twice that which is measured as plant growth (Box et al. 1978). For long-lived woody plants, maintenance respiration increases with stand age, consuming an increasing fraction of the gross photosynthesis and eventually leading to a reduction in the rate of plant growth (Kira and Shidei 1968, Waring and Schlesinger 1985).

Net Primary Production

When we consider plants in the field, we say that

$$\text{Gross primary production} - \text{plant respiration} = \text{net primary production}$$
$$\text{(GPP)} \qquad\qquad \text{(R}_P) \qquad\qquad \text{(NPP)} \qquad (5.4)$$

Net primary production is not directly equivalent to plant growth as measured by foresters, ranchers, and farmers. Some fraction of NPP is lost to herbivores and in the death and loss of tissues, known collectively as litterfall. Ecologists frequently call the NPP that remains the *true increment*, which may add to the accumulation of biomass over many years. When mortality occurs during forest development, true increment is the net increase in the mass of woody tissue in living plants, after subtracting the mass of individuals that die over the same interval.

For the biogeochemist NPP is expressed on an area basis, often in units of kg ha^{-1} yr^{-1}, by multiplying the growth of individual plants by the density of plants in the field. Net primary production can also be expressed in units of energy, by measurements of the caloric content of various plant tissues (Darling 1976). This expression is useful for expressing the efficiency of photosynthesis relative to the receipt of sunlight energy (e.g., Botkin and Malone 1968, Reiners 1972).

The measurement of NPP in the field is not easy, but the methods are well developed and reviewed extensively elsewhere for forests (Whittaker and Marks 1975) and grasslands (Singh et al. 1975). Traditional methods for forests and shrublands involve the harvest of vegetation and calculation of the annual growth of wood and the mass of foliage at the peak of annual leaf display. Independent estimates of the seasonal loss of plant parts can be obtained from collections of litterfall through the year. In grasslands, there is little or no true increment, and estimates of net primary production generally involve the difference in the mass of tissue harvested from small plots at the beginning and the end of the growing season (e.g., Wiegert and Evans 1964, Lauenroth and Whitman 1977). These estimates must be corrected for the consumption and loss of tissues during that interval.

Allocation of net primary production varies with vegetation type and age. In forests 25–35% of aboveground production is found in leaves (Whittaker et al. 1974), with this percentage tending to decrease with stand age. Allocation to foliage in shrublands is generally greater, ranging from 35 to 60% in desert and chaparral shrubs (Whittaker and Niering 1975, Gray 1982). In grassland communities, essentially all net aboveground primary production is found in photosynthetic tissue.

Comparing plant communities in different regions, Jordan (1971) found that the allocation of NPP to wood growth was greater in boreal forests than in the tropics, that is, there is greater wood production per unit foliage in boreal forests. As a result of their massive structure and high environmental temperatures tropical forests may expend a greater percentage of their gross primary production in respiration (Whittaker and Marks 1975), leaving less for wood growth. Webb et al. (1983) found a logarithmic relationship between total aboveground NPP and foliage biomass for a variety of plant communities in North America, with some deserts showing exceptionally high values of this ratio (Fig. 5.4). However, compared to communities with abundant precipitation, desert shrublands show relatively low allocation of NPP to wood production (Jordan 1971), perhaps as a result of a large allocation to roots.

Root growth is difficult to study, and many estimates of NPP include data only from the aboveground tissues. Nevertheless, when roots have been examined carefully, the annual growth and turnover of root tissues comprise a significant fraction of the NPP in most communities. Edwards and Harris (1975) reported that the growth and death of roots delivered 733 g C m^{-2} yr^{-1} to the soil in a forest in Tennessee, where the aboveground production was 685 g C m^{-2} yr^{-1} (Reichle et al. 1973a). Similarly, Vogt et al. (1982) found that roots comprise 59–67% of NPP in coniferous forests in Washington (Table 5.1). An even larger proportion of NPP is allocated to root growth in many grassland ecosystems (Lauenroth and Whitman 1977, Warembourg and Paul 1977). Much of the published data probably underestimates NPP on land by overlooking the importance of root growth. In forests the proportional allocation of photosynthate to root growth varies as an inverse function of site fertility (Waring and Schlesinger 1985), although *total* root growth is greatest on nutrient-rich sites (Raich and Nadelhoffer 1989).

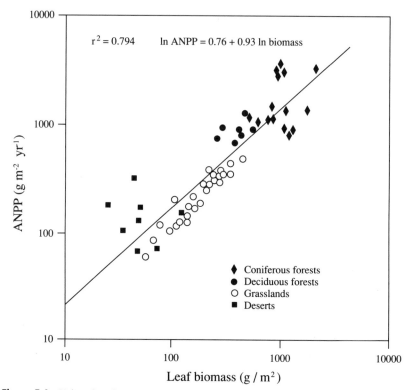

Figure 5.4 Using data from a variety of ecosystems in North America, Webb et al. (1983) found a strong relation between the annual aboveground net primary production (ANPP) and the biomass of foliage.

Remote Sensing of Primary Production

Harvest measurements of NPP are labor intensive and necessarily applied only to small areas. Since the productivity of vegetation may vary greatly over the landscape, regional estimates of productivity by harvest become prohibitively expensive. Nevertheless, for understanding biogeochemistry, regional and global estimates are essential, and various methods using remote sensing to provide integrated estimates of NPP are currently under development.

The most recent LANDSAT satellite gathered information on the reflectance of the Earth's surface, collecting data in discrete portions of the visible and infrared spectrum, labeled TM 1-7 in Fig. 5.5. The LAND-SAT instrument measures an average reflectance for a 30 × 30 m plot or pixel on land. Bare soil shows similar reflectance in the TM4 and TM3 wavelengths, whereas vegetation shows a TM4/TM3 ratio >>1.0 as a

Table 5.1 Net Primary Production in 23- and 180-yr-old *Abies amabilis* Ecosystems in the Cascade Mountains, Washington[a]

	23-yr-old		180-yr-old	
	$kg \cdot ha^{-1} \cdot yr^{-1}$	% of total	$kg \cdot ha^{-1} \cdot yr^{-1}$	% of total
Aboveground				
Biomass increment				
Tree total	4 260		2 320	
Shrub stems	60		< 10	
Total	4 320	18.37	2 320	9.33
Detritus production				
Litterfall	1 510		2 180	
Mortality	300			
Herb layer turnover	320		50	
Total	2 130	9.06	2 230	8.97
Total aboveground	6 450	27.42	4 550	18.30
Roots				
Fine (≤2 mm)	6 500	27.64	12,900	51.87
Fibrous-textured	5 710		11,960	
Mycorrhizal (host tissue)	790		940	
Coarse (>2 mm)	3 580		3 240	
Angiosperm fine root turnover	3 730		440	
Total root turnover	13,810	58.72	16,580	66.67
Mycorrhizal fungal component				
Sporocarps				
Epigeous	30		30	
Hypogeous	< 10		380	
Sclerotia	2 700		2 700	
Mycorrhizal sheath	530		630	
Total	3 260	13.86	3 740	15.04
Total belowground	17,070	72.58	20,320	81.70
Ecosystem total	23,520		24,870	

[a]From Vogt et al. (1982).

result of the absorption of visible light by chlorophyll. The TM4/TM3 ratio was directly correlated to leaf area for 17 coniferous forests studied by harvest measurements in the northwestern United States (Fig. 5.6). Here, leaf area is expressed as a leaf-area index (LAI), in units of $m^2 \, m^{-2}$, the area of leaves above a unit area of ground surface. Since previous studies had shown a direct relation between LAI and NPP in these forests (Fig. 5.7), the potential extrapolation from the satellite measurements to regional estimates of NPP is obvious. Cook et al. (1989) found a good relationship between LANDSAT data and regional estimates of NPP.

Goward et al. (1985) follow a similar approach using data from an advanced very high resolution radiometer (AVHRR) carried on the

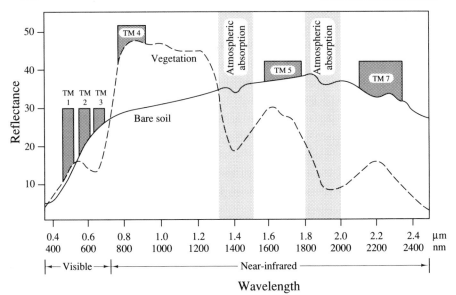

Figure 5.5 A portion of the solar spectrum, showing the typical reflectance from soil (—) and leaf (– – –) surfaces and the portions of the spectrum that are measured by the LAND-SAT satellite.

NOAA-7 satellite. They calculate a normalized difference vegetation index (NDVI):

$$NDVI = (NIR - VIS)/(NIR + VIS) \qquad (5.5)$$

where NIR is reflectance in the near infrared and VIS is reflectance in the visible, respectively. The index minimizes the effects of variations in background reflectance and emphasizes variations in the data that occur because of the density of green vegetation. Their data allow global mapping of a "greenness" index for the Earth's land surface (Plate 1).

Satellite measurements of "greenness" may lead to global estimates of NPP, assuming that "greenness" is directly related to leaf area and that LAI is a good predictor of NPP (Figs. 5.4 and 5.7). Integrations of NDVI measured at frequent intervals over the growing season show a direct correlation to regional average values of NPP measured by harvest methods (Goward et al. 1985, Box et al. 1989). Recently, Fung et al. (1987) have shown that the seasonal patterns of NDVI for the latitudinal bands of the globe are consistent with the magnitude of the seasonal oscillation of atmospheric CO_2 measured at various latitudinal stations (Fig. 3.6) Although the LANDSAT data have finer resolution than those gathered by AVHRR (1.1 km^2), the AVHRR data are more useful in

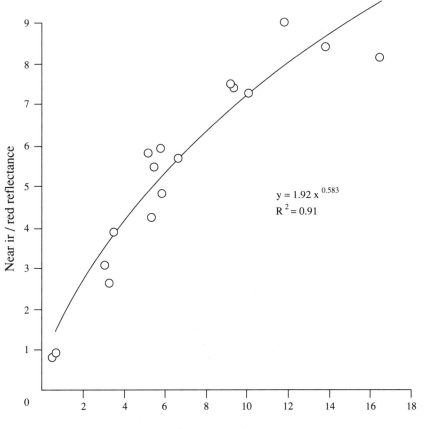

Figure 5.6 The ratio of light reflected in near infrared and red spectral bands (wavebands TM4 and TM3 of the LANDSAT satellite) is related to LAI for forest stands in the Pacific northwest. From Peterson et al. (1987).

global estimates since the number of pixels covering the land surface remains manageable during computer data processing. Running et al. (1989) use AVHRR data to estimate leaf area index for forests in western Montana, and apply the LAI data to a model for forest growth to calculate regional evapotranspiration and NPP.

Global Estimates of Net Primary Production and Biomass

While remote sensing techniques will undoubtedly offer future refinements, current estimates of global net primary production and biomass are largely based on compilations of data from harvest mea-

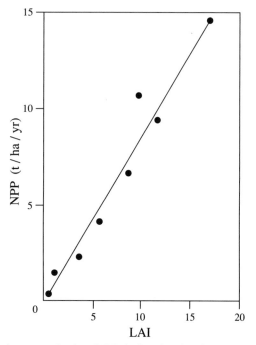

Figure 5.7 Net primary production (NPP) is directly related to leaf area index (LAI) for forests in the northwestern United States. Modified from Gholz (1982).

surements. Among such compilations, the table by Whittaker and Likens (1973) is widely cited (Table 5.2), but alternatives are available (e.g., Atjay et al. 1979). Most of the estimates of NPP are in a range of $45-65 \times 10^{15}$ g C yr^{-1}. Olson et al. (1983) offer the most comprehensive estimate of the total biomass of plants on land; their value is 560×10^{15} gC. The ratio of biomass to NPP is an estimate of the mean residence time (Chapter 3) for an atom of carbon in plant tissues. The global values yield an overall mean residence time of about 9 yr, but this value varies from ~3 in grasslands to >25 in some forests. Of course, we must remember that these are weighted averages. In forests some tissues, such as leaves, may last only a few months, while wood may last for centuries. The carbon content of most plant tissues is 45–50%. When estimates of NPP and biomass are expressed in terms of dry matter, the values are about twice as large as in Table 5.2.

Estimates such as those in Table 5.2 are calculated by classifying the land vegetation into a small number of units and by assigning a mean value to the NPP and biomass of each unit based on data from the widest possible number of field studies. The classification of vegetation is arbi-

Table 5.2 Primary Production and Biomass Estimates for the Biosphere[a]

Ecosystem Type	Area (10^6 km = 10^12 m^2)	Mean Net Primary Productivity (g C/m^2/yr)	Total Net Primary Production (10^9 metric tons C/yr)	Mean Plant Biomass (kg C/m^2)	Total Plant Mass (10^9 metric tons C)
Tropical rain forest	17.0	900	15.3	20	340
Tropical seasonal forest	7.5	675	5.1	16	120
Temperate evergreen forest	5.0	585	2.9	16	80
Temperate deciduous forest	7.0	540	3.8	13.5	95
Boreal forest	12.0	360	4.3	9.0	108
Woodland and shrubland	8.0	270	2.2	2.7	22
Savanna	15.0	315	4.7	1.8	27
Temperate grassland	9.0	225	2.0	0.7	6.3
Tundra and alpine meadow	8.0	65	0.5	0.3	2.4
Desert scrub	18.0	32	0.6	0.3	5.4
Rock, ice, and sand	24.0	1.5	0.04	0.01	0.2
Cultivated land	14.0	290	4.1	0.5	7.0
Swamp and marsh	2.0	1125	2.2	6.8	13.6
Lake and stream	2.5	225	0.6	0.01	0.02
Total continental	149	324	48.3	5.55	827
Open ocean	332.0	57	18.9	0.0014	0.46
Upwelling zones	0.4	225	0.1	0.01	0.004
Continental shelf	26.6	162	4.3	0.005	0.13
Algal bed and reef	0.6	900	0.5	0.9	0.54
Estuaries	1.4	810	1.1	0.45	0.63
Total marine	361	69	24.9	0.0049	1.76
Full total	510	144	73.2	1.63	829

[a]From Whittaker and Likens (1973).

trary, and estimates of the land area in each unit often vary considerably (Golley 1972). Similarly, the NPP data often do not reflect the full range of variation in the field, since ecologists often tend to select mature, well-developed stands for study. Brown and Lugo (1982, 1984) have considered the effect of differences in classification and stand selection on estimates of the biomass of tropical forests. Their data are considerably lower than those reported in Table 5.2. Botkin and Simpson (1990) also report lower, revised estimates of biomass in boreal forest regions. Global estimates of NPP and biomass by remote sensing should help resolve some of these differences.

The data in Table 5.2 suggest that the primary productivity of forests is greatest in the tropics and declines with increasing latitude to low values in boreal forests and shrub tundra. Along a gradient of decreasing precipitation, NPP declines from forests to grasslands, showing very low values in most deserts. Sala et al. (1988) show a direct relation between net primary production and precipitation within the grasslands of the central United States. Wetland vegetation often has high NPP; we will examine wetlands in more detail in Chapter 7.

Ecologists have examined the broad patterns in vegetation data to uncover the main factors controlling global NPP and biomass. Lieth (1975) related NPP in 52 field studies to the mean annual temperature and precipitation recorded in nearby weather stations (Figs. 5.8 and 5.9). He considered temperature to be an index of solar irradiance and to determine the length of the growing season. Each equation was used with local weather data to predict productivity for other areas of the world. Then a map of global productivity was developed using the lower of the two predictions of NPP at each site, to reflect a temperature or moisture limitation on NPP. The global map of NPP (Fig. 5.10) is surprisingly similar to the satellite picture of "greenness" (Plate 1). Leith's (1975) approach suggests that light and moisture are the main factors determining NPP, with available nutrients playing a lesser role. His global map suggests a total terrestrial NPP of about 63×10^{15} g C yr^{-1} (Esser et al. 1982).

Similar evidence for the importance of temperature and moisture is seen in regional comparisons of productivity, especially patterns along gradients of elevation. Whittaker (1975) found that NPP declined with increasing elevation in the forested mountains of the eastern United States, presumably reflecting the influence of declining temperatures. In the southwestern United States, where precipitation is more limited, NPP tends to increase with elevation in communities ranging from desert shrublands to montane forests (Whittaker and Niering 1975). In forests of the northwestern United States, NPP and LAI are directly related to site water balance, which is the difference between precipitation inputs

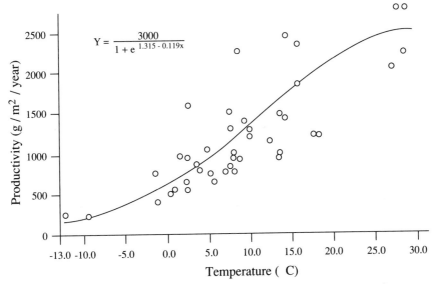

Figure 5.8   Relationship between NPP determined by harvest and mean annual temperature for 52 studies on various continents. From Lieth (1975).

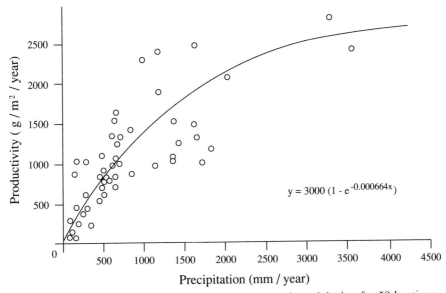

Figure 5.9   Relationship between NPP and mean annual precipitation for 52 locations around the world. From Lieth (1975).

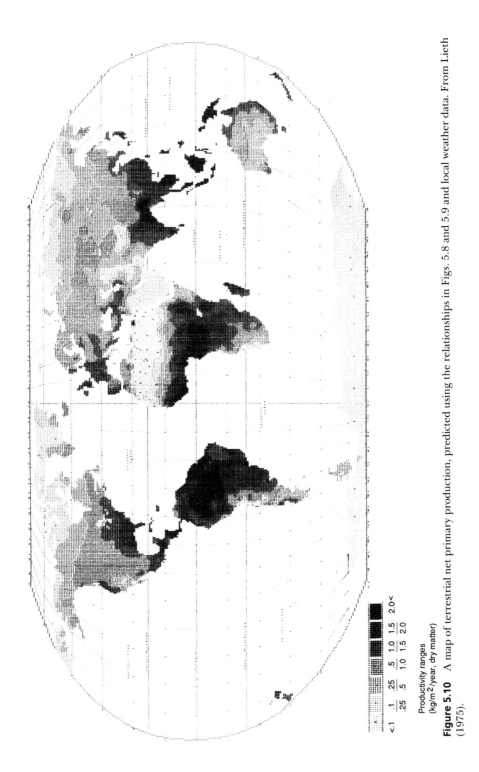

Productivity ranges
(kg/m2/year, dry matter)

<1	.1	.25	.5	1.0	1.5	2.0<
	.25	.5	1.0	1.5	2.0	

Figure 5.10 A map of terrestrial net primary production, predicted using the relationships in Figs. 5.8 and 5.9 and local weather data. From Lieth (1975).

and losses of soil moisture during the growing season (Grier and Running 1977, Gholz 1982). In other studies, light and temperature are combined to calculate actual evapotranspiration, which shows a positive correlation to NPP in temperate zone communities (Rosenzweig 1968, Webb et al. 1978). In tropical forests, where both light and moisture are abundant, the relationship of NPP to these variables is weak, and local soil conditions determining fertility are potentially more important (Brown and Lugo 1982). Nutrient availability also determines differences in net primary productivity among local sites within the same climatic region (e.g., Pastor et al. 1984).

Net Primary Production and Global Change

The direct harvest of plants for food, fuel and shelter accounts for about 3.3×10^{15} g C yr^{-1} or about 6% of the terrestrial productivity worldwide (Vitousek et al. 1986). As a result of inadvertent activities, such as wildfires and pollution, humans may have reduced total net primary production by up to 40%. Vitousek et al. (1986) suggest that this is probably the largest diversion of primary production to support a single species in the history of life on Earth—certainly a provocative implication for the prospects of further human population growth.

Beyond the social aspects, the effect of humans on biomass and net primary productivity is seen in changes in the global cycles of essential elements and in the composition of the atmosphere (Chapter 3). Most of the increase in atmospheric carbon dioxide is due to the burning of fossil fuels, but a significant portion is also due to the destruction of plant biomass, especially in tropical forests (Woodwell et al. 1983). Although fast-growing successional vegetation is found on most areas that are harvested, the rate of carbon accumulation does not equal the rate of carbon lost during harvest, and there is a net transfer of carbon from biomass to atmospheric CO_2 (Harmon et al. 1990). Moreover, the carbon storage in agricultural crops or forest regrowth is less than the carbon contained in the original forest biomass (Table 5.2).

Houghton et al. (1983) attempt to account for changes in world biomass between 1860 and 1980, compiling land use statistics to calculate the rate of agricultural expansion and forest harvest. They suggest that world biomass has been reduced by 21% since 1860; thus, the preindustrial terrestrial biota contained about 827×10^{15} g C. The release of carbon from land, including the release from soils, was estimated to be in the range of $1.8–4.7 \times 10^{15}$ g C yr^{-1} in 1980, compared to a release from fossil fuels of 5×10^{15} g C yr^{-1} (Rotty and Masters 1985). Whereas net primary production and decomposition may have been in balance in the early 1800s, leading to a steady-state in terrestrial biomass, the current net release is approximately equivalent to a 7% imbalance between these processes. Much of the current destruction occurs in tropical forests, which is why an accurate estimate of tropical forest biomass and its harvest is so critical to our understanding of changes in the global carbon cycle (Molofsky et al. 1984, Brown et al. 1986, Palm et al. 1986, Houghton et al. 1987, Detwiler and Hall 1988).

Release of carbon from forest destruction in the tropics could, of course, be balanced by the abandonment of farmland and the permanent regrowth of vegetation elsewhere. Forest regrowth in the southeastern United States has apparently been a sink for atmospheric carbon of about 0.07×10^{15} g yr^{-1} during this century (Delcourt and Harris 1980). Globally, reforestation may sequester between 0.3 and 1.9×10^{15} g C yr^{-1} (Armentano and Ralston 1980, Johnson and Sharpe 1983). We can expect the strength of this regional carbon sink to diminish as reforestation is complete and most forests become mature (Schiffman and Johnson 1989). Of course, the effect of the net sink is lost if these forests are harvested and the wood is converted to short-term products, such as paper, that are burned.

The net destruction of terrestrial vegetation is reflected in changes in the $\delta^{13}C$ of atmospheric CO_2. Since photosynthesis discriminates against $^{13}CO_2$ in favor of $^{12}CO_2$, plant tissues and fossil fuels are depleted in ^{13}C and dilute the atmospheric content of $^{13}CO_2$ when they are burned. In addition, fossil fuels have no ^{14}C content; that radioactive isotope decays away with a half life of 5700 yr. Thus, the burning of fossil fuels also dilutes the atmospheric content of $^{14}CO_2$. Conveniently, tree rings provide a record of the atmospheric content of $^{13}CO_2$ and $^{14}CO_2$ during the last several centuries. When changes in the content of ^{14}C are used to factor out the fossil fuel effect, any change in ^{13}C should be indicative of changes in the net size of the terrestrial biosphere. Records from many trees show a significant decline in $\delta^{13}C$ during the last century, implying a net reduction in terrestrial biomass (Fig. 5.11) (Stuiver 1978, Leavitt and Long 1988). Similar records have been extracted from the bubbles trapped in Antarctic ice (Friedli et al. 1986).

Some workers have suggested that the primary production of land vegetation will increase as the concentration of atmospheric CO_2 rises, stimulating photosynthesis by a greater delivery of CO_2 to ribulose bisphosphate carboxylase. If this effect were significant globally, then increased productivity by undisturbed vegetation could sequester CO_2 released by fossil fuels and forest destruction (Esser 1987). In fact, several recent analyses have suggested increases in the growth rate of trees at high elevations during the past several decades (LaMarche et al. 1984, Jacoby 1986). In most areas, however, the growth of vegetation is limited by other factors, and CO_2 should have little direct effect (Kramer 1981, Billings et al. 1984).

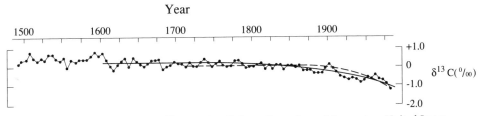

Figure 5.11 Decline in the $\delta^{13}C$ ratio in cellulose of tree rings of the western United States, after correction for changes in ^{13}C due to the burning of fossil fuels (dashed line). The excess decline suggests that the atmospheric content of $^{13}CO_2$ has been affected by a reduction in the net size of the terrestrial biosphere. From Leavitt and Long (1988).

Elevated CO_2 concentrations should increase water use efficiency of vegetation, since stomates show partial closure at high CO_2 concentrations. Higher water use efficiency by terrestrial vegetation could leave greater amounts of moisture in the soil, contributing to increases in the volume of runoff and global riverflow (Probst and Tardy 1987; Chapter 10). Unfortunately, as for the CO_2 response, most of our work has concerned agricultural plants grown in laboratory conditions. There are few studies of the response of whole ecosystems to increasing CO_2, so our ability to predict effects on the storage of organic carbon and riverflow is limited (Strain 1985, Gates 1985, Eamus and Jarvis 1989).

Potentially more significant are changes in the distribution of terrestrial vegetation that may occur as a result of a CO_2-induced global warming. Emanuel et al. (1985a) examine the current distribution of world vegetation types and expected changes in their distribution with global temperature change. In the northern hemisphere, a northward shift in the distribution of productive forests may increase the rate of carbon storage by net primary production in some areas, but other regions may show a decline in biomass and NPP (Solomon and Tharp 1985, Pastor and Post 1988). Changes in the distribution of vegetation and its carbon storage during the climatic warming at the end of the last glacial period may be useful in developing predictions for these parameters in a warmer climate of the future (COHMAP 1988).

Decomposition—The Fate of Organic Carbon

As communities of long-lived plants develop on land, a certain fraction of net primary production is allocated to perennial woody tissues that accumulate as biomass through time. Plant communities achieve a steady state in living biomass when the allocation to woody tissue is balanced by the death and loss of older parts (Fig. 5.12). At that point, there is no true increment in biomass, although dead organic matter may be accumulating in the soil. Odum (1969) summarized these trends in community development, showing that increasing fractions of gross primary production are lost to plant respiration and decomposition through time (Fig. 5.13). His work defines net ecosystem production (NEP) as

$$NEP = NPP - (R_a + R_d) \tag{5.6}$$

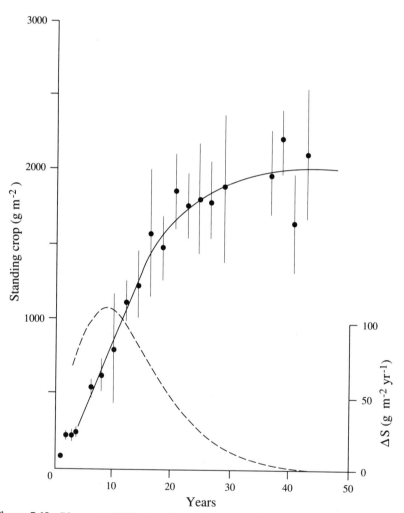

Figure 5.12 Biomass (solid line) and true increment (short dashes) of the aboveground components of a *Calluna* shrubland during 50 yr of recovery after fire. From Chapman et al. (1975).

where R_a is respiration of herbivores and R_d is respiration of decomposers, respectively. Remembering that

$$NPP = GPP - R_p \qquad (5.7)$$

we can say that

$$NEP = GPP - R_t \qquad (5.8)$$

where R_t is the total respiratory loss of CO_2 from the ecosystem.

These relationships suggest that increments in organic matter are possible only during the early stages of plant community development. In older communities, there is no true increment to live biomass, and all NPP is delivered to the soil, where it is decomposed. The role of animals is relatively minor. While herbivory may play a role in controlling forest productivity and nutrient cycling (Chapter 6), its proportion of total respiration in natural ecosystems is nearly always <10% (e.g., Reichle et al. 1973b, Mispagel 1978).

Production of Detritus

Global patterns in the deposition of plant litterfall are similar to global patterns in net primary production. The deposition of litterfall declines with increasing latitude from tropical to boreal forests (Bray and Gorham 1964, Van Cleve et al. 1983, Lonsdale 1988). Leaf tissues comprise about 70% of aboveground litterfall in forests (O'Neill and DeAngelis 1981), but the deposition of woody litter tends to increase with forest development, and fallen logs may be a conspicuous component of the forest floor

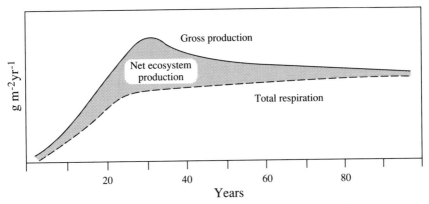

Figure 5.13 Generalized trends in primary production and respiration during ecosystem development. Modified from Odum (1969).

in old-growth forests (Harmon et al. 1986). In grassland ecosystems, where little of the aboveground production is contained in perennial tissues, the annual litterfall is nearly equal to annual net primary production. In most studies, the annual growth and death of fine roots contributes a large amount of detritus to the soil, which has been overlooked by studies that consider only aboveground litterfall. Analogous to relationships for net primary production, Meentemeyer et al. (1982) used actual evapotranspiration to predict global patterns of litterfall and to estimate 54.8×10^{15}g C for the total annual production of plant detritus.

The Decomposition Process

Most detritus, whether from litterfall or root turnover, is delivered to the upper layers of the soil where it is subject to the decomposition by microfauna, bacteria, and fungi (Swift et al. 1979, Schaefer 1990). Decomposition leads to the release of CO_2, H_2O, and nutrient elements, and to the microbial production of highly resistant organic compounds known as humus. Humus compounds accumulate in the lower soil profile (Chapter 4) and comprise the bulk of soil organic matter (Schlesinger 1977). The dynamics of the pool of soil organic carbon is best viewed in two stages—processes leading to rapid turnover of the majority of litter at the surface, and processes leading to the slower production, accumulation, and turnover of humus at depth.

The litterbag approach is widely used to study decomposition at the surface of the soil. Fresh litter is confined in mesh bags that are placed on the ground and collected for measurements at periodic intervals (Singh and Gupta 1977). Simple models of decay are based on an exponential pattern of loss, where the fraction remaining after 1 yr is given by

$$X/X_0 = e^{-k} \tag{5.9}$$

An alternative, the mass-balance approach, suggests that the annual decomposition should equal the annual input of fresh detritus, so that the mass of detritus stays constant. Under these assumptions, a constant fraction k of the detrital mass decomposes, so that

$$\text{Litterfall} = k(\text{detrital mass}) \tag{5.10}$$

or

$$\frac{\text{Litterfall}}{\text{Detrital Mass}} = k. \tag{5.11}$$

When the detritus is in steady state, the values for k calculated from the litterbag and mass-balance approaches should be equivalent, and mean residence time for plant debris is $1/k$ (Olson 1963). Vogt et al. (1983) shows the importance of fine root turnover to the calculation of mean residence times by the mass balance approach. When root turnover was included, mean residence time of organic matter in the forest floor was 8.2–15.6 yr, compared to 31.7–68.6 yr calculated from aboveground litter alone.

With either approach, when decomposition rates are rapid, there is little surface accumulation and values for k are greater than 1.0 (e.g., in tropical rain forests; Cuevas and Medina 1988). In such systems, decomposition has the potential to respire more than the annual input of carbon in litterfall. In contrast, in some peatlands, values for k are very small (e.g., 0.001; Olson 1963). Decomposition in grasslands shows a range of 0.20–0.60 in values for k (Vossbrinck et al. 1979, Seastedt 1988), but values for deserts may be as high as 1.00 due to the action of termites and photooxidation (Schaefer et al. 1985). Esser et al. (1982) suggest a global mean residence time of 3 yr (i.e., $k = 0.33$) for litter on the surface of the soil.

Decomposition rates vary as a function of temperature, moisture, and the chemical composition of the litter material. Microbial activity increases exponentially with increasing temperature (e.g., Edwards 1975). This relation often shows a Q_{10} of 2.0, that is, a doubling in activity per 10°C increase in temperature (Singh and Gupta 1977). For example, Van Cleve et al. (1981) found that the thickness of forest floor layers in black spruce forests in Alaska was inversely related to the cumulative degree days favorable to decomposition each year. In contrast, in areas of arid and semi-arid conditions, soil moisture may limit the rate of decomposition (Wildung et al. 1977, Santos et al. 1984).

Meentemeyer (1978a) compiled data from various decomposition studies to relate surface decomposition to actual evapotranspiration, and to use the resulting equation to predict regional patterns of decomposition in the United States (Fig. 5.14). His predictions are consistent with observations of surface litter in much of the United States (e.g., Lang and Forman 1978). Improvements in his predictions are found when chemical parameters such as lignin and nitrogen are added to his equations (Meentemeyer 1978b, Melillo et al. 1982). We will discuss the dynamics of other elements during decomposition in Chapter 6.

Humus Formation and Soil Organic Matter

Plant litter and soil microbes constitute the cellular fraction of soil organic matter. As decomposition proceeds, there is an increasing content of amorphous organic matter, humus, that appears to result from microbial activity. The structure of humus is poorly known, but it contains numerous aromatic rings with phenolic (—OH) and organic acid (—COOH) groups (Flaig et al. 1975, Stevenson 1986). The humus molecule appears to have no consistent molecular weight or repeating units in its structure.

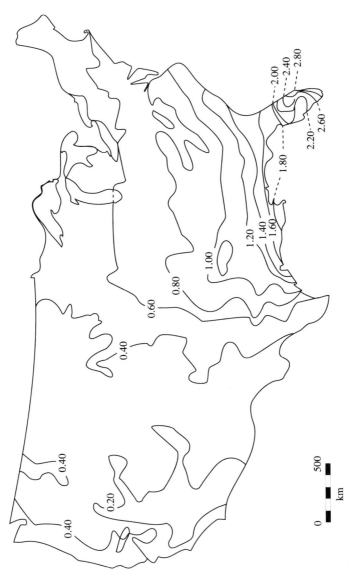

Figure 5.14 Rates of decomposition of fresh litter in the United States predicted by a simulation model using actual evapotranspiration as a predictive variable. Isopleth values are the fractional loss rate (k) of mass from fresh litter during the first year of decay. From Meentemeyer (1978a).

Chemical characterizations of humus are often based on the solubility of humic and fulvic acid components in alkaline and acid solutions, respectively (Fig. 5.15). The water soluble components of humus, primarily fulvic acids, may control the movement of plant nutrients in the soil solution (Schoenau and Bettany 1987), and humus content is a major source of cation exchange capacity in soils (Chapter 4).

During soil development, humus accumulates in the lower horizons (Chapter 4), often complexed with clay mineral components and calcium (Nichols 1984, Oades 1988). Studies in soil chronosequences suggest that humus accumulates at rates of about 1–15 g C m^2 yr (Table 5.3), with the highest rates in cool, wet conditions. Humus is very resistant to microbial attack. Campbell et al. (1967) extracted humic materials from a forest soil in Saskatchewan and measured a weighted mean ^{14}C age of 250–940 yr. Under most vegetation, the mass of humus in the soil profile exceeds the combined content of organic matter in the forest floor and aboveground biomass (Schlesinger 1977). Table 5.4 provides a global inventory of plant detritus and soil organic matter, totaling 1456×10^{15} g C. Similar inventories based on soil groups or climatic regions give values ranging from 1395 to 1477×10^{15} g C (Post et al. 1982, Buringh 1984).

The global estimates of soil organic matter, divided by the estimate of global litterfall, suggest a mean residence time of slightly over 26 yr for the total pool of organic carbon in soils, but the mean residence time

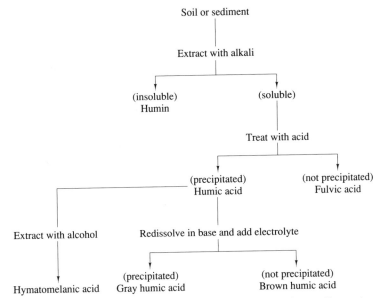

Figure 5.15 Fractionation of fulvic and humic acid components from soil organic matter. From Stevenson (1986).

Table 5.3 Rate of Accumulation of Organic Matter during Soil Development

Ecosystem Type	Vegetation in Terminal State	Soil Origin	Accumulation Interval (yr)	Rate of Accumulation (g C/m²/yr)	Reference
Tundra	Polar desert	Glacial retreat	8 000	0.20	Evans and Cameron (1979)
	Polar desert	Glacial retreat	9 000	0.2	Bockheim (1979)
	Polar desert	Glacial retreat	2 600	2.40	Birkeland (1978)
Boreal forest	Spruce	Glacial retreat	3 500	11.7	Crocker and Dickson (1957)
					Chandler (1942)
	Spruce	Glacial retreat	250	15.3	Ugolini (1968)
			150	13.3	Crocker and Major (1955)
	Spruce–fir	Glacial retreat	2 740	2.2	Protz et al. (1988)
			5 435	0.8	Protz et al. (1984)
Temperate forest	Broadleaf evergreen	Volcanic ash	1 277	12.0	Tezuka (1961)
	Coniferous	Volcanic mudflow	1 200	10.0	Sollins et al. (1983)
				6.8	Dickson and Crocker (1953)
	Deciduous	Alluvium	1 955	5.1	Bilzi and Ciolkosz (1977)
	Podocarpus forest	Dunes	10,000	2.1	Syers et al. (1970)
	Angophora	Dunes	4 200	1.7	Burges and Drover (1953)
	Deciduous	Dunes	10,000	0.7	Franzmeir et al. (1963)
Tropical forest	Metrosideros	Volcanic ash	3 500	2.5	Vitousek et al. (1983)
	Rainforest	Volcanic ash	8 620	2.3	Harris (1971)

Table 5.4 Distribution of Soil Organic Matter by Ecosystem Types[a]

Ecosystem Type	Mean Soil Organic Matter (kg C m^{-2})	World Area (ha × 10^8)	Total World Litter (mt C × 10^9)	Total World Soil Organic Carbon (mt C × 10^9)
Tropical forest	10.4	24.5	255	3.6
Temperate forest	11.8	12	142	14.5
Boreal forest	14.9	12	179	24.0
Woodland and shrubland	6.9	8.5	59	2.4
Tropical savanna	3.7	15	56	1.5
Temperate grassland	19.2	9	173	1.8
Tundra and alpine	21.6	8	173	4.0
Desert scrub	5.6	18	101	0.2
Extreme desert, rock and ice	0.1	24	3	0.02
Cultivated	12.7	14	178	0.7
Swamp and marsh	68.6	2	137	2.5
Totals		147	1456	55.2

[a]From Schlesinger (1977).

varies over several orders of magnitude between the surface litter and the various humus fractions (Fig. 5.16). O'Brien and Stout (1978) used radio-carbon dating to find that 16% of the organic matter in a pasture soil had a minimum age of 5700 yr, while the rest was of recent origin and was concentrated near the surface. Because of different turnover times, decomposition constants k for surface litter cannot be applied to the entire mass of organic matter in the soil profile.

Field measurements of the flux of CO_2 from the soil surface provide an estimate of the total respiration in the soil, and a potential alternative approach for estimating turnover of the humus pool. Most of the production of CO_2 occurs in the surface litter where decomposition is rapid and a large proportion of the fine root biomass is found. Edwards and Sollins (1973) found that only 17% of the annual production of CO_2 in a temperate forest soil was contributed by soil layers below 15 cm. Flux of CO_2 from the deeper soil layers is presumably due to the decomposition of humus substances. Production of CO_2 in the soil leads to the accumulation of CO_2 in the soil pore space, which drives carbonation weathering in the lower profile (Chapter 4).

Unfortunately, the respiration of living roots makes it difficult to use estimates of CO_2 flux in calculations of turnover of the soil organic pool. In a compilation of values, Schlesinger (1977) found that CO_2 evolution exceeded the deposition of aboveground litter by a factor of about 2.5 (Fig. 5.17). The additional CO_2 is presumably derived from root respiration and the decomposition of root detritus (Raich and Nadelhoffer

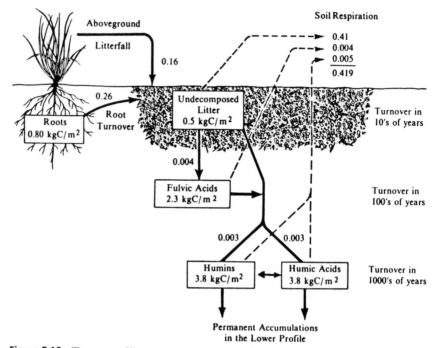

Figure 5.16 Turnover of litter and soil organic fractions in a grassland soil. Note that mean residence time can be calculated for each fraction from measurements of the quantity in the soil and the annual production or loss (respiration) from that fraction. Flux estimates are in kg C m^{-2} yr^{-1}. From Schlesinger (1977).

1989). This comparison speaks strongly for the need to improve our understanding of soil processes as components of ecosystem studies.

Accumulations of soil organic matter show how moisture and temperature control the balance between primary production and decomposition in surface and lower soil layers. Accumulations of soil organic matter are greatest in wetland ecosystems and least in deserts (Table 5.4). Among forests, accumulations increase from tropical to boreal climates. Net primary productivity shows the opposite trend, so the accumulation of soil organic matter is largely due to differences in decomposition. Thus, compared to the process of primary production, soil microbes are more sensitive to regional differences in temperature and moisture (Fig. 5.17). Parton et al. (1987) developed a model based on the differential turnover of soil organic fractions to predict accumulations of soil organic matter in grassland ecosystems. Accurate predictions were achieved when temperature, moisture, soil texture, and plant lignin content were included as variables. Despite relatively low NPP, soils of temperate grasslands contain large amounts of soil organic matter (Sanchez et al. 1982b), due to relatively low rates of decomposition and a larger fraction of detritus

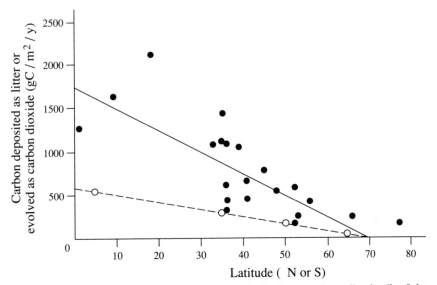

Figure 5.17 Latitudinal trends for carbon dynamics in forest and woodland soils of the world. The dashed line shows the mean annual input of organic carbon to the soil by litterfall. The solid line shows the loss of carbon, measured as the flux of CO_2 from the surface. The difference between these lines represents the loss of CO_2 from root respiration and from the respiration of root detritus and mycorrhizae. From Schlesinger (1977).

inputs from fine root turnover (Oades 1988). In contrast, tropical grass-lands and savannas have relatively low soil organic content (Kadeba 1978, Jones 1973).

Storage of soil organic matter represents the net ecosystem production (NEP) in terrestrial ecosystems. Although many wetland ecosystems may show long-term net accumulations (Chapter 7), the mass of soil organic matter in most upland ecosystems is likely to have been fairly constant before widespread human disturbance of soils. When soils show a steady state in soil organic content, the production of humic compounds must be equal to their degradation and to the removal of soil organic matter by erosion. Thus, an estimate of organic carbon transport in rivers is an upper limit for terrestrial NEP. Recent estimates of the global transport of organic carbon in rivers are 0.4×10^{15} g C yr^{-1} (Schlesinger and Melack 1981, Meybeck 1982), so the terrestrial NEP for the globe is not likely to be more than 0.7% of NPP. Total storage of carbon in soils, 1456×10^{15} g or 121×10^{15} moles, can account for only 0.03% of the O_2 content of the atmosphere, remembering that the storage of organic carbon and the release of O_2 occur on a mole-for-mole basis during photosynthesis. Thus, accumulations of atmospheric O_2 cannot be the result of the storage of organic carbon on land. Despite the stability of humus substances in the soil profile, the limited accumulations of soil organic matter on land speak strongly for the efficiency of decomposers using aerobic metabolic pathways of degradation. Long-term storage of organic carbon appears to be dominated by accumulations in anoxic marine sediments (Chapter 9).

For areas covered by the last continental glaciation, the total accumulation of soil organic matter represents NEP for the last 10,000 years. The maximum extent of the last glacial, covering 29.5×10^6 km^2 of the present land area (Flint 1971), now contains roughly 400×10^{15} gC or about 25% of the carbon contained in all soils of the world (Table 5.4). Thus, for the Holocene period, soil organic matter has accumulated at rates of about 1.35 g C m^{-2} yr^{-1}. At this rate, the current rate of storage in northern ecosystems (0.04×10^{15} g C yr^{-1}) is too small to be a significant sink for human releases of CO_2 to the atmosphere, and it is unlikely to increase in the future (Billings et al. 1982, 1984).

Soil Organic Matter and Global Change

When soils are brought under agriculture, their content of soil organic matter declines (Fig. 5.18). Losses from many soils are typically 20–30% (Schlesinger 1986, Mann 1986, Detwiler 1986). The loss is greatest during the first few years that native land is converted to agriculture. Eventually a new, lower level of soil organic matter is achieved that is in equilibrium with the lower production of plant detritus and the greater rates of decomposition under cropland (Jenkinson and Rayner 1977). Some of the soil organic matter is lost in erosion, but most is probably oxidized to CO_2 and released to the atmosphere. Since about 10% of the world's soil is under cultivation (Table 5.4), losses of organic matter from soils may be a major component of the past increase in atmospheric CO_2 (Schlesinger 1984). The current rate of release from soils, as much as 0.8×10^{15} g C/yr, is largely dependent upon the current rate at which natural ecosystems are being converted to agriculture. Especially large losses of soil carbon are seen when organic soils in wetlands and peatlands are drained (Armentano and Menges 1986).

Dynamics of soil organic matter are illustrated by the pattern of loss after land is converted to agriculture. Soil organic matter consists of a labile and a resistant fraction. The labile fraction is composed of fresh plant materials that are subject to rapid decomposition, whereas the resistant fraction is composed of humic materials that are often complexed with clay minerals. Rather than biochemical fractionations (Fig. 5.15), some workers have used size or density fractionation to quantify the labile and resistant organic matter. Density fractionations are performed by adding soil samples to solutions of increasing specific gravity and collecting the material that floats to the surface (Spycher et al. 1983). In size fractionation, soils are passed through screens of varying mesh

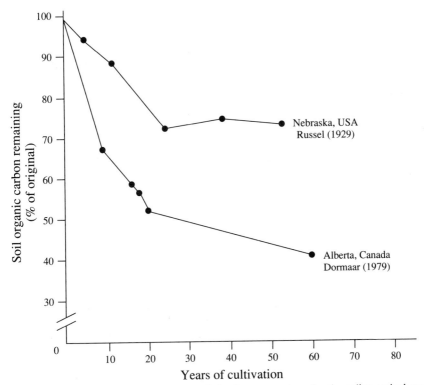

Figure 5.18 Decline in soil organic matter following conversion of native soil to agriculture for two grassland soils.

(Tisdall and Oades 1982, Elliott 1986). Most of the turnover of soil organic matter is in the "light" or large fractions that represent fresh plant materials (Foster 1981, Tiessen and Stewart 1983, Dalal and Mayer 1986a, 1986b). The "heavy" fraction is composed of polysaccharides (sugars) and humic materials that are complexed with clay minerals to form micro-aggregates of relatively high specific gravity (Tisdall and Oades 1982, Tiessen and Stewart 1988). The radiocarbon age of the different size or weight fractions indicates the rate of turnover. Anderson and Paul (1984) reported a ^{14}C age of 1255 yr for the coarse clay fraction in a soil for which the weighted average age was 795 yr. The decline in soil organic matter in agricultural soils is largely due to losses from the light fraction. Successful management of agricultural soils may depend upon the preservation of their organic microaggregate structure.

In addition to changes in the pool of organic carbon with agriculture, soil carbon will change with climatic warming, which should stimulate rates of decomposition in many ecosystems (Schleser 1982). The effect of temperature will interact with other factors. For example, experiments show that warming of organic soils in the tundra will produce large losses of carbon, but the losses will be greater if the water table is also lowered due to melting of permafrost (Billings et al. 1982). Tundra ecosystems appear nutrient-limited, and additions of nitrogen stimulated increased carbon storage in experiments conducted at higher temperatures and ambient CO_2 (Billings et al. 1984). The most accurate predictions of changes in net ecosystem production may be derived from simulatiom models that include these interactive factors (e.g., Pastor and Post 1986, 1988). Changes in soil carbon storage will be closely associated with changes in the distribution and productivity of vegetation discussed earlier.

Summary

Photosynthesis provides the energy that powers the biochemical reactions of life. That energy is captured from sunlight. Globally, net primary production of about 60×10^{15} g C yr^{-1} is available in the terrestrial biosphere. Although that is a large value, NPP typically captures less than 1% of the available sunlight energy falling on a forest annually (e.g., Reiners 1972) and about 0.1% of that received by deserts (Szarek 1979). Most of the remaining energy evaporates water and heats the air, resulting in the global circulation of the atmosphere (Chapters 3 and 10). Thus, the terrestrial biosphere is fueled by a relatively inefficient initial process.

During photosynthesis, plants take up moisture from the soil and lose it to the atmosphere in the process of transpiration. Available moisture appears to be a primary factor determining the display of leaf area and NPP (Figs. 5.4 and 5.7). Among communities with adequate soil moisture, net primary production is determined by the length of the growing season and mean annual temperature—both are an index of the receipt of solar energy. Soil nutrients appear to be of secondary importance to NPP on land, perhaps because plants have various adaptations to obtain and recycle nutrients efficiently when they are in short supply (Chapter 6).

Most net primary production is delivered to the soil, where it is decomposed by a variety of organisms. The decomposition process is remarkably efficient, so only small amounts of NPP are added to the long-term storage of soil organic matter or humus. Soil organic matter consists of a dynamic pool near the surface, in which there is rapid turnover of fresh plant detritus and little long-term accumulation, and a large refractory pool of humic substances that are dispersed throughout the soil profile. Thus, the turnover time of organic carbon in the soil ranges from about 3 yr for the litter to thousands of years for humus.

Humans have altered the processes of net primary production and decomposition on land, resulting in the transfer of organic carbon to the atmosphere, and perhaps a permanent reduction in the global rate of NPP. This disruption has produced global changes in the biogeochemical cycle of carbon, but little change in the atmospheric concentration of O_2.

Recommended Reading

Reichle, D.E., editor. 1981. Dynamic Properties of Forest Ecosystems. Cambridge University Press, Cambridge.

Waring, R.H. and W.H. Schlesinger. 1985. Forest Ecosystems. Academic Press, Orlando.

6

Biogeochemical Cycling on Land

Introduction

Although living tissue is composed of carbon, hydrogen, and oxygen in the approximate proportion of CH_2O, as many as 23 other elements are necessary for biochemical reactions and for the growth of structural biomass. For instance, the proteins found in plants and animals contain about 16% nitrogen by weight. Earlier we saw that the protein ribulose bisphosphate carboxylase is directly related to the rate of carbon fixation during photosynthesis in many plant species (Chapter 5). Thus, the linkage of carbon and nitrogen that is seen in global biogeochemical cycles has a basis at the level of cellular biochemistry (Stock et al. 1990). In other molecules, phosphorus is required for adenosine triphosphate (ATP), the universal molecule for energy transformations, and calcium is a major structural component of plants and animals.

The various elements essential to biochemical structure and function are often found in predictable proportions in living tissues (e.g., wood, leaf, bone, muscle etc.; Reiners 1986). For instance, the ratio of C to N in forest biomass is about 160 (Vitousek et al. 1988). At the global level, our estimate of net primary production, 60×10^{15} g C/yr, implies that *at least* 3.8×10^{14} g of nitrogen must be supplied each year through biogeochemical cycling. As we shall see, the actual amount is much higher since leaf tissues have high concentrations of N. The availability of some elements, such as N and P, is often limited, and these elements may control the rate of net primary production on land. Conversely, for elements that are typically available in greater quantities, such as Ca and S, the rate of net primary production often determines the rate of cycling in the ecosystem and losses to streamwaters. In every case, the biosphere exerts a strong control on the geochemical behavior of the major elements of life. Much less biological control is seen in the cycling of elements such as Na and Cl, which are less important constituents of biomass.

In earlier chapters we saw that the atmosphere is the dominant source of C, N, and S for the growth of land plants, and that rock weathering is the major source for most of the remaining biochemical elements (e.g., Ca, Mg, K, Fe, and P). In any terrestrial ecosystem the receipt of elements from the atmosphere and the lithosphere represents an input of new quantities for plant growth. However, as a result of retention and internal cycling, the annual rate of net primary production is not solely dependent upon new inputs to the system. In fact, the annual circulation of important elements such as N within an ecosystem is often 10 to 20 times greater than the amount received from outside the system (Table 6.1). This large internal or *intrasystem cycle* is achieved by long-term accumulations of elements received from the atmosphere and the lithosphere. Important biochemical elements are retained in terrestrial ecosystems by biotic uptake, whereas nonessential elements pass through these systems under simple geochemical control (Johnson 1971, Vitousek and Reiners 1975).

In this chapter we will examine the cycle of biochemical elements in terrestrial ecosystems. We will begin by examining aspects of plant uptake, allocations during growth, and losses in the death of plants and

Table 6.1 Percentage of the Annual Requirement of Nutrients for Growth in the Northern Hardwoods Forest at Hubbard Brook, New Hampshire, That Could Be Supplied by Various Sources of Available Nutrients[a]

Process	N	P	K	Ca	Mg
Growth requirement (kg/ha/yr)	115.6	12.3	67.3	62.2	9.5
Percentage of the requirement that could be supplied by:					
Intersystem inputs					
Atmospheric	18	0	1	4	6
Rock weathering	0	13	11	34	37
Intrasystem transfers					
Reabsorptions	31	28	4	0	2
Detritus turnover (includes return in throughfall and stemflow)	69	81	86	85	87

[a] From Waring and Schlesinger (1985). Reabsorption data are from Ryan and Bormann (1982). All other data are from Likens et al. (1977) and Wood et al. (1984).

plant tissues. Then, we will examine processes by which elements such as N, P, and S are transformed in the soil, leading to their release for plant uptake or for potential loss from the ecosystem. We will stress interactions between carbon and other biochemical elements and examine how land plants have adapted to the widespread limitations of nitrogen and phosphorus in most ecosystems. A brief examination will be given of how biogeochemical processes may control the distribution of plants and animals on land.

Biogeochemical Cycling in Land Plants

Nutrient Uptake

It is easy to forget the essential, initial role played by plants in all of biochemistry. Plants obtain inorganic minerals from the soil (e.g., NO_3) and incorporate their elements into biochemical molecules (e.g., amino acids). Animals may eat plants, and each other, and synthesize new proteins, but the building blocks of animal proteins are the amino acids originally synthesized in plants. Only in isolated instances, for example, in animals at natural salt licks, do we find a direct transfer of elements from inorganic form to animal biochemistry (Jones and Hanson 1985). There are no vitamin pills in the natural biosphere!

Plant uptake of essential elements begins within the initial constraints set by chemical reactions in the soil, such as ion exchange and solubility (Chapter 4). However, when plant uptake of an element such as phosphorus is rapid, additional phosphorus may dissolve in the soil solution from mineral forms, and we have seen how plants can release various substances that enhance the solubility of nutrient elements from soil minerals.

Delivery of ions to plant roots can occur by several pathways (Barber 1962). The concentration of some elements in the soil solution is such that their passive uptake with water is adequate for plant nutrition. In some cases, such delivery is excessive, and ions are actively excluded from uptake. For example, it is not unusual to see accumulations of Ca, as $CaCO_3$, surrounding the roots of desert shrubs growing in calcareous soils (Klappa 1980, Wullstein and Pratt 1981). In contrast, for N, P, and K the concentration in the soil solution is much too low for adequate delivery in the transpiration stream, and the uptake is mediated by active transport by enzymes located in root membranes (Ingestad 1982, Robinson 1986, Chapin 1988).

Enzymes involved in root membranes yield increasing rates of uptake as a function of increasing concentrations in the soil solution until the activity of the enzyme system is saturated (Fig. 6.1). Chapin and Oechel (1983) found that the arctic sedge *Carex aquatilis* from colder habitats had higher rates of uptake than those from warmer habitats, presumably

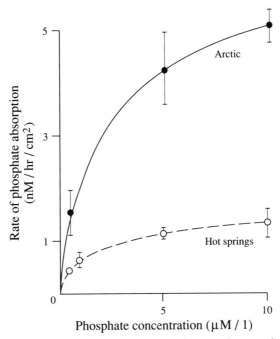

Figure 6.1   Rate of phosphate absorption per unit of root surface area in populations of *Carea aquatilis* from cold (Arctic) and warm (hot springs) habitats measured at 5°C. From Chapin (1974).

reflecting enzymatic adaptation to the lower availability of phosphorus in cold environments as a result of reduced microbial activity.

The uptake of nitrogen and phosphorus is so rapid and the concentrations in the soil solution are typically so low that these elements are effectively absent in the soil solution surrounding roots, and the rate of uptake is determined by diffusion to the root from other areas (Nye 1977). Phosphate is particularly immobile in most soils, and the rate of diffusion strongly limits its supply to plant roots (Robinson 1986). Although adaptations for more efficient root enzymes are seen in some species, the most apparent response of plants to low nutrient concentrations is an increase in the root/shoot ratio, which increases the volume of soil exploited and decreases diffusion distances (Chapin 1980, Clarkson and Hanson 1980). In many species the relative growth rate of roots determines the uptake of nitrogen and phosphorus (Newman and Andrews 1973) (Fig. 6.2), and roots show a rapid response to added nutrients (Jackson et al. 1990).

Higher plants and soil microbes release enzymes to the soil that can mineralize inorganic phosphorus from organic matter. These extracellular enzymes are known as phosphatases, and separate forms active in acid and alkaline soils are known (Malcolm 1983, Tarafdar and Claassen

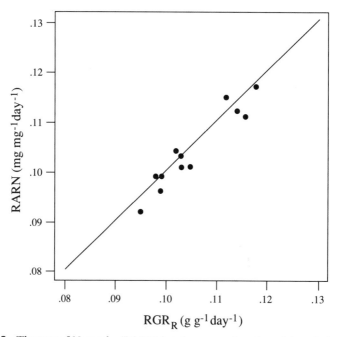

Figure 6.2 The rate of N uptake (RARN) in tobacco as a function of the relative growth rate of roots (RGR$_R$) From Raper et al. (1978).

1988). In many cases, root phosphatase activity is inversely proportional to available soil P (McGill and Cole 1981). Phosphatase activity associated with root surfaces is particularly significant in plants from phosphorus-poor habitats, and it may provide up to 65% of the annual phosphorus demand by some tundra plants (Kroehler and Linkins 1988).

Nutrient Balance

In addition to an adequate supply of nutrient elements, plant growth is affected by the balance of nutrients (Shear et al. 1946). For seedlings of several tree species, Ingestad (1979a) has shown that an appropriate ratio of nutrient elements is supplied in a solution containing 100 parts N, 15 parts P, 50 parts K, 5 parts Ca and Mg, and 10 parts S. However, unless the supply of a nutrient reaches very low levels, plants do not show deficiency symptoms, they simply grow more slowly (Clarkson and Hanson 1980). Inherent, slow growth rate is a characteristic of plants adapted to infertile habitats (Chapin et al. 1986a), and it often persists even when nutrients are added experimentally (e.g., Schlesinger et al. 1989).

Because more soil nutrients occur as positive ions than negative ions, one might expect that plant roots would develop a charge imbalance as a result of ion uptake. When ions such as K^+ are removed from the soil solution in excess of the uptake of negatively charged ions, the plant releases H^+ to maintain an internal balance of charge. This H^+ may, in turn, replace K^+ on a cation exchange site, driving another K^+ into the soil solution. The high concentration of N in plant tissues causes the form in which N is taken up to dominate this process (Table 6.2). Nye (1981) has shown how plants that use NH_4^+ as a N source tend to acidify the immediate zone around their roots. The uptake of NO_3^- has the opposite effect as a result of plant releases of HCO_3^- and organic acids to balance the negative charge (Hedley et al. 1982a).

Table 6.2 Chemical Composition and Ionic Imbalance for Perennial Ryegrass[a]

Elements	N	P	S	Cl	K	Na	Mg	Ca
Percent in leaf tissue	4.00	0.40	0.30	0.20	2.50	0.20	0.25	1.00
Equivalent weight (g)	14.00	30.98	16.03	35.46	39.10	22.99	12.16	20.04
Meq % present	285.7	12.9	18.7	5.6	63.9	8.8	20.6	49.9
Sum of meq %	±285.7		−37.2			+143.1		

Imbalance in meq %	(a) where ammoniacal nitrogen is taken up: $285.7 + 143.1 - 37.2 = +391.6$
	(b) where nitrate nitrogen is taken up: $143.1 - 285.7 - 37.2 = -179.8$

[a] From Middleton and Smith (1979).

Nitrogen Assimilation

Among various habitats, the availability of soil nitrogen as NH_4^+ or NO_3^- differs largely depending upon the environmental conditions that affect the conversion of NH_4^+ to NO_3^- in the microbial process known as nitrification [equation (2.15)]. For example, in the water-logged soils of the tundra, almost all nitrogen is found as NH_4^+ (Barsdate and Alexander 1975), whereas in deserts and some forests, NO_3^- is important (Virginia and Jarrell 1982, Nadelhoffer et al. 1984). Many species show a preference for NO_3^-, although species occuring in sites where nitrification is slow or inhibited often tend to show superior growth with ammonium (Haynes and Goh 1978, Adams and Attiwill 1982). A few unusual, insectivorous plants obtain their N by digesting captured organisms. Dixon et al. (1980) found that 11–17% of the annual uptake of N in *Drosera erythrorhiza* could be obtained from captured insects.

Inside the plant, both forms of inorganic N are converted to amino groups ($-NH_2$) that are attached to soluble organic compounds. In many woody species these conversions occur in the roots and N is transported as amides, amino acids, and ureide compounds through the xylem stream (Andrews 1986). However, in some species the reduction of NO_3^- to $-NH_2$ occurs in leaf tissues and N is found as NO_3^- in the xylem stream (Smirnoff et al. 1984). Eventually, most plant N is incorporated into protein.

The conversion of NO_3^- to $-NH_2$ is a biochemical reduction reaction that requires metabolic energy and is catalyzed by the enzyme, nitrate reductase, containing Mo. One might puzzle why most plants do not show a clear preference for NH_4^+, which is assimilated more easily. Several explanations have been offered. Remembering that NH_4^+ interacts with soil cation exchange sites, whereas NO_3^- is highly mobile in most soils, the rate of delivery of NO_3^- to the root by diffusion or mass flow is much higher than that of NH_4^+ in otherwise equivalent conditions. Plants that utilize NH_4^+ may have to compensate for the differences in diffusion by having a greater investment in root growth (Gijsman 1990). Uptake of NO_3^- avoids the competition that occurs in root enzyme carriers between NH_4^+ and other positively charged nutrient ions. For example, the presence of large amounts of K^+ in the soil solution can reduce the uptake of NH_4^+ (Haynes and Goh 1978). Finally, relatively low concentrations of NH_4^+ are potentially toxic to plant tissues. These potential disadvantages in the uptake of NH_4^+ may explain why many plants take up NO_3^- when thermodynamic calculations suggest that metabolic costs of reducing NO_3^- are about 8–17% greater than for plants that assimilate NH_4^+ directly (Middleton and Smith 1979, Gutschick 1981).

It is unclear why so many species concentrate nitrate reductase in their roots, when the same reaction performed in leaf tissues, where it can be coupled to the photosynthetic reaction, is much less costly (Gutschick

1981, Andrews 1986). Additions of NO_3^- induce the production of root enzymes for NO_3^- uptake and the synthesis of nitrate reductase in plant tissues (Lee and Stewart 1978). There is some evidence that the proportion of nitrate reductase in the shoot increases at high levels of available NO_3^- (Andrews 1986).

Nitrogen Fixation

Several types of bacteria and blue-green algae possess the enzyme nitrogenase, which converts atmospheric N_2 to NH_4^+. Some of these exist as free-living forms (asymbiotic) in soils, but others, such as *Rhizobium* and *Frankia*, form symbiotic associations with the roots of higher plants. Symbiotic bacteria reside in root nodules that can be recognized in the field.

Nitrogen that enters terrestrial ecosystems by fixation is a "new" input in the sense that it is derived from outside the ecosystem from the atmosphere. The reduction of N_2 to NH_4^+ has large metabolic costs that are seen in the respiration of organic carbon. Nevertheless, Gutschick (1981) suggests that symbiotic fixation in higher plants is not greatly less efficient than the uptake of NO_3^- for those species in which the nitrate reductase activity is concentrated in plant roots. Only a few land plants show symbiotic nitrogen fixation, and it is interesting to speculate why nitrogen fixation is not more widespread, when nitrogen limitations of net primary production are so frequent.

The energy cost of nitrogen fixation links this biogeochemical process to the availability of organic carbon, provided by net primary production. In plants with symbiotic nitrogen fixation, the rate is often related to the efficiency of net primary production (Bormann and Gordon 1984). Heterotrophic bacteria conducting asymbiotic nitrogen fixation are usually found in soils with high levels of organic matter for decomposition (Granhall 1981). Nitrogen fixation that is observed in fallen logs (Roskoski 1980, Silvester et al. 1982) is probably due to anaerobic cellulolytic bacteria that may be widespread in most natural ecosystems (Leschine et al. 1988). In both symbiotic and asymbiotic forms, nitrogen fixation is generally inhibited at high levels of available nitrogen (Cejudo et al. 1984). Added phosphorus stimulates asymbiotic N fixation in prairie soils, and the rate of fixation appears to be controlled by the N/P ratio in the soil (Fig. 6.3). In bacteria the regulation of nitrogen fixation by phosphorus is seen at the level of molecular biology (Stock et al. 1990). Requirements for Mo, Co, and Fe also link nitrogen fixation to the biogeochemical cycles of these elements. Some N-fixing species appear to acidify their rooting zone to make Fe and P more available (Ae et al. 1990, Raven et al. 1990). Silvester (1989) suggests that low availability of Mo may limit asymbiotic N fixation in forests of the northwestern United States.

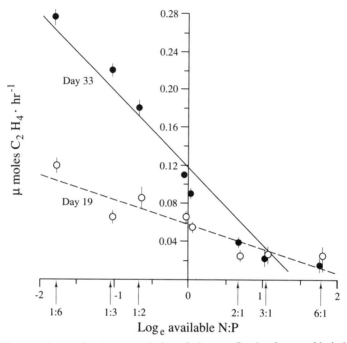

Figure 6.3 Acetylene reduction as an index of nitrogen fixation by asymbiotic N-fixing bacteria as a function of the N : P ratio in soil. From Eisele et al. (1989).

Nitrogenase activity can be measured using the acetylene-reduction technique, which is based on the observation that this enzyme also converts acetylene to ethylene under experimental conditions. Plants or nodules are placed in small chambers or small chambers are placed over field plots, and the conversion of injected acetylene to ethylene over a known time period is measured using gas chromatography. The conversion of acetylene in moles is not exactly equivalent to the potential rate of fixation of N_2 because the enzyme has different affinities for these substrates. However, appropriate conversion ratios can be determined using other techniques. For instance, investigators have applied $^{15}N_2$, the heavy stable isotope of N, in chambers and measured the increase in organic ^{15}N in test plants or field plots through time.

The natural isotopic ratio of N in plant tissues is expressed as $\delta^{15}N$, using a calculation analogous to that which we saw for the isotopes of carbon in Chapter 5. In the case of nitrogen, the standard is the atmosphere, which contains 99.63% ^{14}N and 0.37% ^{15}N. Nitrogenase shows a discrimination between the isotopes of N—that is, between $^{15}N_2$ and $^{14}N_2$. Differences in the isotopic ratio of nitrogen among plant species can be used to suggest which species may be involved in nitrogen fixation in the field (Virginia and Delwiche 1982). Showing depletion in ^{15}N from

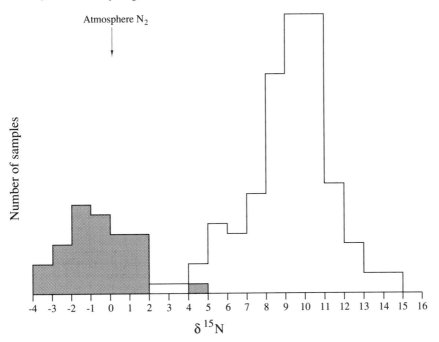

Figure 6.4 Frequency distribution of $\delta^{15}N$ values in the plant tissues of 20 nitrogen-fixing species and in the organic matter of 124 soils from throughout the United States. From Shearer and Kohl (1988, 1989).

inputs of fixed nitrogen, these species have negative ratios, whereas species depending on nitrogen uptake from the soil typically show positive values, which are typical of the soil pool (Fig. 6.4) (Shearer and Kohl 1988, 1989). Shearer et al. (1983) used the difference in isotopic ratio between *Prosopis* grown in the laboratory without added N (i.e., all nitrogen was derived from fixation), and the same species in the field, to estimate that the field plants derived 43–61% of their nitrogen from fixation. Of course, when nitrogen fixing plants die, their nitrogen content is available for other species in the ecosystem (Huss-Danell 1986). Lajtha and Schlesinger (1986) found that the desert shrub, *Larrea tridentata*, growing adjacent to nitrogen-fixing *Prosopis* had lower $\delta^{15}N$ than when *Larrea* were growing in isolation.

Asymbiotic bacteria and blue-green algae are widespread, and their nitrogen fixation can be an important source of N for some terrestrial ecosystems. Exceptionally high rates of fixation have been recorded in blue-green algae crusts that cover the soil surface in some desert ecosystems (Rychert et al. 1978); however, in most cases the total input from asymbiotic fixation is in the range of 1–5 kg N ha^{-1} yr^{-1} (Boring et al.

1988, Cushon and Feller 1989). This input rivals the deposition of nitrogen in wetfall and dryfall from the atmosphere.

The importance of fixation in higher plants varies widely depending upon the presence of species that harbor symbiotic bacteria. Some of the greatest rates of fixation are seen in species that invade after disturbance. For example, in the recovery of Douglas fir forests after fire, Youngberg and Wollum (1976) found that the nodulated shrub *Ceanothus velutinus* contributed up to 100 kg N ha^{-1} yr^{-1} on some sites. Invasion of the exotic nitrogen-fixing tree *Myrica faya* in Hawaii provides important inputs of nitrogen (18 kg ha^{-1} yr^{-1}) on volcanic ashflows (Vitousek et al. 1987). In most cases the importance of plants with symbiotic nitrogen fixation declines with the recovery of mature vegetation, and their occurrence in undisturbed communities is limited. The sporadic occurrence of symbiotic nitrogen fixation in space and time makes it difficult to extrapolate from studies in local areas to provide a global estimate of its importance. The widespread distribution of leguminous species in mature tropical forests is deserving of further study.

Mycorrhizae

Symbiotic associations between fungi and higher plants are found in most ecosystems (Harley and Smith 1983). The symbiosis is important for the nutrition of plants, and may have even determined the origin of land plants (Pirozynski and Malloch 1975). There are several forms of symbiosis. In temperate regions, many trees are infected by ectotrophic mycorrhizae. These fungi form a hyphal sheath around the active fine roots and extend additional hyphae into the surrounding soil. In many areas, especially the tropics, plants are infected by endotrophic mycorrhizae in which the fungal hyphae actually penetrate cells of the root cortex. By virtue of their large surface area and efficient absorption capacity, mycorrhizal fungi are able to obtain soil nutrients and transfer these to the higher plant root. In addition, recent work suggests that these fungi are directly involved in the decomposition of soil organic materials through the release of extracellular enzymes such as cellulases and phosphatases (Antibus et al. 1981, Dodd et al. 1987) and in the weathering of soil minerals through the release of organic acids (Bolan et al. 1984; see also Chapter 4). In return, mycorrhizal fungi depend upon the host plant for supplies of carbohydrate.

The importance of mycorrhizae in infertile sites is well known. Many species of pine require ectotrophic mycorrhizae, which perhaps accounts for their success in nutrient-poor soils. Most tropical trees appear to require endotrophic mycorrhizal associations for proper growth (Janos 1980), and mycorrhizal fungi are widespread among the *Eucalyptus* species growing in the low-phosphorus soils of Australia. Berliner et al.

(1986) report complete exclusion of *Cistis incanus* from basaltic soils in Israel due to a failure of mycorrhizal development. The same species grows well on adjacent calcareous soils, or in basaltic soils supplied with fertilizer.

Mycorrhizal fungi are most important in the transfer of those soil nutrients with low diffusion rates in the soil. A large number of studies document the importance of mycorrhizae in P nutrition, but absorption of N and other nutrients is also known (Bowen and Smith 1981). Some plants with mycorrhizal fungi show higher levels of various nutrients in foliage, but frequently the enhanced uptake of nutrients results in higher rates of growth (Schultz et al. 1979). Rose and Youngberg (1981) provide an insightful experiment with *Ceanothus velutinus* growing in nitrogen-deficient soils with and without mycorrhizae and symbiotic nitrogen-fixing bacteria (Table 6.3). Greatest rates of growth were seen in the presence of both of these symbiotic associations, which also allowed a decrease in the root/shoot ratio. Nitrogen fixation enhanced the uptake of phosphorus by mycorrhizae. These results illustrate the interaction between N, P, and C in the nutrition of higher plants.

In conditions of nutrient deficiency, plant growth usually slows whereas photosynthesis continues at relatively high rates (Chapin 1980), and the content of soluble carbohydrate in the plant increases. Marx et al. (1977) found that high concentrations of carbohydrate in root tissues of

Table 6.3 Effects of Mycorrhizae and N-Fixing Nodules on Growth and Nitrogen Fixation in *Ceanothus velutinus* Seedlings.[a]

	Control	+Mycorrhizae	+Nodules	+Mycorrhizae and Nodules
Mean shoot dry weight (mg)	72.8	84.4	392.9	1028.8
Mean root dry weight (mg)	166.4	183.4	285.0	904.4
Root/shoot	2.29	2.17	0.73	0.88
Nodules per plant	0	0	3	5
Mean nodule weight (mg)	0	0	10.5	44.6
Acetylene reduction (mg/nodule/h)	0	0	27.85	40.46
Percent mycorrhizal colonization	0	45	0	80
Nutrient contents (% ODW in shoot)				
N	0.32	0.30	1.24	1.31
P	0.08	0.07	0.25	0.25
Ca			1.07	1.15

[a] From Rose and Youngberg (1981).

loblolly pine stimulated mycorrhizal infections (Fig. 6.5). Thus, internal plant allocation of carbohydrates to roots may result in increased nutrient uptake by mycorrhizae and an alleviation of nutrient deficiencies.

Mycorrhizae use a fraction of the fixed carbon of the host plant and represent a drain on net primary production that might otherwise be allocated to growth. That the cost of symbiotic fungi is significant is underscored by experiments in which the degree of colonization declined and plant growth increased when plants were fertilized (e.g., Blaise and Garbaye 1983). Vogt et al. (1982) found that mycorrhizal biomass was only 1% of the ecosystem total in a fir forest, but the growth of mycorrhizae utilized about 15% of the net primary production (see Table 5.1). Again, we have few data from which a global estimate of the effect of mycorrhizae on net primary production might be calculated.

Acid Rain: Effect on Plant Nutrient Uptake

Recent studies suggest that forest growth has declined in areas that are downwind of air pollution. In addition to direct effects of ozone, nitric oxide and other gaseous pollutants on plant growth, plants in these areas

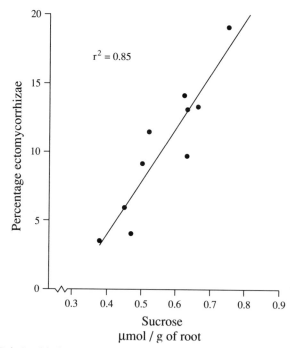

Figure 6.5 Relationship between infection of the roots of loblolly pine by ectomycorrhizal fungi and the sucrose concentration in the root. From Marx et al. (1977).

are subject to "acid rain." Acid rain is characterized by low pH, as a result of NO_3 and SO_4 that are derived from the incorporation of gaseous pollutants in raindrops (Chapter 3). The chemical input in acid rain may affect several aspects of the mineral nutrition of plants, leading to changes in their growth rate.

Inputs of H^+ in acid rain increase the rate of weathering of soil minerals, increase the release of cations from cation exchange sites, and increase movement of Al^{3+} in the soil solution (Chapter 4). High concentrations of Al^{3+} may reduce the uptake of Ca and other cations (Godbold et al. 1988, Bondietti et al. 1989), and in the northeastern United States, forest growth appears to decline as a result of an increased Al/Ca ratio in the soil solution (Shortle and Smith 1988). Depending upon the underlying parent rocks, the soil exchange capacity may be depleted of various nutrient cations. Bernier and Brazeau (1988a, 1988b) link dieback of sugar maple to deficiencies of K on areas of low-K rocks and to deficiences of Mg on low-Mg granites in southeastern Quebec. Magnesium deficiencies are also seen in forests of central Europe (Schulze 1989).

Inputs of N in acid rain may act as fertilizer in areas where forest growth is limited by nitrogen. However, excessive N may lead to the loss of mycorrhizal fungi (Littke et al. 1984), which may exacerbate P deficiency. Along three air pollution gradients in southern California, Zinke (1980) showed that N content in the foliage of Douglas fir increased from 1% to more than 2%, while the P content decreased abruptly, changing the ratios of N/P from about 7 in relatively pristine areas to 20–30 in polluted areas. Such an imbalance in leaf N/P ratios is also seen in the Netherlands, in areas of excessive inputs of NH_4^+ from the atmosphere (Mohren et al. 1986). Excessive uptake of nitrogen leads to the loss of fine root biomass and deficiencies of other nutrients (Schulze 1989, Aber et al. 1989, van Dijk et al. 1990). Greater rates of nitrification in the soil lead to the leaching of cations from the soil profile. Current hypotheses for forest decline as a result of such nutrient imbalances should be tested by the response to experimental additions of P in affected areas (Waring 1987).

Nutrient Allocations and Cycling in Land Vegetation

The Annual Intrasystem Cycle

The uptake of nutrients from the soil is allocated to the growth of new plant tissues. Although short-lived tissues (leaves and fine roots) comprise a small fraction of total plant biomass, they receive a large proportion of the annual nutrient uptake. Growth of leaves and roots received 87% of N and 79% of P allocated to new tissues in a deciduous forest in England (Cole and Rapp 1981, p. 404). In a perennial grassland dominated by *Bouteloua gracilis*, current growth of aboveground tissues sequestered 67% of the annual uptake of N (Woodmansee et al. 1978).

When leaf buds break and new foliage begins to grow, the leaf tissues often have high concentrations of N, P, and K. As the foliage matures, these concentrations often decrease, while concentrations of Ca, Mg, and Fe usually increase (Van den Driessche 1974). Some of these changes are due to increasing accumulation of photosynthetic products and to leaf thickening during development. Leaf mass per unit area (mg cm^{-2}) may increase as much as 50% during the growing season and then decline as the leaf senesces (Smith et al. 1981). The initial concentrations of N and P are diluted as the leaf tissues accumulate carbohydrates and cellulose. Increases in calcium concentration with leaf age result from secondary thickening, including calcium pectate deposition in cell walls, and from increasing storage of calcium oxalate in cell vacuoles.

Although there are variations among species, nutrient concentrations in mature foliage are related to plant growth (e.g., Tilton 1978), and analysis of foliage is often used as an index of site fertility (Van den Driessche 1974). Vitousek et al. (1988) found that C/N and C/P ratios in foliage varied systematically with higher values among species in colder habitats than in the tropics. Among tropical forests, concentrations of major nutrients in leaves are significantly higher on more fertile soils (Vitousek and Sanford 1986). Leaf concentrations of trace metals often reflect the content of the underlying soil, such that leaf tissues are useful for mineral prospecting in some areas.

Upon fertilization with a specific nutrient, the concentrations of other leaf nutrients can show unpredictable patterns of change. Leaf N increased when Miller et al. (1976) fertilized Corsican pine with N, but in the same samples, concentrations of P, Ca, and Mg declined. Apparently, N fertilization of N-deficient stands stimulates photosynthesis such that the concentrations of other nutrients in foliage are diluted by accumulations of carbohydrates (Timmer and Stone, 1978). In these cases, uptake of P from the soil may fall behind the rates needed for growth at the newly established level of N availability. In other cases, improvements in plant nitrogen status enhance the uptake of other elements as well (e.g., Table 6.3). As in the response to acid rain, plant responses to single-element fertilizations suggest that the normal balance of nutrient availability and uptake has been disturbed.

Once leaves are fully expanded, seasonal changes in the nutrient content per unit of leaf area indicate the pattern of nutrient movements between the foliage and the stem. For scarlet oak, Woodwell (1974) found that leaf N accumulated rapidly during the early summer, presumably as a component of photosynthetic enzymes. The leaf content of N, P, and K was relatively constant at high levels during the growing season, but strongly removed from leaves in autumn. Such losses often represent active withdrawal of nutrients from foliage for reuse during the next year. Some trace micronutrients are withdrawn before leaf fall (Killingbeck

1985), but usually reabsorption of foliar Ca and Mg is limited. Fife and Nambiar (1984) observed that reabsorption of N, P, and K was not just related to leaf senescence in Radiata pine, but these nutrients could also move from the early to the later tissues produced during the same growing season.

Leaf nutrient contents are also affected by rainfall that leaches nutrients from the leaf surface (Tukey 1970, Parker 1983). In particular, seasonal changes in the content of K, which is highly soluble and especially concentrated in cells near the leaf surface, may represent leaching. The losses of nutrients in leaching often follow the order:

$$K >> P > N > Ca \hspace{2cm} (6.1)$$

Leaching rates generally increase as foliage senesces before abscission; thus, care must be taken to recongize changes due to leaching versus changes due to active withdrawals (Ostman and Weaver 1982).

Nutrient losses by leaching differ among leaf types. Luxmoore et al. (1981) calculated lower rates of leaching loss from pines than from broadleaf deciduous species in a forest in Tennessee. Such differences may be due to differences in leaf nutrient concentration, surface-area-to-volume ratio, surface texture, and leaf age. Among the trees of the humid tropics, the smooth surface of broad sclerophylls may be an adaptive response to reducing leaching by minimizing the length of time that rainwater is in contact with the leaf surface. Species-specific differences in rates of leaching may explain differences in epiphyte loads of forest species (Schlesinger and Marks 1974).

Rainwater that passes through a vegetation canopy is called throughfall, which is usually collected in funnels or troughs placed on the ground. Throughfall contains nutrients leached from leaf surfaces and is important in the cycling of nutrients such as K (Parker 1983). In forests, rainwater that travels down the surface of stems is called stemflow. The concentrations of nutrients in stemflow waters are high, but usually much more water reaches the ground as throughfall. The annual nutrient return in throughfall typically accounts for 90% of the nutrient movement by leaching of plant tissues. Stemflow is significant to the extent that it returns highly concentrated nutrient solutions to the soil at the base of plants (Gersper and Holowaychuk 1971).

Leaching varies seasonally depending on forest type and climate. Not surprisingly, in temperate deciduous forests, the greatest losses are during the summer months (Lindberg et al. 1986). In some cases the canopy appears to accumulate nutrients from rainfall, particularly soluble forms of N (Carlisle et al. 1966, Miller et al. 1976, Olson et al. 1981, Lang et al. 1976). The leaching of nutrients from vegetation makes it difficult to use nutrient concentrations in the rainfall collected under a canopy to calculate dry deposition of nutrients on leaf surfaces (Chapter 3). Lindberg and Garten (1988) found that about 85% of the flux of sulfate from a forest canopy was due to dry deposition on leaf surfaces.

Litterfall

When the biomass of vegetation is not changing, the annual production of new tissues is balanced by the senescene and loss of plant parts (Chapter 5). In the intrasystem cycle, plant litterfall is the dominant pathway for nutrient return to the soil, especially for N and P. Root death also makes a major contribution to nutrient return to the soil each year (Cox et al. 1978, Vogt et al. 1983).

The nutrient concentrations in litterfall differ from the nutrient concentrations in mature foliage by the reabsorption of constituents during leaf senescene. In the tundra shrub, *Eriophorum vaginatum*, Chapin et al. (1986b) found that all organic N and P compounds decreased to a similar extent during leaf senescence, suggesting that reabsorption is not limited to certain biochemical compounds that are susceptible to hydrolysis. Nutrient reabsorption potentially confers a second type of nutrient-use efficiency on vegetation (see Chapter 5 for nutrient-use efficiency in photosynthesis). Nutrients that are reabsorbed can be used in net primary production in future years, increasing the carbon fixed per unit nutrient uptake. In a wide range of species in the boreal forest, Chapin and Kedrowski (1983) found a mean fractional reabsorption of 52% N and 43% P. Somewhat lower values are seen in a California shrubland (Table 6.4), in the Hubbard Brook forest (Table 6.1), and in grassland ecosystems (Woodmansee et al. 1978). Lajtha (1987) found exceptionally high values for P reabsorption (72–86%) in the desert shrub *Larrea tridentata*, growing in calcareous soils in which P availability is limited due to the formation of calcium phosphates (see Fig. 4.3).

Comparing several temperate forests in Poland, Zimka and Stachurski (1976) found that species with high rates of reabsorption of foliar nutrients tended to dominate nutrient-poor sites, which resulted in an efficient intrasystem cycle of nutrients in these ecosystems. Other studies have found similar (Miller et al. 1976, Tsutsumi et al. 1983, Shaver and Melillo 1984) or opposite results (e.g., Chapin and Kedrowski 1983, Lennon et al. 1985). Chapin (1988) states that plants grown at low nutrient availability or occurring on infertile sites tend to have low nutrient concentrations in mature leaves and litter; they generally reabsorb a smaller *amount* but a larger *proportion* of the nutrient pool in senescing leaves compared to individuals of the same species in conditions of greater nutrient availability (e.g., Pastor et al. 1984, Boerner 1984). Failure to distinguish between the total amount of reabsorption and reabsorption as a fraction of the canopy pool has led to different conclusions about the response of species to fertility gradients in natural ecosystems.

Table 6.4 Nutrient Cycling in a 22-yr-old Stand of the Chaparral Shrub *Ceanothus megacarpus* near Santa Barbara, California.[a]

	Biomass	N	P	K	Ca	Mg
Atmospheric input (g/m^2/yr)						
Deposition		0.15		0.06	0.19	0.10
N fixation		0.11				
Total input		0.26		0.06	0.19	0.10
Compartment pools (g/m^2)						
Foliage	553	8.20	0.38	2.07	4.50	0.98
Live wood	5929	32.60	2.43	13.93	28.99	3.20
Reproductive tissues	81	0.92	0.08	0.47	0.32	0.06
Total live	6563	41.72	2.89	16.47	33.81	4.24
Dead wood	1142	6.28	0.46	2.68	5.58	0.61
Surface litter	2027	20.5	0.6	4.7	26.1	6.7
Annual flux (g/m^2/yr)						
Requirement for production						
Foliage	553	9.35	0.48	2.81	4.89	1.04
New twigs	120	1.18	0.06	0.62	0.71	0.11
Wood increment	302	1.66	0.12	0.71	1.47	0.16
Reproductive tissues	81	0.92	0.08	0.47	0.32	0.07
Total in production	1056	13.11	0.74	4.61	7.39	1.38
Reabsorption before abscission		4.15	0.29	0	0	0
Return to soil						
Litter fall	727	6.65	0.32	2.10	8.01	1.41
Branch mortality	74	0.22	0.01	0.15	0.44	0.02
Throughfall		0.19	0	0.94	0.31	0.09
Stemflow		0.24	0	0.87	0.78	0.25
Total return	801	7.30	0.33	4.06	9.54	1.77
Uptake (=increment + return)		8.96	0.45	4.77	11.01	1.93
Streamwater loss (g/m^2/yr)		0.03	0.01	0.06	0.09	0.06
Comparisons of turnover and flux						
Foliage requirement/total requirement (%)		71.3	64.9	61.0	66.2	75.4
Litter fall/total return (%)		91.1	97.0	51.7	84.0	79.7
Uptake/total live pool (%)		21.4	15.6	29.0	32.6	45.5
Return/uptake (%)		81.4	73.3	85.1	86.6	91.7
Reabsorption/requirement (%)		31.7	39.0	0	0	0
Surface litter/litter fall (yr)	2.8	3.1	1.9	1.2	3.3	4.8

[a] Modified from Gray (1983) and Schlesinger et al. (1982).

In a compilation of data from various forest ecosystems of the world, Vitousek (1982) found that the C/N ratio of leaf litterfall varied by a factor of 4, declining as an inverse function of the apparent nutrient availability of the site. Since the nutrient concentrations in mature foliage seldom vary by more than a factor of 2, his correlation suggests that species in nutrient-poor conditions reabsorb a greater proportion of leaf N before leaf fall. Nutrient-rich sites are associated with high productivity and abundant nutrient circulation, but low nutrient use efficiency. In a later study, he found a similar pattern for phosphorus in tropical forests

(Vitousek 1984). As a result of mycorrhizal associations and internal conservation of P, it appears that tropical trees are adapted to P-deficient soils, which are widespread in these regions (Cuevas and Medina 1986).

Differences in nutrient use efficiency in reabsorption between nutrient-rich and nutrient-poor sites are not as likely to be due to a direct response of plants, as to the tendency for species with higher inherent capabilities for nutrient reabsorption to dominate nutrient-poor sites (Chapin et al. 1986a, Birk and Vitousek 1986, Schlesinger et al. 1989). Net primary production is positively correlated to N availability in both coniferous and deciduous forests (Cole and Rapp 1981), but differences in nutrient reabsorption tend to reduce the correlation, so that light and moisture are the primary determinants of net primary production on a global basis (Chapter 5).

Mass Balance of the Intrasystem Cycle

The annual circulation of nutrients in land vegetation, the intrasystem cycle, can be modeled using the mass-balance approach. Nutrient requirement is equal to the peak nutrient content in newly produced tissues during the growing season (Tables 6.1 and 6.4). Nutrient uptake cannot be measured directly, but uptake must equal the annual storage in perennial tissues such as wood plus the replacement of losses in litterfall and leaching. Uptake is less than the annual requirement by the amount reabsorbed from leaf tissues before abscission. The requirement is the nutrient flux needed to complete a mass balance; it should not be taken as indicative of biological requirements, and in fact it can be calculated for nonessential elements such as Na.

As an example, the mass-balance approach has been used to analyze the internal storage and the annual transfers of nutrients in a California shrubland (Table 6.4). These data serve to summarize many aspects of the intrasystem cycle. Note that 71% of the annual requirement of N is allocated to foliage, whereas much less is allocated to stem wood. However, total nutrient storage in short-lived tissues is small compared to storage in wood, which has lower nutrient concentrations than leaf tissues but has accumulated during 22 yr of growth. For most nutrients in this ecosystem, the storage in wood increases by about 5% each year. In this community the nutrient flux in stemflow is unusually large, but the total annual return in leaching is relatively small, except for K. Despite substantial reabsorption of N and P before leaf abscission, litterfall is the dominant pathway of return of these elements to the soil from the aboveground vegetation. It appears that Ca is actively exported to the leaves before abscission (i.e., requirement < uptake). In this shrubland, annual uptake is 16–46% of the total storage in vegetation, but 73–92% of the uptake is returned each year. As in most studies, some of these calculations would be revised if belowground transfers were better understood.

Nutrient cycling changes during the development of vegetation, as the allocation of net primary production changes. During forest regrowth after disturbance, the leaf area develops rapidly, and the nutrient movements dependent upon leaf area (i.e., litterfall and leaching) are quickly reestablished (Marks and Bormann 1972, Boring et al. 1981). Gholz et al. (1985) found that the proportion of the annual requirement met by internal cycling (i.e., nutrient reabsorption from leaves) increased with time during the development of pine forests in Florida. Nutrients are accumulated most rapidly during the early development of forests, and more slowly as the aboveground biomass reaches a steady-state (Gholz et al. 1985, Pearson et al. 1987). For a forest in Tennessee, the mass-balance approach was used to show that accumulations of Ca and Mg in vegetation were directly related to decreases in the content of exchangeable Ca and Mg in the soil during 11 yr of growth (Johnson et al. 1988). Percentage turnover in vegetation declines as the mass and nutrient storage in vegetation increase. In mature forests, leaf biomass is <5% of the total, and leaves contain only 5–20% of the total nutrient pool in vegetation (Waring and Schlesinger 1985).

Vitousek et al. (1988) have compiled data showing the proportions of biomass (i.e., carbon) and major nutrient elements in various types of mature forest (Table 6.5). The nutrient ratios vary over a surprisingly small range, so the global pattern of element stocks in vegetation is similar to that for biomass: that is, tropical > temperate > boreal forests (Table 5.2). It is important to remember that these ratios are calculated for the total plant biomass; the content of nutrients in leaf tissues is higher and C/N and C/P ratios are correspondingly smaller. Thus, nutrient ratios increase with time as the vegetation becomes increasingly dominated by structural tissues with lower nutrient concentrations (Vitousek et al. 1988).

Nutrient-Use Efficiency in Growth

A mass balance for the intrasystem cycle of vegetation allows us to calculate an integrated measure of nutrient use efficiency by vegetation—net primary production per unit nutrient uptake. This measure is affected by various factors that we have examined individually, including the rate of photosynthesis per unit leaf nutrient (Chapter 5), respiration, root uptake capacity, nutrient reabsorptions during leaf senescence, differences in leaching, and inherent differences in the rate at which photosynthate is incorporated into plant growth. As a result of changes in these various factors, net primary production per unit of nitrogen or phosphorus taken from the soil increased by factors of 5 and 10, respectively, during the growth of pine forests in central Florida (Gholz et al. 1985).

Table 6.5 Biomass and Element Accumulation in Biomass of Mature Forests

Forest Biome	Number of Stands	Total Biomass (t/ha)	Percent of Total Biomass					Mass Ratio		
			Leaf	Branch	Bole	Roots	C/N	C/P	N/P
Northern/subalpine conifer	12	233	4.5	10.2	62.8	22.6	143	1246	8.71
Temperate broadleaf deciduous	13	286	1.1	16.2	63.1	19.5	165	1384	8.40
Giant temperate conifer	5	624	2.5	10.2	66.4	20.8	158	1345	8.53
Temperate broadleaf evergreen	15	315	2.7	14.7	66.2	16.5	159	1383	8.73
Tropical/subtropical closed forest	13	494	1.9	21.8	59.8	16.4	161	1394	8.65
Tropical/subtropical woodland and savanna	13	107	3.6	19.1	60.4	16.9	147	1290	8.80

[a] From Vitousek et al. (1988).

Differences in nutrient use efficiency among terrestrial ecosystems might be due to species differences among sites, with vegetation on poor sites being dominated by species that use nutrients efficiently. Differences in nutrient use efficiency might also appear within a species as a result of responses to differing nutrient availability. These differences have been examined in laboratory experiments. Although there were few differences in root uptake capacity or reabsorption from senescing foliage, Birk and Vitousek (1986) found that net production per unit N in pine seedlings was significantly higher in N-limited plants, accounting for greater nutrient use efficiency in pine forests with low nutrient availability. Ingestad (1979b) found that the growth of birch seedlings increased in response to additions of N, but N-use efficiency (dry matter production per unit N) declined sharply over the same range.

In temperate regions, the annual circulation of nutrients in coniferous forests is much lower than the circulation in deciduous forests, largely as a result of lower leaf turnover in coniferous forest species (Cole and Rapp 1981). Leaching losses are also lower in coniferous forests (Parker 1983), and in many cases evergreen species reabsorb a greater proportion of their leaf nutrient content during leaf senescence (Vitousek 1982). These mechanisms result in greater nutrient-use efficiency in coniferous forests compared to deciduous forests of the world (Table 6.6). Higher nutrient-use efficiency in coniferous species may explain their frequent occurrence on nutrient-poor sites and in boreal climates with slow nutrient turnover in the soil (Schlesinger et al. 1989). These findings may also extend to the occurrence of broad-leaf evergreen vegetation on nutrient-poor soils in other climates (Monk 1966, Beadle 1966, Goldberg 1982, 1985). Significantly, larch, one of the few deciduous species in the boreal forest, has exceptionally high fractional reabsorption of foliar nutrients (Carlyle and Malcolm 1986).

For biogeochemical cycling in vegetation, we have seen that the leaves and fine roots contain only a small portion of the nutrient content in biomass, but the growth, death, and replacement of these tissues largely determine the annual intrasystem cycle of nutrients. Net primary produc-

Table 6.6 Net Primary Production (kg/ha/yr) per Unit of Nutrient Uptake Used as an Index of Nutrient-Use Efficiency to Compare Deciduous and Coniferous Forests[a]

Forest Type	Production per Unit Nutrient Uptake				
	N	P	K	Ca	Mg
Deciduous	143	1859	216	130	915
Coniferous	194	1519	354	217	1559

[a] From Cole and Rapp (1981).

tion is partially dependent upon the soil nutrient pool, but the coupling is weakened by nutrient reabsorptions before leaf drop and other attributes that confer nutrient-use efficiency. When nutrient concentrations in litter are low, as might be expected after reabsorption of nutrients, decomposition is slower. Thus, intrasystem cycling contains a positive feedback to the extent that an increase in nutrient-use efficiency by vegetation may reduce the future availability of soil nutrients for plant uptake (Shaver and Melillo 1984).

Biogeochemical Cycling in the Soil

Despite new inputs from the atmosphere and from rock weathering and plant adaptations to minimize the loss of nutrients, most of the annual nutrient requirements by land vegetation are supplied from the decomposition of dead materials in the soil (Table 6.1). Decomposition of dead organic matter completes the intrasystem cycle by releasing nutrient elements for plant uptake. *Decomposition* is a general term to refer to the breakdown of organic matter. *Mineralization* is a more specific term that refers to processes that release carbon as CO_2 and nutrients in inorganic form, such as P as PO_4^{3-}. A variety of soil animals, including earthworms, fragment and mix fresh litterfall (Swift et al. 1979); however, the main biogeochemical transformations are performed by fungi and bacteria in the soil. Most of the mineralization reactions are the result of the activity of extracellular degradative enzymes, released by soil microbes (Burns 1982). During the course of decomposition, humus compounds are synthesized by microbial activity (Chapter 5).

Soil microbes typically comprise up to 5% of the organic carbon found in soils (Anderson and Domsch 1980). Fungi dominate over bacteria in most well-drained upland soils. Determination of microbial biomass is usually performed by one of several techniques involving fumigation with chloroform (Jenkinson and Powlson 1976a, 1976b). For instance, in a subdivided soil sample, respiration (CO_2 evolution) is measured before and after fumigation with chloroform. The higher rate of respiration in the fumigated sample is assumed to result from the decay of microbes that were killed by chloroform (Stevenson 1986). A correction factor K_c must be applied, since a portion of the dead microbial biomass does not decay immediately (Voroney and Paul 1984). Recently, the chloroform fumigation technique has also been performed using K_2SO_4 to extract soluble organic carbon that is released from dead microbial biomass (Brookes et al. 1985, Vance et al. 1987, Tate et al. 1988). Following a similar approach, extractable N or PO_4^{3-} is measured in a soil sample before and after fumigation, and the higher content after fumigation is assumed to derive from dead microbes (Brookes et al. 1985). Microbial biomass is calculated assuming a standard nitrogen content in microbial tissue and a correction factor K_n to account for microbial N that is not released immediately after fumigation. The technique seems justified by the observation that the C/N and C/P ratio in soil microbial biomass is rather constant over a broad range of values (e.g., Fig. 6.6).

Bacteria and fungi have high concentrations of N and P that are sequestered during their growth. Accumulation of N, P, and other constituents by soil microbes is known as *immobilization*. As a result of high nutrient concentrations, microbial biomass contained 2.5–5.6% of the organic carbon, but up to 19.2% of the organic phosphorus in tropical soils of central India (Srivastava and Singh 1988). Compared to microbial tissue,

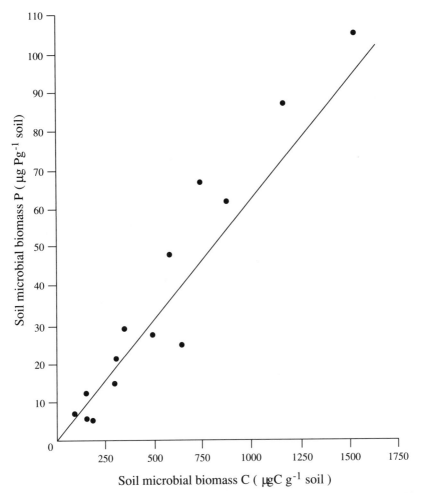

Figure 6.6 Relationship between the phosphorus and carbon contained in microbial biomass of 14 soils. From Brookes et al. (1984).

plant litter has lower concentrations of N and P (i.e., high C/N and C/P ratios). During decomposition, soil microbes respire organic carbon as CO_2, while N and P are retained in microbial biomass. When the decomposition of fresh litter is observed in litterbags (Chapter 5), the C/N and C/P ratios decline as decomposition proceeds and as the remaining materials are progressively dominated by microbial biomass that has colonized and grown on the substrate (Table 6.7).

Immobilization is most significant for N and P, which are limiting to microbial growth, and much less significant for Mg and K that are usually available in excess quantities (Jorgensen et al. 1980, Staaf and Berg 1982, O'Connell 1988). In the process of immobilization, soil microbes not only can retain nutrients released from their substrate, but also can accumulate nutrients that are available in the soil solution from other sources (Berg 1988). Microbial uptake of NH_4^+ is rapid, sequestering available NH_4^+ from plant uptake or from use by nitrifying bacteria (Jackson et al. 1989, Schimel and Firestone 1989). In cases of net accumulation, the apparent *total* content of the substrate increases during the initial phases of decomposition (e.g., Aber and Melillo 1980, Schlesinger 1985a).

When microbial growth slows, there is little further nutrient immobilization. As the microbial populations die, N is released as NH_4^+ from dead microbial tissue (Ladd et al. 1981, Van Veen et al. 1987). This net mineralization of N often begins with C/N ratios near 30:1, but this can vary depending on the substrate and the assimilation efficiency of the decomposer (Rosswall 1982). Using ^{15}N as a tracer, Marumoto et al. (1982) have shown that much of the N mineralized in the soil is released from dead microbes and not directly from soil organic matter. The presence of soil

Table 6.7 Ratios of Nutrient Elements to Carbon in the Litter of Scots Pine (*Pinus sylvestris*) at Sequential Stages of Decomposition[a]

	C/N	C/P	C/K	C/S	C/Ca	C/Mg	C/Mn
				Needle litter			
Initial	134	2630	705	1210	79	1350	330
After incubation of:							
1 yr	85	1330	735	864	101	1870	576
2 yr	66	912	867	ND	107	2360	800
3 yr	53	948	1970	ND	132	1710	1110
4 yr	46	869	1360	496	104	704	988
5 yr	41	656	591	497	231	1600	1120
				Fungal biomass			
Scots pine forest	12	64	41	ND	ND	ND	ND

[a] Some values for fungal tissues are also given. Note that C/N and C/P ratios decline, which indicates retention of these nutrients as C is lost, whereas C/Ca and C/K ratios increase, which indicates that these nutrients are lost more rapidly than carbon. From Staaf and Berg (1982).

animals that feed on bacteria and fungi can increase the rates of release of N and P from microbial tissues (Cole et al. 1978, Anderson et al. 1983).

Litter Decomposition

Litter with higher concentrations of plant nutrients decomposes more rapidly, and net mineralization is likely to begin earlier. Fallen logs, on the other hand, have low N contents and the long-term immobilization of N is especially evident during log decay (Lambert et al. 1980, Fahey 1983, Schimel and Firestone 1989). Ecologists have long used the C/N ratio of litterfall as an index of its potential rate of decomposition (Taylor et al. 1989). More recently, Melillo et al. (1982) have used the lignin/nitrogen ratio in litterfall as a predictor of the rate of decomposition in various ecosystems (Fig. 6.7).

Immobilization of nutrients predominates in the layer of fresh litter on the soil surface, while mineralization of N, P, and S is usually greatest in the lower forest floor (Federer 1983). During soil development, nutrient-rich fulvic acids with low C/N, C/P, and C/S ratios are transported to the lower soil horizons (Schoenau and Bettany 1987). Sollins et al. (1984) found that the "light" fraction of soil organic matter, representing fresh plant residues, had a higher C/N ratio and lower mineralization than the "heavy" fraction, comprised of humic substances (Chapter 5). Release of

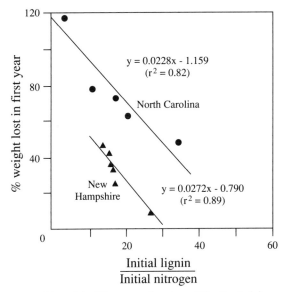

Figure 6.7 Decomposition of leaf litter as a function of the lignin/nitrogen ratio in fresh litterfall of various forest species in New Hampshire and North Carolina. From Melillo et al. (1982).

N, P, and S from soil organic matter is likely to occur at different rates (McGill and Cole 1981). Nitrogen is largely bound directly to C in amino groups ($-C-NH_2$). Thus, N is mineralized as a result of the balance between the degradation of organic substances for energy and the synthesis of protein by microbes. While some S is also bound directly to C, much of the S and P is found in ester linkages (i.e., $C-O-S$ and $C-O-P$). These groups may be mineralized by the release of extracellular enzymes (e.g., phosphatases) in response to specific microbial demand for nutrients. Release of acid phosphatases by soil microbes is directly related to levels of soil organic matter (Tabatabai and Dick 1979). For P, organic transformations are increasingly important as soils age and inorganic P is complexed into secondary minerals (Chapter 4).

Differential losses of nutrients and nutrient immobilizations mean that the loss of mass from litterbags cannot be directly equated with the proportional release of its original nutrient contents (Jorgensen et al. 1980). Table 6.8 shows the mean residence time for organic matter and its nutrient content in the surface litter of various ecosystems. Some nutrients such as K are easily leached from litter and may show mineralization rates in excess of the loss of litter mass. Others such as N turn over more slowly due to immobilization in microbial tissues. Vogt et al. (1986) suggest that immobilization of N is greatest in temperate and boreal forests, whereas immobilizations of P are more important in tropical forests.

In Chapter 5 we saw that the pool of soil organic matter greatly exceeds the mass of living tissue in most ecosytems. Generally, the ratio of C, N, P, and S in humus is close to 140 : 10 : 1.3 : 1.3 (Stevenson 1986). As a result

Table 6.8   Mean Residence Time (yr) for Organic Matter and Nutrients in the Surface Litter of Forest and Woodland Ecosystems[a]

	Mean Residence Time (yr)					
Region	Organic Matter	N	P	K	Ca	Mg
Boreal forest	353	230	324	94	149	455
Temperate forest						
Coniferous	17	17.9	15.3	2.2	5.9	12.9
Deciduous	4	5.5	5.8	1.3	3.0	3.4
Mediterranean	3.8	4.2	3.6	1.4	5.0	2.8
Tropical rain forest	0.4	2.0	1.6	0.7	1.5	1.1

[a] Values are calculated by dividing the forest floor mass by the mean annual litterfall. Boreal and temperate values are from Cole and Rapp (1981), tropical values are from Edwards and Grubb (1982) and Edwards (1977, 1982), and Mediterranean values are from Gray and Schlesinger (1981).

of its high nutrient content, humus also dominates the storage of biogeochemical elements in most ecosystems. In temperate forests, for example,
the aboveground biomass contains only 4–8% of the total quantity of N
within the ecosystem (Cole and Rapp 1981). Slightly higher percentages
are found in tropical forests (Edwards and Grubb 1982), since the pool of
nutrients in humus declines from boreal to tropical regions, whereas
vegetation biomass increases over the same gradient (Tables 5.2 and 5.4).
The global pool of soil N, 95×10^{15} g (Post et al. 1985), dwarfs the pool of
nitrogen in vegetation, 5.2×10^{15} g [calculated using the global biomass
of 827×10^{15} g C (Table 5.2) and a C/N ratio in vegetation of 160 (Table
6.5)]. The stability of humus substances in the mineral soil means that this
large nutrient pool turns over very slowly.

Simple measurements of extractable nutrients, such as NH_4^+ or PO_4^{3-},
are unlikely to give a good index of nutrient availability in terrestrial
ecosystems. These nutrients are subject to active uptake by plant roots,
immobilization by soil microbes, and a variety of other processes that
rapidly remove available forms from the soil solution. At any moment, the
quantity extractable from a soil sample may be only a small fraction of that
which is made available by mineralization during the course of a growing
season. Thus, studies of biogeochemical cycling in the soil are based on
measurements that record the dynamic nature of nutrient turnover.

Nitrogen Cycling

The mineralization of N from decomposing materials begins with the
release of NH_4^+ by heterotrophic microbes (Fig. 6.8). This process is
known as ammonification. Subsequently, a variety of processes affect the
concentration NH_4^+ in the soil solution, including uptake by plants,
immobilization by microbes, and fixation in clay minerals (Chapter 4).
Some of the remaining NH_4^+ may undergo nitrification, in which oxidation of NH_4^+ to NO_3^- is coupled to the fixation of carbon by chemoautotrophic bacteria in the genera *Nitrosomonas* and *Nitrobacter* [equations
(2.15) and (2.16)]. In some cases NH_4^+ is also oxidized in heterotrophic
nitrification to NO_3^- (Schimel et al. 1984). Nitrate is subject to plant
uptake, and loss from the ecosystem in runoff waters or by denitrification.
Nitrate is also subject to immobilization (assimilatory reduction), but
soil microorganisms often show a distinct preference for NH_4^+ (Jones
and Richards 1977, Vitousek and Andariese 1986, Jackson et al. 1989).
Extractable quantities of NH_4^+ and NO_3^- at any time represent the net
result of these processes. A low concentration of NH_4^+ is not necessarily
an indication of low mineralization rates, because it can also indicate rapid
nitrification or plant uptake (Rosswall 1982).

Various techniques are available to study the individual transformations of nitrogen (Binkley and Hart 1989). Many workers have used

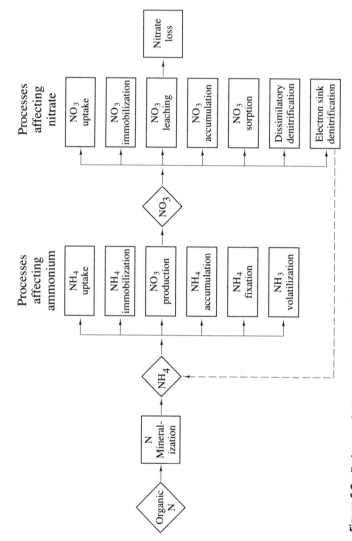

Figure 6.8 Pathways of mineralization and nitrification of organic nitrogen in the soil, and potential fate of the products, NH_4^+ and NO_3^-. From Vitousek and Melillo (1979). Reprinted from *Forest Science*, published by the Society of American Foresters, 5400 Grosvenor Lane, Bethesda, MD 20814-2198.

the "buried-bag" approach to examine net mineralization. A soil sample is subdivided and part is extracted immediately, usually with KCl, to measure the available NH_4^+ and NO_3^-. The remaining portion is replaced in the field in a polyethylene bag, which is permeable to O_2 but not to H_2O. After a short period, usually 30 days, the second bag is retrieved and analyzed for the forms of available N. An increase in the quantity of available N is taken to represent net mineralization, that is, the mineralization in excess of microbial immobilization, in the absence of plant uptake. Repeated samples taken through an annual cycle allow an estimate of annual net mineralization, which can be correlated with plant uptake and cycling (Pastor et al. 1984). Although this technique has proven useful in a variety of studies, it is not without problems. During the course of incubation, soil moisture content in the buried bag does not fluctuate as it does in the natural ecosystem, and the original soil sample inevitably contains fine root material that is severed during collection.

Field measurements can also be performed in tubes (Raison et al. 1987) or trenched plots (Vitousek et al. 1982). In the latter, a block of soil, often 1 m², is isolated on all sides by trenching and the trenches are lined with plastic to prevent the ingrowth of roots. Plants rooted in this plot are removed, but the area is not otherwise disturbed. Periodic measurements of NH_4^+ and NO_3^- indicate rates of mineralization and nitrification in the absence of plant uptake. Since trenching also eliminates the plant uptake of water, this approach measures microbial activity at artifically high soil moisture content, and with potential losses from the ecosystem due to leaching and denitrification.

An expensive but improved approach involves the use of ^{15}N to label the initial pool of available N (Van Cleve and White 1980). After a period of time, the pool is remeasured for ^{15}N content, and the decline in ^{15}N is taken to represent turnover of the available pool by microbial mineralization. This gives a measure of gross mineralization in natural field conditions. Net nitrification can be studied by measuring changes in the concentration of NH_4^+ and NO_3^- after application of compounds that specifically inhibit nitrification, including nitrapyrin (Bundy and Bremner 1973) or acetylene (Berg et al. 1982).

Mineralization and nitrification have been studied in a wide variety of ecosystems (Vitousek and Melillo 1979, Robertson 1982b, Vitousek and Matson 1988). Generally net mineralization is directly related to the total content of organic nitrogen in the soil (e.g., Marion and Black 1988), but mineralization is also closely linked to the availability of carbon. Vegetation with a high C/N ratio in litterfall often shows low rates of mineralization in the soil (Gosz 1981, Vitousek et al. 1982). When field plots are fertilized with sugar, net mineralization slows as immobilization of NH_4^+ by soil microbes increases. Fertilization of Douglas fir with sugar resulted in lower N content in leaves and greater nutrient reabsorption before leaf fall (Turner and Olson 1976), showing a direct link between microbial processes in the soil and nutrient-use efficiency by vegetation.

Although soil microbial populations may adapt to a wide variety of field conditions, nitrification is generally lower at low pH, low O_2, and high litter C/N ratios (Rosswall 1982, Robertson 1982a). Vitousek and Matson (1988) found high rates of mineralization and nitrification in most tropical forests, but Marrs et al. (1988) reported that net mineralization and nitrification were inhibited by the high soil water content in montane tropical forests in Costa Rica. Nitrification rates are high when NH_4^+ is readily available (Robertson and Vitousek 1981), but the concentrations of other nutrients generally have little effect (Robertson 1982b, 1984, Christensen and MacAller 1985).

A large amount of effort has been directed toward understanding the control of nitrification following disturbances, such as forest harvest or fire (Vitousek and Melillo 1979, Vitousek et al. 1982). When vegetation is removed, soil temperature and moisture contents are generally higher, and rapid ammonification increases the availability of NH_4^+. Subsequently, nitrification may be so rapid that uptake by vegetation and immobilization by soil microbes are insufficient to prevent large losses of NO_3^- in stream water following disturbance. However, not all sites show large losses of NO_3^- upon disturbance. In pine forests in the southeastern United States, microbial immobilization in harvest debris accounted for 83% of the uptake of ^{15}N that was applied as an experimental tracer following forest harvest (Vitousek and Matson 1984). Microbial immobilization also retards the loss of nitrate following burning of tallgrass prairie (Seastedt and Hayes 1988). In general, nitrification and losses of NO_3^- in stream water are greatest in forests with high nitrogen availability prior to disturbance (Vitousek et al. 1982). Rates of nitrification decline during the early recovery of vegetation, and only minor differences are seen between early and late successional forests (Robertson and Vitousek 1981, Christensen and MacAller 1985). There is some evidence that nitrification is inhibited by terpenoid and tannin compounds released by some types of vegetation (Olson and Reiners 1983, White 1986, 1988), but little evidence for a direct inhibition of nitrification by mature vegetation, as predicted by Rice and Pancholy (1972).

Increases in nitrification following disturbance affect other aspects of ecosystem function. Since nitrification generates acidity [equation (2.15)], losses of NO_3^- in stream water are often accompanied by increased losses of cations, which are removed from cation exchange sites in favor of H^+ (Likens et al. 1970). Stream-water losses of nearly all biogeochemical elements increased following harvest at the Hubbard Brook Forest in New Hampshire; however, sulfate was a curious exception (Fig. 6.9). Nodvin et al. (1988) have now shown that the decline in stream-water SO_4^{2-} concentrations is related to an increase in soil anion absorption

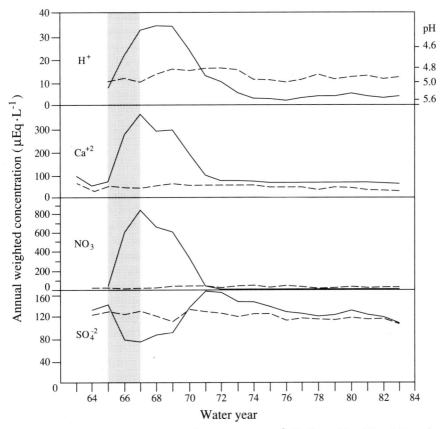

Figure 6.9 Concentrations of H^+, Ca^{2+}, NO_3^-, and SO_4^{2-} in the Hubbard Brook Experimental forest for the years 1964–1984. Streams draining undisturbed forest are shown with the dashed line. The solid line depicts the concentrations in a stream draining a single watershed that was disturbed between 1965 and 1967 (shaded). Losses of Ca and NO_3^- increased strongly during the period of disturbance, and then recovered to normal values as the vegetation regenerated. The budget for SO_4^{2-} shows greater retention during and after the period of disturbance, presumably as a result of increased acidity and anion absorption capacity in the soil. Modified from Nodvin et al. (1988).

capacity as a result of the acidity generated from nitrification (Mitchell et al. 1989; Chapter 4). These observations are a good example of the linkage between the biogeochemical cycles of N and S in terrestrial ecosystems.

Denitrification

Nitrate is converted to N_2O and N_2 in the process of denitrification (Knowles 1982, Firestone 1982). This reaction [equation (2.17)] is performed by soil bacteria that are aerobic heterotrophs in the presence of O_2, but facultative anaerobes in the absence of O_2. Heterotrophic activity continues with nitrate serving as a terminal electron acceptor in the absence of O_2. Since the NO_3^- is reduced, but not incorporated into microbial biomass, denitrification is also known as dissimilatory nitrate reduction. Bacteria in the genus *Pseudomonas* are the best-known denitrifiers, but many others are reported (Knowles 1982, Tiedje et al. 1989).

For a long time, denitrification was thought to occur only in flooded, anaerobic soils (Chapter 7), and its importance in upland ecosystems was overlooked. Now, soil scientists have shown that oxygen diffusion to the center of soil aggregates is so slow that anaerobic microzones are common (Tiedje et al. 1984, Sexstone et al. 1985a). Thus, denitrification is widespread in terrestrial ecosystems, especially those in which organic carbon and nitrate are readily available. Denitrification returns N_2 to the atmosphere, completing the biogeochemical cycle of nitrogen (Bowden 1986). Although N_2O comprises only a small portion of denitrification, recent changes in the production of N_2O in soils are potentially important, given the role of N_2O in the destruction of stratospheric ozone and in greenhouse warming (Chapters 3 and 12).

Field measurements of denitrification are usually based on the observation that acetylene blocks the conversion of the intermediate product, N_2O, to N_2 (Fig 6.10) (Yoshinari and Knowles 1976, Tiedje et al. 1989). Since it is much easier to measure N_2O without contamination from the atmosphere, application of acetylene in laboratory incubations or field plots is followed by the collection of gas for N_2O determinations by gas chromatography. The incubations must be short, since acetylene also blocks nitrification, and long-term incubations could be affected by a decline in the concentration of NO_3^-, which is needed for denitrification. Denitrification can also be measured by the application of $^{15}NO_3^-$ to field plots, and measurement of the release of $^{15}N_2$ or the decline in $^{15}NO_3^-$ remaining in the soil (Parkin et al. 1985, Mosier et al. 1986).

Denitrification usually proceeds most rapidly when organic carbon and

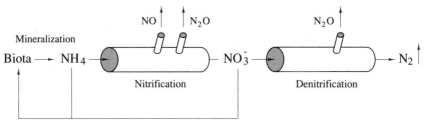

Figure 6.10 Transformations producing nitrogen gases during nitrification and denitrification. Based on an unpublished diagram of M. Firestone.

nitrate are readily available (Burford and Bremner 1975); however, the C/N ratio of the organic carbon must be low enough that microbial immobilization is minimal. Davidson and Swank (1987) found that additions of NO_3^- stimulated denitrification in undisturbed forest soils in western North Carolina, but additions of organic carbon were significant only in the mineral soil. Additions of NO_3^- stimulate the release of N_2O in tropical forests of the Amazon basin, implying that denitrification is inhibited by low levels of NO_3^- in natural conditions (Livingston et al. 1988, Keller et al. 1988). In Costa Rica, however, additions of organic carbon had a much greater effect than NO_3^- (Matson et al. 1987) and denitrification is inhibited when NO_3^- concentrations are very high (Fig. 6.11). Rainfall stimulates denitrification, as the diffusion of oxygen is slow in wet soils (Sexstone et al. 1985b, Smith and Tiedje 1979).

During decomposition, soil microbes mineralize ^{14}N in favor of ^{15}N, which increases in the undecomposed residue (Nadelhoffer and Fry 1988). Denitrifying bacteria further fractionate among the isotopes of nitrogen, that is, between $^{14}NO_3^-$ and $^{15}NO_3^-$. Preference for $^{14}NO_3^-$ leads to positive $\delta^{15}N$ in most soils (Fig. 6.4), as $^{14}N_2$ is lost from the soil by denitrification (Shearer and Kohl 1988).

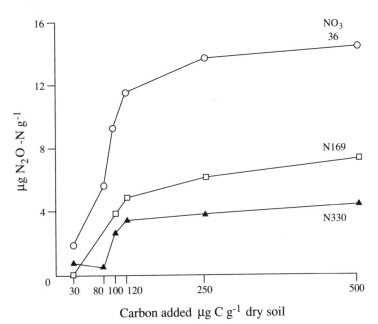

Figure 6.11 Effects of organic carbon and available NO_3 on the rate of denitrification, measured as the accumulation of N_2O after application of acetylene. Modified from Lalisse-Grundmann et al. (1988). Nitrate levels are $\mu g/g$ soil.

The measurement of denitrification is complicated by the observation that N_2O is also released as a byproduct during nitrification and perhaps by other soil nitrogen transformations (Robertson and Tiedje 1987, Davidson et al. 1986, Tortoso and Hutchinson 1990). The relative importance of nitrification and denitrification as a source of N_2O varies among ecosystems. Matson and Vitousek (1987) found a direct relation of N_2O production and nitrogen mineralization in comparisons of various tropical forests (Fig. 6.12), but in the Amazon, N_2O appeared to be mostly from denitrification (Livingston et al. 1988). On the other hand, nitrification was the major source of N_2O lost from a shortgrass prairie ecosystem, in which 2.5–9.0% of the annual input of nitrogen from rainfall was lost to the atmosphere (Parton et al. 1988a). Factors affecting the relative loss of N_2O and N_2 by nitrification and denitrification are poorly understood, but include soil pH and the relative abundance of NO_3^- and O_2 as oxidants and organic carbon as a reductant (Firestone et al. 1980). When NO_3^- is abundant relative to the supply of organic carbon, N_2O can be an important product. The loss of N_2O is minimal in flooded organic soils when pH > 7.0 (Weier and Gilliam 1986). Recent evidence also suggests that soils release NO, perhaps at a rate greater than the release of N_2O, but the processes leading to the release of NO are poorly understood (Anderson and Levine 1987, Johansson et al. 1988, Kaplan et al. 1988).

Loss of N_2O to the atmosphere increases greatly when agricultural

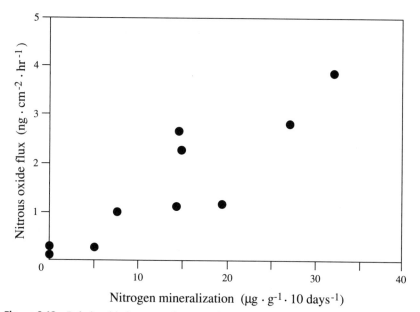

Figure 6.12 Relationship between nitrogen mineralization measured in laboratory incubations and loss of N_2O from 10 tropical forest soils. From Matson and Vitousek (1987).

fields are fertilized with nitrogen (Bremner and Blackmer 1978, Conrad et al. 1983, Slemr et al. 1984), suggesting that the increasing use of commercial fertilizers might be a cause of increasing concentrations of N_2O in the atmosphere (Fig. 3.9). When nitrification increases after fire or forest harvest, higher losses of N_2O are also seen (Bowden and Bormann 1986, Anderson et al. 1988, Robertson and Tiedje 1988, Luizão et al. 1989). High concentrations of NO_3^- in stream water may be reduced by denitrification in streambed sediments (Swank and Caskey 1982).

Regional and global estimates of denitrification are complicated by high spatial variability. At the local scale, a large portion of the total variability is found at distances of <10 cm, which Parkin et al. (1987) link to the local distribution of soil aggregates that provide anaerobic microsites. Parkin (1987) found that 85% of the total denitrification in a 15-cm diameter soil core was located under a 1-cm^2 section of a decaying pigweed (*Amaranthus*) leaf! In desert ecosystems, soil nitrogen content and nitrification rates are greatest under shrubs (Charley and West 1977, Klopatek 1987, Burke 1989, Burke et al. 1989), and denitrification is largely confined to those areas (Virginia et al. 1982).

Robertson et al. (1988) have documented the pattern of mineralization, nitrification, and denitrification in a field in Michigan. All these processes showed large variation, but the coefficient of variation for denitrification, 275%, was the largest measured. Significant autocorrelation was seen among these processes. Soil respiration and potential nitrification explained 37% of the variation in denitrification, presumably due to the dependence of this process on organic carbon and NO_3^- as substrates. The high variability of these processes makes it difficult to use measurements from a few sample chambers to calculate a mean or total flux from an ecosystem. Groffman and Tiedje (1989) suggest that correlations of denitrification to soil texture may allow the most accurate extrapolations of laboratory measurements to regional estimates of gaseous N loss.

At a larger scale, high rates of denitrification are often confined to particular landscape positions where conditions are favorable. For example, Peterjohn and Correll (1984) suggest that the runoff of nitrate from agricultural fields was largely denitrified in streamside forests, minimizing the losses in rivers (Davidson and Swank 1986, Lowrance et al. 1984, Jacobs and Gilliam 1985). In calculating regional averages for denitrification, investigators must weight the contributions from local areas of high and low activity.

Mean values for loss of N by denitrification are typically <2 kg ha^{-1} yr^{-1} in forests and grasslands (Robertson and Tiedje 1984). However, denitrification losses as high as 19 kg N ha^{-1} yr^{-1} in deserts are necessary to balance an internal nitrogen budget in some sites (West and Skujins 1977). Melillo et al. (1983) found that N_2O was the only significant product of denitrification in the soils of four forests in New Hampshire.

Although large values are reported from the tropics (Keller et al. 1986), N_2O flux is generally $<<2$ kg ha^{-1} yr^{-1} in most ecosystems (Goodroad and Keeney 1984, Bowden et al. 1990). Anderson and Levine (1987) measured a loss of 0.5 kg N ha^{-1} yr^{-1} as NO from an unfertilized field in Virginia.

Global estimates of denitrification range from 7 to 16×10^{12} g yr^{-1} for N_2O and 13 to 233×10^{12} g yr^{-1} for $N_2 + N_2O$ (Bowden 1986; Chapter 12). At least half of this flux is from wetlands, which have high rates of denitrification (Chapter 7). Slemr and Seiler (1984) estimate the global loss of NO from soils may be as large as 11×10^{12} g N yr^{-1}. Further refinement of these estimates will require extensive field work, combining local measurements with regional extrapolations.

Ammonia Volatilization

In soils of high pH, ammonium is converted to NH_3, which is lost to the atmosphere. The reaction is

$$NH_4^+ + OH^- \rightarrow NH_3 \uparrow + H_2O \tag{6.2}$$

The reaction is favored in deserts where accumulations of $CaCO_3$ in the soil maintain alkaline pH, and dry, permeable soils with low cation exchange capacity maximize the conversion and loss of NH_4 (Nelson 1982). Ammonia volatilization is also greater under conditions where nitrification is slow (Fleisher et al. 1987). The highest rates of ammonia volatilization are associated with the application of nitrogen fertilizer and with cattle grazing, where NH_3 is derived from the mineralization of urea (Terman 1979, Freney et al. 1983). Denmead et al. (1974) found losses of 0.26 kg ha^{-1} day^{-1} from a grazed pasture in Australia. Extremely high NH_3 volatilization from feedlots results in high deposition of NH_4^+ in precipitation in the Netherlands (Van Breeman et al. 1982), but in most cases volatile losses from natural ecosystems are rather small (Schimel et al. 1986). During volatilization of ammonia, isotopic fractionation occurs, leaving soils with high $\delta^{15}N$ (Mizutani et al. 1986, Mizutani and Wada 1988).

The flux of NH_3 to the atmosphere is especially significant as ammonia is the only substance capable of generating alkalinity in rainfall (Chapter 3). Dawson (1977) used a model of soil nitrogen transformations to estimate a flux of 47×10^{12} g NH_3 yr^{-1} from undisturbed land. The total flux may be as large as 150×10^{12} g yr^{-1} globally (Bowden 1986), although Warneck (1988) balances the atmospheric budget with a flux of 50×10^{12} g yr^{-1}.

Phosphorus Cycling

Transformations of organic phosphorus in the soil are difficult to study because of the reactions of inorganic phosphorus with soil mineral forms (Fig. 6.13) (Chapter 4). A few workers have examined phosphorus mineralization using the buried bag approach (e.g., Pastor et al. 1984), but in many cases there is no apparent mineralization because of complexation of P with soil minerals. Thus, most studies of phosphorus cycling have followed the decay of radioactively labeled plant materials (Harrison 1982) or measured the dilution of ^{32}P that is applied to the soil pool as a radioisotope (Walbridge and Vitousek 1987).

With the isotope dilution technique, one must assume that ^{32}P equilibrates with all the chemical pools in the soil, and that the only dilution of its concentration is by the mineralization of organic phosphorus. Unfortunately, these assumptions are not always valid, making the technique difficult to apply in many instances (Walbridge and Vitousek 1987).

Recognizing the shortcomings of using simple extractions to measure available P, most workers have followed a sequential extraction scheme to quantify phosphorus availability in the soil (Hedley et al. 1982b, Stevenson 1986). Extraction with 0.5 M NaHCO$_3$ is a convenient index of labile

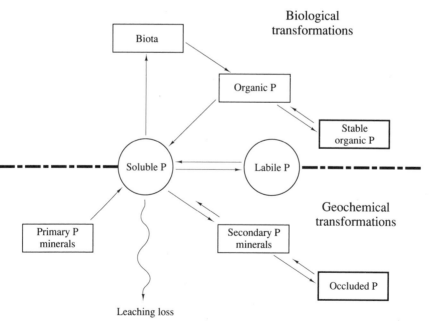

Figure 6.13 Phosphorus transformations in the soil. From Smeck (1985).

inorganic and soluble organic phosphorus in many soils (Olson et al. 1954, Sharpley et al. 1987). Organic P is often determined as the difference between PO_4 in an "ashed" sample and an untreated sample (Stevenson 1986), and microbial P by the change in extractable phosphorus after fumigation with chloroform (Brooks et al. 1982). Extraction with various acids and bases indicates the quantities that are held in Fe, Al, Ca, and primary minerals (Tiessen et al. 1984, Lajtha and Schlesinger 1988).

In most ecosystems phosphorus available for biogeochemical cycling is held in organic forms (Chapin et al. 1978, Wood et al. 1984a, but see Lajtha and Schlesinger 1988). Mineralization of organic P usually begins at C/P ratios that are <200, although we have earlier discussed the ability of microbes to release phosphatase enzymes and organic acids that mineralize P from organic and inorganic forms.

Despite fokelore to the contrary, the production of phosphine gas (PH_3) is impossible in natural soils, requiring extremely low redox potential (Bartlett 1986; Chapter 7). There are scattered reports that such conditions exist in sewage treatment ponds (Dévai et al. 1988), but the movement of phosphorus as a gas is of negligible importance in its regional or global cycle.

Sulfur Cycling

Like the phosphorus cycle, the cycle of sulfur in the soil is also affected by both chemical and biological reactions. Sulfur is derived from the weathering of pyrite and gypsum in rocks and from atmospheric deposition (Chapter 3). In many soils most SO_4^{2-} is held on anion absorption sites on Fe and Al minerals in the lower profile (Chapter 4). For instance, in the study of a forest in Tennessee, Johnson et al. (1982) found that the pool of adsorbed SO_4^{2-} was larger than the total pool of S in vegetation by a factor of 15. In these soils, sulfate available for plant uptake exists in equilibrium with sulfate adsorbed on soil minerals. In other soils, sesquioxide minerals are of limited importance, and the major pool of S is organic (Watwood et al. 1988, Bartel-Ortiz and David 1988).

Most of the sulfur in plants is found in protein. Plant uptake of SO_4^{2-} is followed by assimilatory reduction, and incorporation of carbon-bonded sulfur in the amino acids, cysteine and methionine, that are constituents of protein (Johnson 1984). A small quantity of sulfur in plants is found in ester-bonded sulfates (—C—O—SO_4), and when soil sulfate concentrations are high, plants may also accumulate SO_4 in leaf tissues.

Decomposition of plant tissues is accompanied by microbial immobilizations (Saggar et al. 1981, Staaf and Berg 1982, Fitzgerald et al. 1984), and mineralization of SO_4^{2-} generally begins at C/S ratios <200 (Stevenson 1986). In most cases, the majority of microbial S is found in carbon-bonded form (David et al. 1982, Watwood et al. 1988, S. C. Schindler et al.

JUNE 1982 GREEN LEAF DENSITY
BY NASA/GSFC GIMMS

Plate 1 Normalized Difference Vegetation Index (NDVI) for July 1982 as obtained from the Advanced Very High Resolution Radiometer on the NOAA satellite. Note that the greatest vegetation density is colored blue, whereas green and yellow indicate lower leaf area. The Northern Hemisphere is in mid-summer. From NASA, 1987, Moderate-Resolution Imaging Spectrometer, Instrument Report, Washington, D.C.

Plate 2 Distribution of chlorophyll in the western North Atlantic Ocean during May 1981, as recorded by the Coastal Zone Color Scanner (CZCS) on the Nimbus-7 satellite. Areas rich in phytoplankton are shown as red (> 1mg chlorophyll/m^3); blue and purple areas have lower chlorophyll concentrations (< 0.01 mg/m^3). Note the high productivity of coastal areas, especially from North Carolina to Maine. From NASA, 1987, High-Resolution Imaging Spectrometer, Instrument Panel Report, Washington, D.C.

1986, Dhamala et al., 1990). However, at the Cowetta Experimental Forest in North Carolina, a large portion of the immobilization of sulfur by soil microbes formed ester sulfates (Fitzgerald et al. 1985, Watwood and Fitzgerald 1988), conferring a significant sink for SO_4^{2-} derived from the atmosphere in this region (Swank et al. 1984). Downward movement of fulvic acids appears to transport ester sulfate to the lower soil profile (Schoenau and Bettany 1987). Sulfur in soil humus shows lower $\delta^{34}S$ than soil sulfate, suggesting that soil microbes discriminate against the rare, heavy isotope in favor of ^{32}S during the synthesis of humic substances (Fuller et al. 1986).

To maintain a charge balance, plant uptake and reduction of SO_4^{2-} consume H^+ from the soil, whereas the mineralization of organic sulfur returns H^+ to the soil solution, producing no net increase in acidity (Binkley and Richter 1987). In contrast, reduced inorganic sulfur is found in association with some rock minerals, and the oxidative weathering of reduced sulfide minerals accounts for highly acidic solutions draining mine tailings. This oxidation is performed by chemoautotrophic bacteria, generally in the genus *Thiobacillus* [equation (2.14)].

Production of reduced sulfur gases, such as H_2S, COS (carbonyl sulfide), and $(CH_3)_2S$ (dimethylsulfide), is largely confined to wetland soils, since highly reducing, anaerobic conditions are required (Chapter 7). However, Adams et al. (1981) have measured the release of several of these compounds from upland soils. They suggest that 59% of the global production of reduced sulfur gases may arise from upland areas. Delmas and Servant (1983) also argue for a large release of H_2S from soils, based on a regional study in the humid tropical forests of the Ivory Coast. The global estimate of Adams et al. (1981) is higher than that given by most other workers (Warneck 1988, Goldan et al. 1987), but there is no doubt that the emission of reduced S gas from land is a significant component of the global sulfur budget (Chapter 13).

Transformations in Fire

During fires, nutrients are lost in gases and in the particles of smoke, and soil nutrient availability increases with the addition of ash to soil (Raison 1979, Woodmansee and Wallach 1981). Following fire, there is often increased runoff and erosion from bare, ash-covered soils. Before human intervention, fires were a natural part of the environment in many regions; thus, these nutrient losses from ecosystems occurred at infrequent but somewhat regular intervals. Using the mass-balance approach we can estimate the length of time that it takes to replace the nutrients that are lost in a single fire. For instance, 11–40 kg/ha of N is lost in small ground fires in southeastern pine forests (Richter et al. 1982), equivalent to 3–12 times the annual deposition of N from the atmosphere in this region (Swank and Henderson 1976).

When leaves and twigs are burned in laboratory conditions, up to 85% of their N content can be lost, presumably as N_2 or as one or more forms of nitrogen oxide gases (DeBell and Ralston 1970, Hegg et al. 1990, Lobert et al. 1990). Forest fires volatilize N in proportion to the heat generated and the organic matter consumed (DeBano and Conrad 1978, Raison et al. 1985). Typically N losses in forest fires range from 100 to 300 kg/ha, or 10–40% of the amount in aboveground vegetation and surface litter. Grier (1975) reported a volatile N loss of 855 kg/ha, 39% of the vegetation pool, in an intense wildfire in a montane coniferous forest in Washington.

Air currents and updrafts during fire carry particles of ash that remove other nutrients from the site. These losses are usually much smaller than N losses. Expressed as a percentage of the amount present in above-ground vegetation and litter before fire, the losses often follow the order N >> K > Mg > Ca > P > 0%. Differential loss changes the balance of nutrients available in the soil after fire (Raison et al. 1985), and nutrient losses to the atmosphere in fire may result in added atmospheric deposition in adjacent locations (Clayton 1976, Lewis 1981).

Depending on intensity, fire kills aboveground vegetation and transfers varying proportions of its mass and nutrient content to the soil as ash. There are a large number of changes in soil chemical and biological properties as a result of fire and additions of ash to the soil (Raison 1979). Cations and P may be readily available in ash, which usually increases soil pH (Woodmansee and Wallach 1981). DeBano and Klopatek (1988) found that burning increased extractable P, but reduced the levels of organic P and phosphatase activity in the soils of pinyon–juniper woodlands. Nitrogen may be released from ash by mineralization and nitrification that is stimulated by fire (Christensen 1973, 1977, Dunn et al. 1979, Matson et al. 1987). Thus, available NH_4^+ and NO_3^- increase after fire, even though total soil N may be lower. The increase in available nutrients as a result of ashfall is usually short-lived, as nutrients are taken up by vegetation or lost to leaching and erosion (Lewis 1974, Christensen 1977, Uhl and Jordan 1984). High rates of nitrification can stimulate the loss of NO and N_2O from burned soils (Anderson et al. 1988, Levine et al. 1988); thus, increased clearing and burning of tropical forests may be responsible for part of the observed increase in the atmospheric concentration of N_2O globally. Losses of N from surface soils may be overestimated by increases in N and other nutrients that are transferred to the lower soil profile (Mroz et al. 1980, Grier 1975).

Stream-water runoff is often greater after fire because of reduced water losses in transpiration. High nutrient availability in the soil coupled with greater runoff can lead to large nutrient losses from the ecosystem. These losses depend on many factors, including the season, rainfall pattern, and the growth of postfire vegetation. Wright (1976) noted significant increases in the loss of K and P from burned forest watersheds in Minnesota. These losses were greatest in the first 2 yr after fire; by the third year there was actually less P lost from burned watersheds than from adjacent mature forests, presumably due to uptake by regrowing vegetation (McColl and Grigal 1975). Percentage losses of Ca, Mg, Na, and K in runoff waters often exceed those of N and P, but there are exceptions to this pattern.

The Role of Land Animals

Discussions of terrestrial biogeochemistry center on the role of plants and soil microbes. Having seen that animals harvest only a small fraction of net primary production (Chapter 5), it is legitimate to ask if they might play a significant role in nutrient cycling. Certainly an impressive nutrient influx is observed below roosting birds (Gilmore et al. 1984, Mizutani and Wada 1988). Various workers have also suggested that animals grazing aboveground vegetation, especially insects, stimulate the intrasystem cycle of nutrients and might even be advantageous for terrestrial vegetation (Owen and Wiegert 1976). Risley and Crossley (1988) noted that significant premature leaf fall in a forest was related to insect grazing. These leaves delivered large quantities of nutrients to the soil, since nutrient reabsorption had not yet occurred. In the same forest, Swank et al. (1981) noted an increase in streamwater nitrate when trees were defoliated by grazing insects.

Trees that are susceptible to herbivory are often those that are deficient in minerals or otherwise stressed (Waring and Schlesinger 1985). Periodic herbivory may stimulate nutrient return to the soil and alleviate nutrient deficiencies (Mattson and Addy 1975). In extreme cases, defoliations may be the dominant form of nutrient turnover in the ecosystem (Hollinger 1986). Usually, however, the role of grazing animals in terrestrial ecosystems is rather minor (Gosz et al. 1978, Woodmansee 1978, Pletscher et al. 1989), and certainly of limited benefit to plants (Lamb 1985). In fact, plants often show marked allocation of net primary production to defensive compounds (Coley et al. 1985) and increases in net primary production when they are relieved of insect herbivores (Morrow and LaMarche 1978).

The role of animals in litter decomposition is much more significant (Swift et al. 1979, Hole 1981, Seastedt and Crossley 1980). Nematodes, earthworms, and termites are particularly widespread and important in the initial breakdown of litter and the turnover of nutrients in the soil. Schaefer and Whitford (1981) found that termites are responsible for the turnover of 8% of litter N annually in a desert ecosystem (Fig. 6.14). An additional 2% of the pool of nitrogen in surface litter was transported

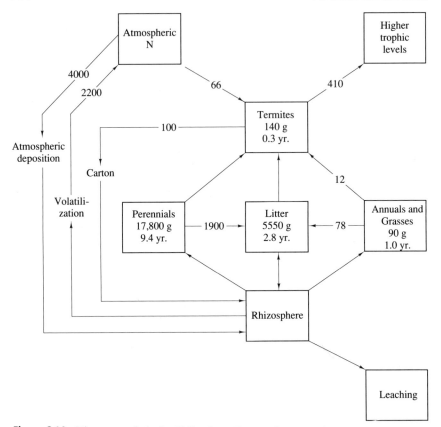

Figure 6.14 Nitrogen cycle in the Chihuahuan Desert of New Mexico, showing the role of termites in nitrogen transformations. Annual flux of nitrogen is shown along arrows in g N/m^2; nitrogen pools are shown in boxes with turnover time in years. From Schaefer and Whitford (1981).

belowground by their burrowing activities. When termites are excluded by applications of pesticides, decomposition is slowed and surface litter accumulates. Since soil animals have short lifetimes, their nutrient content is rapidly decomposed and returned to the intrasystem cycle (Seastedt and Tate 1981).

It is interesting to view the biogeochemistry of animals from another perspective: What is the role of biogeochemistry in determining the distribution and abundance of animals? The death of ducks and cattle feeding in areas of high soil selenium (Se) suggests that such interactions might be of widespread significance.

Plants have no essential role for sodium in their biochemistry, and naturally have low Na contents due to limited uptake and exclusion at the root surface (Smith 1976). On the other hand, sodium is an important, essential element for all animals. The wide ratio between the Na content of herbivores and that in their foodstuffs suggests that Na might limit mammal populations generally. Observations of Na deficiency are supported by the interest that many animals show in natural salt licks (Jones and Hanson 1985) and Na-rich plants (Botkin et al. 1973; but see also Risenhoover and Peterson 1986). Weir (1972) suggested that the distribution of elephants in central Africa was at least partially dependent on sodium in seasonal waterholes, and McNaughton (1988) found that the abundance of ungulates in the Serengeti area was linked to Na, P, and Mg in plant tissues available for grazing. Thus, animal populations may be affected by the biogeochemical cycling of Na in natural ecosystems. Aumann (1965) found high rodent populations in areas of Na-rich soils, and speculated that the increased abundance of rodents in the eastern United States during the 1930s might have been due to an increased deposition of Na-rich soil dust derived from the prairies during the "Dust Bowl." Such a case would link the abundance of animals to the biogeochemistry and global transport from a distant region.

An enormous literature exists on the characteristics of plant tissues that are selected for food. Many studies report that herbivory is centered on plants with high nitrogen contents (Mattson 1980, Lightfoot and Whitford 1987), suggesting that animal populations might also be limited by N. However, the preference for such tissues may be related more to the high water contents (Scriber 1977) and low phenolic contents (Jonasson et al. 1986) that are found in those tissues than to a specific search for leaves with high amino acid content to support the protein requirement of animals. Grazing often reduces plant photosynthesis while nutrient uptake continues, resulting in high nutrient contents in the aboveground tissues that remain (McNaughton and Chapin 1985). Grazing may also enhance nitrogen uptake in some species (Jaramillo and Detling 1988). In both cases, consumers increase the nutritional quality of the forage available for future consumption, although the quantity of defensive compounds may also increase (White 1984, Seastedt 1985).

Integrative Models of Terrestrial Nutrient Cycling

Interactions between plants, animals and soil microbes link the internal biogeochemisty of terrestrial ecosystems. In conditions of high nutrient availability, plants have high nutrient contents and low nutrient reabsorption before leaf fall, reflecting a lower nutrient-use efficiency by vegetation (Fig. 6.15). In some cases these characteristics can be induced by experimental treatments that alter nutrient availability. For instance, when Douglas fir were fertilized with sugar, which increases the C/N ratio in the soil and the immobilization of N by microbes, reabsorption of foliar N increased, implying greater nutrient-use efficiency by the trees (Turner

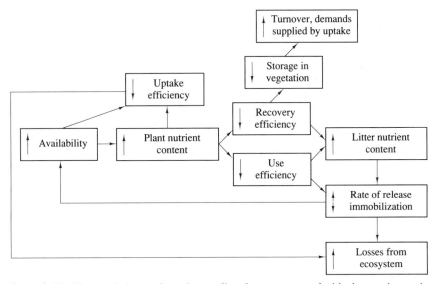

Figure 6.15 Changes in internal nutrient cycling that are expected with changes in nutrient availability. From Shaver and Melillo (1984).

and Olson 1976). Internal cycling by the vegetation may partially alleviate nutrient deficiencies, but decomposition of nutrient-poor litterfall is slow, further exacerbating the low availability of nutrients in the soil. Thus, nutrient-poor sites are likely to be occupied by vegetation specially adapted for long-term persistence in such conditions (Chapin et al. 1986a).

The role of biogeochemistry in controlling the distribution and characteristics of vegetation is seen at varying scales. Continental distributions of vegetation, such as the widespread dominance of conifers in the boreal regions, are likely to be related to the higher nutrient-use efficiency of evergreen vegetation in conditions of limited nutrient turnover in the soil. Regional distribution of vegetation is seen in the occurrence of evergreen vegetation on nutrient-poor hydrothermally altered soils in arid and semi-arid climates (Billings 1950, Goldberg 1982, 1985). Fine-scale spatial heterogeneity in soil properties, as recorded by Robertson et al. (1988) for a field in Michigan, has been linked to the maintenance of diversity in land plant communities (Tilman 1982, 1985), and several early studies show the importance of local soil conditions to the distribution of forest and grassland herbs (Snaydon 1962, Pigott and Taylor 1964).

Linkages among components of the intrasystem cycle suggest that an integrative index of terrestrial biogeochemistry might be derived from the measure of a single component, such as the chemical characteristics

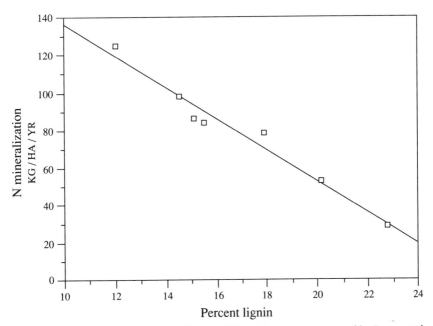

Figure 6.16 Nitrogen mineralization in seven Wisconsin forests, measured by Pastor et al. (1984) using buried bags, as a function of canopy lignin, measured from an airborne reflectance spectrometer by Wessman et al. 1988a. (Reprinted by permission from Nature vol. 335 p. 155, copyright (c) 1988 by Macmillan Magazines Ltd.).

of the leaf canopy. Changes in canopy characteristics might provide an index of the effects of acid rain or other pollutants on nutrient cycling. Variations of leaf C/N ratio across sites might also provide a convenient index of many characteristics of the intrasystem cycle of nutrients. Wessman et al. (1988b) have measured the nitrogen and lignin content of foliage by analyzing the spectral reflectance of tissues in the laboratory, as a first step toward developing an index of forest canopies by remote sensing. Their data show a strong correlation between nitrogen and lignin measured by reflectance properties and by traditional laboratory analyses. An aircraft-borne spectrophotometer was then used to obtain reflectance spectra of forest canopies in Wisconsin. Canopy lignin, calculated by applying the laboratory calibration to the remote-sensing images, was highly correlated to nitrogen mineralization that had been measured in these stands in earlier studies (Fig. 6.16) (Pastor et al. 1984). Recognizing that decomposition is frequently controlled by the lignin and nitrogen content of litter (Fig. 6.7), these data suggest that remote sensing of canopy characteristics has potential for comparative regional studies of nutrient cycling in different plant communities. Myrold et al. (1989)

found that a variety of soil properties were related to canopy characteristics that could be measured by remote sensing, and Reiners et al. (1989) used LANDSAT map images (Chapter 5) to classify landscape units for regional estimates of nitrogen cycling. Studies such as these reinforce the linkage between vegetation and soil characteristics, as outlined in Fig. 6.15.

Various models demonstrate other linkages between plant and soil processes in terrestrial biogeochemistry. Walker and Adams (1958) suggested that the level of available phosphorus during soil development was the primary determinant of terrestrial net primary production, since nitrogen-fixing bacteria depend on a supply of organic carbon and available phosphorus. They use the level of organic carbon in the soil as an index of terrestrial productivity and suggest that organic carbon will peak midway during soil development and then decline as an increasing fraction of the phosphorus is rendered unavailable by precipitation with secondary minerals (Fig. 4.4).

Numerous workers have examined the Walker and Adams (1958) hypothesis in various ecosystems. Tiessen et al. (1984) found that available phosphorus explained 24% of the variability of organic carbon in a collection of 168 soils from eight different soil orders. Roberts et al. (1985) found a similar relationship between bicarbonate-extractable P and organic carbon in several grassland soils of Saskatchewan. Thus, available phosphorus explains some, but not all, of the variation in soil organic carbon, which is ultimately derived from the production of vegetation. The linkage of phosphorus and carbon is likely to be strongest during early soil development, when both organic phosphorus and carbon are accumulating. The importance of organic phosphorus increases during soil development, and through the release of phosphatase enzymes, vegetation interacts with the soil pool to control the mineralization of P. Thus, in mature soils, net primary production is more likely to be limited by nitrogen.

Parton et al. (1988b) present a model linking the cycling of C, N, P, and S in grassland ecosystems. The flow of carbon is shown in Fig. 6.17. The nitrogen cycling submodel has similar structure, since the model assumes that most nitrogen is bonded directly to carbon in amino groups (McGill and Cole 1981). Lignin controls decomposition rates, and nitrogen is mineralized from soil pools when critical N/C ratios are achieved during the respiration of carbon. Phosphorus availability is controlled by a modification of a model first presented by Cole et al. (1977), which includes P/C control over mineralization of organic pools and geochemical control over availability of inorganic forms as in Fig. 6.13. However, unlike N, P/C ratios in plant tissues and soil organic matter are allowed to vary widely as a function of P availability.

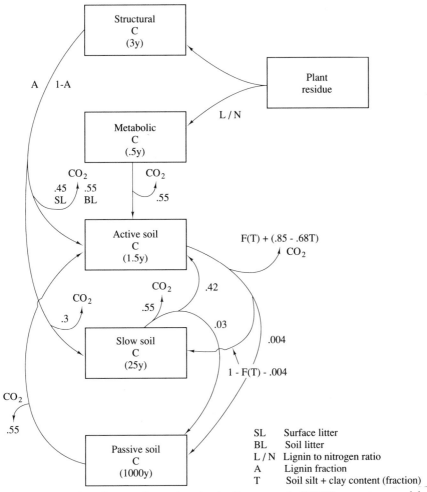

Figure 6.17 Flow diagram for carbon in the Parton et al. (1988b) ecosystem model. Turnover times for each compartment are shown in years.

The complete model was used to predict patterns of primary production and nutrient mineralizations during 10,000 yr of soil development (Fig. 6.18). Net primary production and accumulations of soil organic matter are strongly linked to P availability during the first 800 yr, after which increases in plant production are related to increases in soil N mineralization. Organic P increases throughout the 10,000-yr sequence. In simulations of the response of native soils to cultivation, the model predicted a correlated decline in the native levels of organic carbon and nitrogen in the soil, but a relatively small decline in P. Validation of the

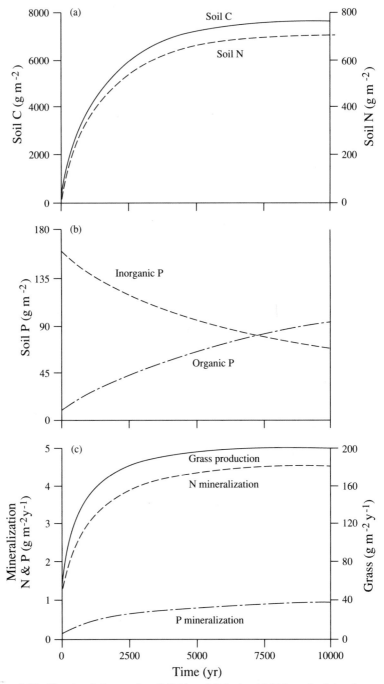

Figure 6.18 Simulated changes in soil C, N, and P during 10,000 yr of soil development in a grassland, using the Parton et al. (1988b) model.

model is seen in the data of Tiessen et al. (1982), who found declines of 51% for C and 44% for N, but only 30% for P in a silt loam soil cultivated for 90 yr in Saskatchewan.

Summary: Calculating Landscape Mass Balance

Elements are retained in terrestrial ecosystems when they play an essential functional role in biochemistry or are incorporated into organic matter. The pool of nutrients held in the soil and vegetation is many times larger than the annual receipt of nutrients from the atmosphere and rock weathering. Turnover times (mass/input) range from 21 yr for Mg to 103 yr for P in the vegetation and forest floor of the Hubbard Brook Experimental Forest in New Hampshire (Likens et al. 1977). In contrast, for nonessential elements, such as Na, the turnover time is rapid (1.2 yr), since these elements are not retained by biota or incorporated into humus. Some nonessential, and toxic, elements such as lead (Pb) bind to soil organic matter and may accumulate in an ecosystem (Smith and Siccama 1981, Lindberg and Turner 1988). Even though these elements are not involved in biochemistry, their retention in the ecosystem is the result of biotic activity. Thus, study of their movement on the surface of the Earth is also in the realm of biogeochemistry.

Annual mineralization, plant uptake, and plant death result in a large internal cycle of elements in most ecosystems. Annual nitrogen inputs are typically 1–5 kg/ha/yr, while mineralization of soil nitrogen is 50–100 kg/ha/yr (Bowden 1986). It is perhaps surprising that nutrient losses to stream waters are so limited, in the face of such large movements of available nutrients within the ecosystem. The minor losses to stream water speak strongly for the efficiency of biological processes that retain elements essential to biochemistry.

Stream-water losses of nonessential elements, and elements that are not in short supply, are useful in calculating the rate of rock weathering on land (Table 4.4). On the other hand, most ecosystems accumulate N and P (Table 6.9), and

Table 6.9 Annual Chemical Budgets for Undisturbed Forests (Total Stream-Water Losses Minus Atmospheric Deposition)[a]

Location and Reference	Precipitation (cm)	Chemical (kg/ha/yr)			
		Ca	Cl	N	P
British Columbia (Feller and Kimmins, 1979)	240	15.8	2.9	−2.6	0
Oregon (Martin and Harr, 1988)	219	41.2	—	−1.2	0.3
New Hampshire (Likens et al., 1977)	130	11.7	−1.6	−16.7	0
North Carolina (Swank and Douglas, 1977)	185	3.9	1.7	−5.5	−0.1
Venezuela (Lewis et al., 1987; Lewis 1988)	450	14.2	−1.4	8.5	0.32

[a] Negative values indicate net accumulations in the ecosystem.

stream-water losses of these elements are often negligible. The incorporation of N and P in biomass should be greatest when structural biomass and soil organic matter are accumulating—that is, when there is positive net ecosystem production (Chapter 5). Vitousek and Reiners (1975) hypothesized that the losses of N and P would increase when the total storage of organic matter was not changing, as in old growth forests. Using the mass-balance approach, where

$$\text{Input} - \text{output} = \Delta\text{storage}, \tag{6.3}$$

they hypothesized that losses would eventually equal inputs in mature vegetation. In support of their hypothesis, Vitousek (1977) found greater losses of available N from old-growth forests than from younger sites in New Hampshire. Losses of N, P, and K are minor during the growing season, and increase during the winter period of plant dormancy. Across the same sites, there were few differences in the loss of Na and Cl, which pass through the system under simple geochemical control (Juang and Johnson 1967, Johnson et al. 1969). Similarly, Lewis (1986) suggested that relatively high losses of N and P from the Caura River in Venezuela were due to the mature vegetation covering most of the watershed (Table 6.9).

Stream-water losses give terrestrial ecosystems the appearance of being "leaky," but it is important to recognize that outputs represent the excess of inputs after uptake by vegetation and immobilization by soil microbes and humus. Comparing forests from Oregon, Tennessee, and North Carolina, Henderson et al. (1978) noted strong N retention in each, despite a 10-fold difference in N input from the atmosphere. The data suggest that plant growth is limited by N in each region. In contrast, losses of Ca were always a large percentage of the annual amount cycling in the system. Especially on limestone soils, ample supplies of Ca were derived from rock weathering and Ca was not in short supply.

Among elements in short supply to biota, nitrogen is unique in that it is only derived from the atmosphere (Table 4.6). Net primary production in some temperate forests appears to show a correlation to N inputs in precipitation (Cole and Rapp, 1981). In addition to uptake by vegetation and incorporation in organic matter, use of available N and S by soil microbes can result in gaseous losses from the ecosystem. Denitrification may explain why the retention of N applied in fertilizer is often much lower than that of other elements (e.g., P and K) when very little N is observed in stream waters (Stone and Kszystyniak 1977, Keeney 1980, Melin et al. 1983). Most future studies of ecosystem mass balance will need to include denitrification in their budgets. Globally, denitrification may explain the tendency for the growth of most vegetation to be N-limited, despite efficient uptake of N from the soil and only minor losses in stream water.

While most studies of ecosystem mass balance have considered single watersheds, Robertson and Rosswall (1986) have compiled an input–output budget for nitrogen in all of West Africa, south of the Sahara Desert. They found that N fixation and precipitation dominated the sources of nitrogen in this region, while fire (8.3×10^{12} g N yr^{-1}), rivers (1.5×10^{12} g N yr^{-1}), and denitrification (1.1×10^{12} g N yr^{-1}) dominated the losses. By including volatile losses, the nitrogen budget was balanced within 1% for the entire region (Fig. 6.19). Similarly, Riggan et al. (1985) developed a nitrogen budget for the Los Angeles Basin,

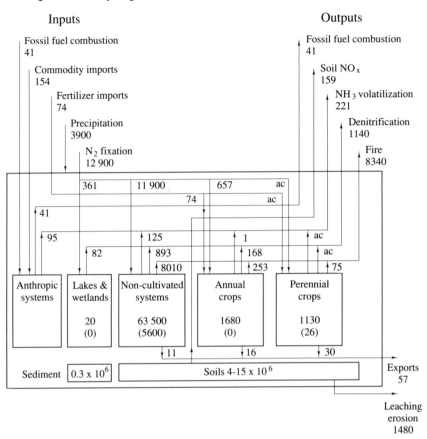

Inputs

| Fossil fuel combustion 41 | Fossil fuel combustion 41 |

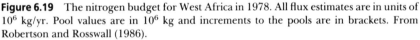

Figure 6.19 The nitrogen budget for West Africa in 1978. All flux estimates are in units of 10^6 kg/yr. Pool values are in 10^6 kg and increments to the pools are in brackets. From Robertson and Rosswall (1986).

in which volatilization of NO_2 in automobile exhaust and its subsequent deposition in chaparral ecosystems were major cycling processes.

Many of the transformations in biochemistry involve oxidations and reductions that generate or consume acidity (H^+). For instance, H^+ is produced during nitrification and consumed in the plant uptake and reduction of NO_3^- [equation (2.15)]. Binkley and Richter (1987) review these processes and demonstrate how ecosystem budgets for H^+ may be useful as an index of net change in ecosystem function as soils acidify during ecosystem development (Chapter 4). Hydrogen ion budgets are also useful as an index of human impact. For example, a net increase in acidity is expected when excess NH_4^+ deposition is subject to nitrification, with the loss of NO_3^- in stream water (Van Breeman et al. 1982). The H^+ budgets are are analogous to measurements of human body temperature. When we see a change, we suspect that the ecosystem is stressed, but we must look carefully within the system for the actual diagnosis.

Recommended Reading

Andreae, M.O. and D.S. Schimel. (eds.). 1990. Exchange of Trace Gases between Atmosphere and Biosphere. Wiley, New York.

Hobbs, R.J. and H.A. Mooney. 1990. Remote Sensing of Biosphere Functioning. Press, San Diego.

Simkiss, K. and K. M. Wilbur. 1989. Biomineralization. Academic Press, San Diego.

Stevenson, F.J. 1986. Cycles of Soil. John Wiley and Sons, New York.

Swift, M.J., O.W. Heal and J.M. Anderson. 1979. Decomposition in Terrestrial Ecosystems. University of California Press, Berkeley.

7

Biogeochemistry in Freshwater Wetlands and Lakes

Introduction

Oxygen is only sparingly soluble in water and diffuses about 10^4 times more slowly in water than in air. Biological activity in lakes and flooded soils often must tolerate relatively low concentrations of oxygen. When heterotrophic respiration occurs in these environments, oxygen may be entirely depleted. For instance, within a few millimeters of depth, the environment of sediments often changes from one dominated by oxidation reactions to one dominated by reduction reactions that occur in anaerobic conditions.

Nutrient cycling in lakes and freshwater wetlands is controlled by redox potential and the microbial transformations of nutrient elements that

195

occur in conditions in which O_2 is not always abundant. The availability of phosphorus in lakes differs strongly between the upper layers of water that are aerobic and the lower layers in which O_2 is depleted. Anaerobic microbial processes—denitrification, sulfate reduction, and methanogenesis—are responsible for the release of N_2, H_2S, $(CH_3)_2S$, and CH_4 to the atmosphere. Other anaerobic microbial processes are coupled to changes in the oxidation state of iron and manganese in wetland soils.

Mitsch and Gosselink (1986) estimate that 6.4% of the world's land area is classified as wetland, but the present area of wetlands has been significantly reduced by human activities in the last 100 yr. The unique environment of wetlands and their role in elemental transformations means that their importance to global biogeochemistry is much greater than their proportional surface area on Earth would suggest. Anaerobic decomposition is often incomplete, so many wetlands store significant amounts of organic carbon—net ecosystem production—in their sediments (Schlesinger 1977, Post et al. 1982). Vast deposits of coal represent the net ecosystem production of swamps during the Carboniferous Period (Berner 1984).

In this chapter we will examine and compare various freshwater wetlands, using the status of oxidation and reduction reactions as a unifying theme. We will begin with a brief review of the concept of redox potential. Then we will discuss the relationship between redox potential and the microbial reactions that occur in the saturated soils and sediments of swamps and lakes. Much of this discussion will also apply to the vast area of arctic tundra, known as muskeg, that is dominated by saturated organic soils known as Histosols. A second section of the chapter treats net primary production and nutrient cycling in lakes. We will compare the nutrient budgets of various wetlands to those of terrestrial ecosystems (Chapter 6).

The biogeochemistry of wetlands is linked to the reactions occurring in the surrounding terrestrial environment by the movements of surface runoff and groundwater (Likens and Bormann 1974). Although rivers are the explicit subject of Chapter 8, we will make frequent reference in this chapter to the importance of terrestrial runoff. Bogs are an interesting wetland type with respect to linkage to the surrounding land. True bogs are ombrotrophic and depend entirely on the atmosphere for inputs of water and nutrients. Taken literally, ombrotrophic means "to feed on rain" (Du Rietz 1949, Gorham 1957). Other bogs, more correctly termed minerotrophic fens, receive at least a portion of their water and nutrient inputs as runoff from the surrounding watershed.

Redox Potential: The Basics

Just as pH expresses the concentration of H^+ in solution, redox potential is used by chemists to express the tendency of an environment to receive or supply electrons. Aerobic systems are said to have a high redox potential because O_2 is available as an electron acceptor. For instance, iron (Fe) oxidizes when it shares the electrons of its outer shell with O_2 to become Fe_2O_3 (rust). Heterotrophic organisms in aerobic environments capitalize on the use of O_2 as a powerful acceptor of electrons to form H_2O. The electrons are derived by the metabolism of reduced or electron-rich organic compounds obtained from the environment and oxidized to CO_2. In eukaryotes, the electrons flow across the internal membrane of the mitochondria, allowing an especially efficient capture of energy for biochemistry.

The oxidation state of the environment, or redox potential, is determined by the particular suite of chemical species that are present. Redox potential is measured as a voltage, E_h, that is necessary to prevent the flow of electrons between the environment in question and a standard electrode. Figure 7.1 illustrates a hypothetical situation in which two containers hold iron chloride in different oxidation states, Fe^{2+} and Fe^{3+}.

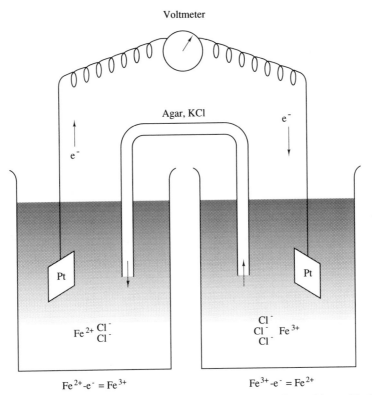

Figure 7.1 A hypothetical chemical cell, connecting two solutions of iron chloride in different oxidation states. The flow of electrons (e^-) can be prevented by the application of 771 mV at the voltmeter; Cl^- ions diffuse through the agar–salt bridge. From Jenny (1980).

The containers are connected by a wire, which passes through a voltmeter and ends in inert platinum electrodes that are placed in each container. A salt bridge allows Cl^- to diffuse between the containers so as to maintain neutral charge. One might expect that electrons would flow from left to right until an equilibrium was established:

$$Fe^{2+} - e^- \rightleftarrows Fe^{3+} \tag{7.1}$$

The voltmeter measures the resistance necessary to prevent this reaction. Experimentally, if the two containers hold equimolar concentrations of Fe and the pH is <3.0, the voltage at equilibrium will be $+771$ mV. We would say that the container on the right is an oxidizing environment, since it draws electrons from Fe^{2+}, the more reduced or electron-rich species on the left. If the container on the right has a higher concentration of Fe^{3+} than the concentration of Fe^{2+} on the left, a greater voltage will be required to prevent the flow of electrons. In aerobic soils, O_2 acts as a powerful electron acceptor, and iron is maintained as Fe^{3+}, which precipitates as one or more forms of oxyhydroxide.

Natural environments are not isolated into separate containers, nor do they contain such a simple mixture of constituents. In practice, we can measure the redox potential of an entire chemical system by expressing the disequilibrium of the mix relative to a standard electrode, which contains H_2 gas overlying a solution of known H^+ concentration. We connect the environment to the standard electrode using an inert platinum electrode, which takes on the potential of the soil, without altering the tendency for electrons to move among soil constituents. When a voltmeter is placed in this circuit, the redox potential is measured as the voltage required to prevent the interconversion of H^+ and H_2 at the standard electrode. In practice, a standard hydrogen electrode is difficult to maintain in the field, so investigators often use other reference electrodes that are calibrated against a hydrogen electrode (Bricker 1982, Faulkner et al. 1989).

When O_2 is present, it accepts electrons at the platinum electrode:

$$O_2 + 4e^- + 4H^+ \rightarrow 2H_2O \tag{7.2}$$

The electrons are generated at the hydrogen electrode:

$$2H_2 \rightarrow 4H^+ + 4e^- \tag{7.3}$$

and the voltmeter records a high voltage or redox potential. Since equation (7.2) is more likely to proceed in acid conditions, a higher redox potential will be found at low pH, assuming all other factors are the same. In the absence of oxygen, other constituents, such as Fe^{3+}, may accept electrons, following equation (7.1), but a lower voltage will be recorded. Thus, Fe^{3+} can accept electrons from more reduced substances, such as Fe^{2+}, but not from O_2, which is more strongly oxidizing. In the absence of strongly oxidizing substances, Fe^{2+} persists in the environment.

The pH of the environment affects the redox potential established by Fe^{3+}. At pH < 3.0 an equilibrium between Fe^{2+} and Fe^{3+} is found at a redox potential of 771 mV, with the underlying reaction being

$$Fe^{3+} + H_2 \rightleftarrows Fe^{2+} + 2H^+ \tag{7.4}$$

Since this reaction is more likely to proceed to the right at higher pH, an equilibrium between these forms will be established at a lower redox potential. At pH 5.0, Fe^{2+} persists only below a redox potential of about 400 mV (Fig. 7.2). Thus, oxidation proceeds more readily, and at lower redox potentials, in neutral or alkaline environments. Fe^{2+} is likely to persist in the acid waters of peatbogs.

Much of the recent work expresses redox in units of pe, which is derived from the equilibrium constant of the oxidation–reduction reaction. Thus, for a reaction

$$\text{Oxidized species} + e^- + H^+ \rightarrow \text{reduced species} \tag{7.5}$$

$$\log K = \log[\text{reduced}] - \log[\text{oxidized}] - \log[e^-] - \log[H^+] \tag{7.6}$$

or

$$\log K = \text{pe} + \text{pH} \tag{7.7}$$

Thus, pe is the negative log of the electron activity and expresses the energy of electrons (Bartlett 1986). Since the sum of pe and pH is constant, if one goes up, the other must decline. When a given reaction occurs at lower pH, it will occur at higher redox potential, expressed as pe. Measurements of redox potential that are expressed as voltage, E_h, can be converted to pe following

$$\text{pe} = \frac{E_h}{(RT/F)2.3} \tag{7.8}$$

where R is the universal gas constant, F is Faraday's constant, T is temperature, and 2.3 is a constant to convert natural to base-10 logarithms.

Environmental chemists use E_h–pH or pe–pH diagrams to predict the oxidation state of various consituents in natural environments (e.g., Fig. 7.2). All diagrams are bounded by two lines. If redox potentials were ever to fall above the upper line, even water would be oxidized, following the

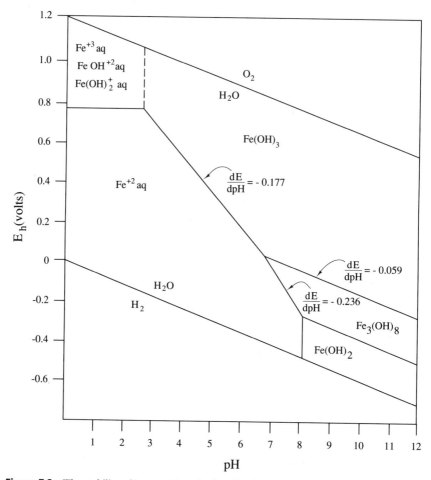

Figure 7.2 The stability of iron and iron hydroxides in soils relative to E_h and pH at 25°C. All conditions refer to 1 mM Fe^{2+} solution. From Ponnamperuma et al., © by Williams & Wilkins, 1967.

reverse of equation (7.2). While photolysis of water occurs during photo-synthesis (Chapter 5) and in the upper atmosphere (Chapter 3), we do not find such strongly oxidizing conditions in the natural waters at the surface of the Earth. For example, an environment dominated by Cl_2 would be more oxidizing than water, so as long as the Earth has contained abundant liquid water, Cl has been dissolved in ocean waters as Cl^- (Bohn et al. 1985). Similarly, any conditions above the lower line allow the reaction

$$H_2 + OH^- \rightarrow H_2O + 2e^-$$ (7.9)

but the reverse of this reaction—the reduction of water—is also rarely seen in the natural environment. Elemental Na reduces water:

$$Na + 2H_2O \rightarrow Na^+ + 2OH^- + H_2 \tag{7.10}$$

which is why sodium exists in ionic form at the surface of the Earth (Bohn et al. 1985). These boundary conditions vary with pH, with E_h declining by 59 mV with each unit of pH increase, reflecting that oxidation requires a higher redox potential in acid conditions.

Figure 7.2 shows the expected forms of iron in natural environments of varying pH and E_h. Note that most transition lines slope downward, indicating that iron is more likely to occur in oxidized form in neutral or alkaline conditions. In interpreting such diagrams, it is important to remember that E_h and pH are properties of the environment, determined by the total suite of chemical species present, and not simply by the conversion of iron from one oxidation state to another. In most cases, organic matter contributes a large amount of "reducing power" that lowers the redox potential in flooded soils and sediments (Bartlett 1986). High concentrations of Fe^{2+} will be found in flooded, low-redox environments, where impeded decomposition leaves undecomposed organic matter in the soil and humic substances impart acidity to the soil solution. The tendency for iron to precipitate in oxidized form at high redox potential or high pH underlies the use of aeration and liming as techniques for ameliorating lakes that are affected by acid mine drainage.

Soils and sediments that resist changes in their redox potential are said to be highly poised. Conceptually, poise is to redox potential as buffer capacity is to pH (Bartlett 1986). As long as O_2 is present, aerobic soils appear to be highly poised, since O_2 will maintain a high redox potential in nearly all conditions. However, in the absence of O_2, these soils may show a rapid decline in redox potential as various weakly oxidizing constituents, NO_3^-, Mn^{4+}, Fe^{3+}, and SO_4^{2-}, sequentially accept electrons from organic matter, which exists as an enormous store of reducing power. Redox potential will fall less rapidly—more poise—when concentrations of Mn^{4+} and Fe^{3+} are high.

Redox Potential in Natural Environments

Most aerobic environments seldom show redox potentials less than +400 mV. However, the diffusion of oxygen in flooded soils and sediments is so slow that redox potentials decline rapdily with increasing depth in wetland soils (Stolzy et al. 1981). Data collected from wetland ecosystems account for much of the total variation in the redox potential that has been reported for environments at the surface of the Earth (Fig.

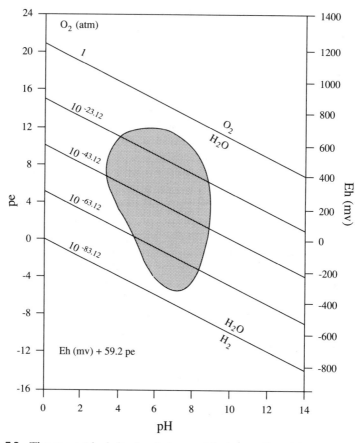

Figure 7.3 The range (shaded) of redox potentials that are found in natural aqueous environments. From Lindsay (1979), based on the compilation of Baas Becking et al. (1960).

7.3). A strong gradient of redox potential may develop in sediments over a depth as short as 2 mm (e.g., Howeler and Bouldin 1971). The progressive decrease in redox potential after aerobic soils are flooded (Fig. 7.4) is analogous to the change in redox potential with depth.

The results of many studies suggest that a sequence of reactions is expected as progressively lower redox potentials are achieved (Table 7.1) (Ponnamperuma 1972). After O_2 is depleted by aerobic respiration, denitrification begins when the redox falls to 421 mV (Cho 1982). Denitrifying bacteria use nitrate as an alternative electron acceptor during the oxidation of organic matter in anaerobic conditions (Chapters 2 and 6). When nitrate is depleted, reduction of Mn^{4+} begins below a redox of +396 mV, followed by reduction of Fe^{3+} at $E_h < -182$ mV. These reactions are catalyzed by various bacteria that use fermentation to obtain energy, with

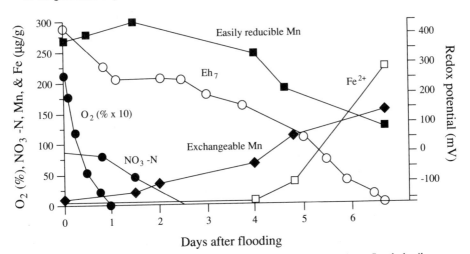

Figure 7.4 Changes in the chemical composition of the waters overlying a flooded soil as a function of time after flooding. Note that the reduction of iron does not begin until fully anaerobic conditions are achieved. Redox potential is expressed at pH 7, that is, E_{h7}. From Turner and Patrick (1968).

Mn^{4+} and Fe^{3+} acting indirectly as electron acceptors (Lovley and Phillips 1989), for example,

$$\text{Pyruvate} \rightarrow \underset{\text{(ethanol)}}{CH_3CH_2OH} + 2CO_2 + e^- \tag{7.11}$$

$$Fe^{3+} + e^- \rightarrow Fe^{2+} \tag{7.12}$$

Table 7.1 Thermodynamic Sequence for Reduction of Inorganic Substances at pH 7.0[a]

Reaction	E_h
Disappearance of O_2	
$O_2 + 4H^+ + 4e \rightleftharpoons 2H_2O$	0.816 V
Disapperance of NO_3^-	
$NO_3^- + 2H^+ + 2e \rightleftharpoons NO_2^- + H_2O$	0.421 V
Formation of Mn^{2+}	
$MnO_2 + 4H^+ + 2e^- \rightleftharpoons Mn^{2+} + 2H_2O$	0.396 V
Reduction of Fe^{3+} to Fe^{2+}	
$Fe(OH)_3 + 3H^+ + e^- \rightleftharpoons Fe^{2+} + 3H_2O$	−0.182 V
Formation of H_2S	
$SO_4^{2-} + 10H^+ + 8e^- \rightleftharpoons H_2S + 4H_2O$	−0.215 V
Formation of CH_4	
$CO_2 + 8H^+ + 8e^- \rightleftharpoons CH_4 + 2H_2O$	−0.244 V

[a] From Stevenson (1986).

Recently, bacteria have also been isolated that couple the reduction of Mn and Fe directly to the oxidation of simple organic substances that diffuse upward from more reducing conditions at lower depth (Myers and Nealson 1988, Lovley and Phillips 1988a). There is some overlap between the zone of denitrification and the zone of Mn reduction in sediments (Klinkhammer 1980), and most of the microbes in this zone are facultative anaerobes that can tolerate periods of aerobic conditions. There is little overlap between the zone of Mn reduction and Fe reduction, since soil bacteria show an enzymatic preference for Mn^{4+}, and Fe reduction will not begin until Mn^{4+} is depleted (Lovley and Phillips 1988b).

Below the zone of Mn^{4+} reduction, most redox reactions are performed by obligate anaerobes. Our earlier emphasis on the redox state of iron (Fig. 7.2) reflects the widespread use of Fe^{2+} as an index of the transition from mildly oxidizing to strongly reducing conditions. Iron is a convenient indicator in the field, since oxidized iron is easily recognized by its red color in soils, whereas reduced iron is grayish. Soil layers with reduced iron are called gley.

Obligate anaerobes such as *Clostridium* use the energy derived from fermentation/Fe^{3+} reduction to engage in nitrogen fixation (Ottow 1971). Such nitrogen fixation is probably essential to augment the meager supplies of nitrogen that are available below the zone of denitrification. Below the depth of iron reduction, the redox potential progressively drops to -215 mV, where sulfate reduction commences, and to -244 mV where methanogenesis occurs (Lovley and Phillips 1987). These reactions are performed by obligate anaerobic bacteria, some of which also engage in nitrogen fixation (Postgate et al. 1988).

The environment of flooded soils and sediments exists as a dynamic equilibrium that is maintained by the availability of oxygen at the surface and buried organic carbon as a source of reducing power at depth. If the surface of such a soil is exposed to the air, as might occur with seasonal fluctuations of the water table, the position of each redox reaction will shift downward in the profile. Products of previous reduction reactions become substrates for oxidizing bacteria. The total rate of denitrification is enhanced when seasonal periods of aerobic conditions stimulate the mineralization and nitrification of organic nitrogen, which makes NO_3^- available for denitrifiers when the water rises (Reddy and Patrick 1975, 1976). In continuously flooded soils, nitrate must diffuse downward from aerobic layers supporting nitrification to anaerobic layers supporting denitrification (Patrick and Tusneem 1972).

Sulfate Reduction

In Chapter 6 we examined the reduction of sulfate that accompanies the assimilation of sulfur by soil microbes and plants. In contrast, sulfate reduction in anaerobic soils is *dissimilatory* sulfate reduction, analogous to denitrification, in which SO_4^{2-} acts as an alternative electron acceptor during the oxidation of organic matter by bacteria in the genera *Desulfovibrio* and *Desulfotomaculum*, for example,

$$2H^+ + SO_4^{2-} + 2(CH_2O) \rightarrow 2CO_2 + H_2S + 2H_2O \qquad (7.13)$$

These bacteria produce a variety of sulfur gases, including hydrogen sulfide (H_2S), dimethylsulfide [$(CH_3)_2S$)], and dimethyldisulfide [$(CH_3)_2S_2$)]. In an analogous reaction, anaerobic bacteria reduce selenium compounds (e.g., SeO_4) to Se, which is often toxic to wildlife (Oremland et al. 1989). Before widespread industrial emissions, the release of biogenic gases from wetland soils was the dominant source of sulfur gases in the atmosphere (Möller 1984, Warneck 1988).

Since H_2S can react with other soil constituents and is oxidized by sulfur bacteria in the overlying sediments and water, many workers once believed that organic sulfur gases might be the dominant form of emission from wetland soils. This belief was reinforced by the lack of good methods for measuring low concentrations of H_2S in the field, and in 1974 Rasmussen was successful in identifying dimethylsulfide as an emission from a temperate pond. However, using new techniques, most investigators now find that H_2S accounts for most of the emission from wetland soils (Adams et al. 1981). Castro and Dierberg (1987) report a flux of H_2S containing $1-110$ mg S m^{-2} yr^{-1} for various wetlands in Florida.

Brown and MacQueen (1985) found that only 0.3% of the sulfate added to peat soils was subsequently recovered as H_2S, as a result of various reactions between H_2S and soil constituents. Hydrogen sulfide can react with Fe^{2+} to precipitate FeS, which gives the characteristic black color to anaerobic soils. H_2S also reacts with elemental iron when steel is used for construction in flooded environments, and the structures corrode rapidly, leading to the formation of FeS:

$$Fe + H_2S \rightarrow FeS + H_2 \uparrow \qquad (7.14)$$

When H_2S diffuses upward through the zone of Fe^{3+}, pyrite (FeS_2) is precipitated following

$$2FeOOH + 2H_2S + 2H^+ \rightarrow FeS_2 + 4H_2O + Fe^{2+} \qquad (7.15)$$

Low concentrations of iron in many wetlands limit the accumulation of reduced iron compounds (Berner 1984). However, hydrogen sulfide also reacts with organic matter to form carbon-bonded sulfur that accumulates in peat and lake sediments (Brown 1985, Rudd et al. 1986a). In a West Virginia bog, Wieder and Lang (1988) found that 81% of the sulfur was carbon bonded, and only small amounts were found in reduced inorganic forms (H_2S, FeS, and FeS_2) in the peat. However, when $^{35}SO_4^{2-}$ was added to the peat, 87% underwent sulfate reduction and the remainder was immobilized by microbes in the surface layers (cf. Brown 1986). Their data suggest that the reduced forms of sulfur are subsequently reoxidized, potentially allowing high rates of sulfate reduction to continue in the soil, even in wetlands in which SO_4^{2-} is not abundant (cf. Wieder et al. 1990). Apparently the carbon-bonded forms—from the reaction of H_2S with organic matter or direct immobilization of SO_4—are more stable and accumulate in the soil.

Precipitation of pyrite in organic muds accounts for some of the sulfur in coals; the remainder is largely derived from the carbon-bonded sulfur content in sedimentary organic matter (Casagrande et al. 1977, Altschuler et al. 1983). Coal containing pyrite sulfur that is the result of dissimilatory sulfate reduction shows negative values for $\delta^{34}S$, as a result of bacterial discrimination against the rare, heavy isotope $^{34}SO_4^{2-}$ in favor of $^{32}SO_4^{2-}$ during sulfate reduction (Hackley and Anderson 1986). Similarly, SO_4^{2-} in rainfall in Ontario shows a lower $\delta^{34}S$ value during the summer, when there are large microbial releases of dimethylsulfide from nearby wetlands, than during the winter (Nriagu et al. 1987). A least a portion of the SO_4^{2-} content in this rain must be derived from the oxidation of dimethylsulfide released to the atmosphere.

Methanogenesis

Since the concentration of SO_4^{2-} in most freshwater wetlands is not high, the zone of sulfate reduction is closely underlaid by a zone in which various methanogenic bacteria are active. Methanogenesis can occur via several metabolic pathways (Chapter 2). Methane production in freshwater environments is dominated by acetate splitting:

$$CH_3COOH \rightarrow CO_2 + CH_4 \tag{7.16}$$

which produces a $\delta^{13}C$ of -65 to $-50\%o$ in CH_4 (Woltemate et al. 1984, Whiticar et al. 1986, Cicerone and Oremland 1988). Acetate-type compounds are produced from cellulose by fermentive bacteria that coexist at the same depths. Methane is also produced by CO_2 reduction:

$$CO_2 + 4H_2 \rightarrow CH_4 + 2H_2O \tag{7.17}$$

where the hydrogen is available as a byproduct of fermentation:

$$CH_2O + H_2O \rightarrow 2H_2 + CO_2 \tag{7.18}$$

In this reaction CO_2, found as HCO_3^-, serves as an electron acceptor, in an analogous role to NO_3^- and SO_4^{2-} in denitrification and sulfate reduction. Methanogenesis by CO_2 reduction accounts for the limited release of H_2 from wetland soils (Schütz et al. 1988). This methane is highly depleted in ^{13}C, with $\delta^{13}C$ of -60 to $-100\%o$ (Whiticar et al. 1986). Recently, Daniels et al. (1987) reported methanogenesis from CO_2 reduction, in which anaerobic bacteria use elemental iron as a source of electrons. This specialized pathway can also account for the corrosion of metals in anaerobic environments.

Methanogenic bacteria can use only certain organic substrates for acetate splitting, and in many cases there is evidence that sulfate-reducing bacteria are more effective competitors for the same compounds (Schönheit et al. 1982). Similarly, Kristjansson and Schönheit (1983) found that sulfate-reducing bacteria had a greater affinity for H_2 than methanogens engaging in CO_2 reduction. Thus, in most environments there is little or no overlap between the zone of methanogenesis and the zone of sulfate reduction in sediments. Methanogenesis begins when sulfate is depleted (Lovley and Klug 1986, Kuivila et al. 1989). In marine environments, methanogenesis by acetate splitting is much less important than CO_2 reduction, because the high concentration of SO_4^{2-} in seawater allows the complete consumption of acetate (Sansone and Martens 1981, Crill and Martens 1986, Whiticar et al. 1986).

In freshwater wetlands, the release of methane is strongly dependent upon the soil moisture content (Moore and Knowles 1989). Sebacher et al. (1986) reported that the flux increased linearly with soil moisture content for tundra bog, fen, and marsh habitats in Alaska (Fig. 7.5). Harriss et al. (1982) found that methane flux from the Great Dismal Swamp (Virginia) was $0.001-0.02$ g $CH_4/m^2/day$ in the wet season, but the swamp became a sink for methane during the dry season, when methane oxidizers were active. Baker-Blocker et al. (1977) found that methane flux from lakes in Michigan was positively related to air temperature. Other workers have also reported temperature-dependent seasonal fluctuations in methanogenesis (King and Wiebe 1978, Kelly and Chynoweth 1981, Yavitt et al. 1988, Crill et al. 1988, Schütz et al. 1989b), but it is surprising to note that the rate of methanogenesis showed little relation to temperature among the Alaskan sites studied by Sebacher et al. (1986).

As a result of a variety of processes that oxidize methane in sediments and surface waters, the flux of methane from the surface is less than the rate of production at depth. Yavitt et al. (1988, 1990a) found that methane oxidation consumed $11-100\%$ of the methane production in some peatlands in West Virginia. Flux to the atmosphere is greatest when the sediment releases large bubbles of gas that quickly pass to the surface

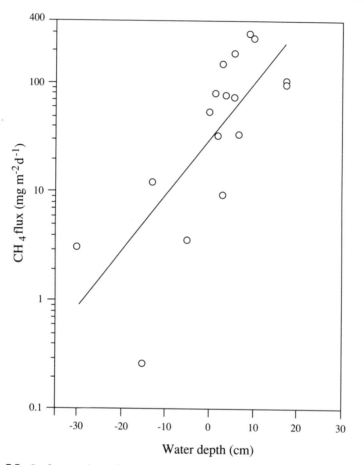

Figure 7.5 Surface methane flux versus water depth for bog, fen, and marsh sites in Alaska. From Sebacher et al. (1986).

(Devol et al. 1988, Wilson et al. 1989). In some areas, wetland plants with hollow stems enhance the movement of methane to the surface without oxidation (Dacey 1981, Sebacher et al. 1985). Rice paddies account for about half of the global production of methane from wetlands (Aselmann and Crutzen 1989). Methane oxidation in the upper sediment consumed about half of the methane generated in the deeper sediments of Lake Washington, with most of the remainder being oxidized in the water column (Kuivila et al. 1988). Only 2% of the carbon entering the system in net primary production is returned to the atmosphere as CH_4. In other lakes, oxidation in the water column is thought to dominate the consumption of methane (Rudd and Taylor 1980), and in highly stratified lakes, large concentrations of methane can accumulate in deep anoxic waters (Tietze et al. 1980).

Biomethylations

Microbial reactions in sediments are responsible for the methylation of a wide variety of metallic elements, some of which are toxic to biota and more rapidly assimilated in methyl form (Ridley et al. 1977, Craig 1980). For instance, the methylation of mercury proceeds as follows:

$$Hg^{2+} \rightarrow CH_3Hg^+ \tag{7.19}$$

Methylated forms of some metals are volatile, allowing escape to the atmosphere and contributing to global biogeochemical movement (Lindqvist and Rodhe 1985).

Biogeochemistry of "Terrestrial" Wetlands

Most areas of shallow water are dominated by emergent vegetation, resembling uplands, but the biogeochemical processes in these ecosystems are mediated by sediments with low redox potential. Swamps, marshes, and bogs comprise the "terrestrial" wetlands, which are important wildlife habitat. These areas are often found at the interface between upland and lake ecosystems (Fig. 7.6), and the nutrients received from the adjacent landscapes are often transformed during their passage through wetlands (e.g., Hooper and Morris 1982). Net primary productivity in these ecosystems varies widely, depending upon nutrient supply (Brinson et al. 1981). Swamp forests that receive seasonal nutrient inputs in floodwaters often have high productivity, since aerobic conditions exist for nutrient turnover in the soil during the remainder of the year (Fig. 7.7). In contrast, bogs that receive little or no nutrient input from runoff usually have very low productivity. Since emergent plants dominate the vegetation of these wetlands, net primary production is usually measured using the harvest approaches outlined briefly in Chapter 5.

As we have seen, decomposition is impeded in flooded and saturated soils, so freshwater wetlands show large accumulations of soil organic matter (Table 7.2). For bogs, Clymo (1984) proposed a model for peat accumulation, which predicts that peatlands will eventually attain a steady state when the input of detritus from primary production at the peat surface is balanced by the loss of organic matter by decomposition throughout the peat profile. The maximum depth to which a peatland soil will accumulate is determined by the rate of decomposition in the aerobic upper levels (the acrotelm) and in the lower levels (the catotelm)

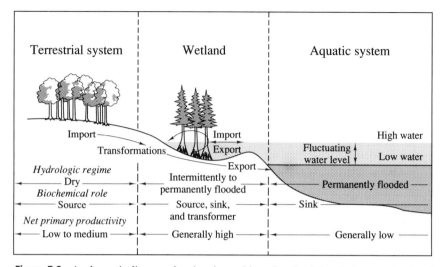

Figure 7.6   A schematic diagram showing the position of wetlands in relation to upland and aquatic ecosystems and the biogeochemical linkages between these landscape components. From Mitsch and Gosselink (1986).

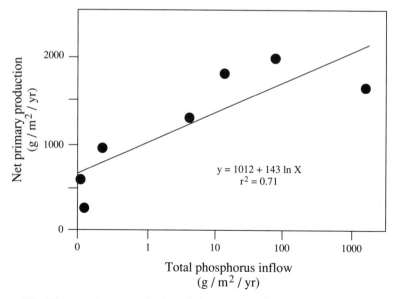

Figure 7.7   The net primary production of cypress swamp forests in relation to the annual input of phosphorus. From Brown (1981).

Table 7.2 Rate of Peat Accumulation in Some Peatland Ecosystems

Location	Vegetation	Accumulation Interval (ybp)[a]	Accumulation Rate[b] (g/m²/yr)	Reference
Alaska	*Picea* and *Sphagnum*	0–4790	22–122	Billings (1987)
Alaska	*Eriophorum vaginatum*	0–7000	53.2	Viereck (1966)
Manitoba	*Picea* and *Sphagnum*	0–2960	52	Reader and Stewart
		0–7939	27	(1972)
Wisconsin	*Sphagnum*	0–8260	34–75	Kratz and DeWitt
				(1986)
Massachusetts	*Sphagnum*	0–132	180	Hemond (1980)
North Carolina	Mixed forest	0–27,700	15	Whitehead (1981)
Georgia	*Taxodium*	0–6500	45	Cohen (1974)

[a] ybp, Years before present.
[b] When data are incomplete, calculated rate assumes bulk density of 0.1 g/m³ and carbon content of 50%.

of the deposit. Losses from the lower layers become more significant over time. The saturated soils of tundra and boreal forest regions contain about 24% of the total storage of organic matter in soils of the world (Table 5.4). Many of these ecosystems have accumulated soil carbon since the retreat of the last continental glacier (Chapter 5). As these areas are subject to drainage and warmer climatic conditions, the rate of carbon storage will decline, decomposition will increase, and wetlands could become a significant source of CO_2 and CH_4 for the atmosphere (Tate 1980, Hutchinson 1980, Armentano and Menges 1986, Matthews and Fung 1987).

Depending upon landscape position, wetland ecosystems are often limited by phosphorus or nitrogen. For instance, many bogs receive little or no runoff from the surrounding land, so it is not surprising that phosphorus is in short supply for plant growth and decomposition (Chapin et al. 1978, Damman 1988). As seen for terrestrial ecosystems in Chapter 6, the nutrient cycle of bogs is characterized by large nutrient storages in vegetation and peat, and small annual turnover through the pool of available nutrients in the soil. In the tundra of Alaska, Chapin et al. (1978) found that the pool of phosphorus in soil organic matter contained 64% of the total phosphorus in the ecosystem and had a mean residence time of 220 yr, while phosphorus available in the soil solution comprised 0.3% of the total with a residence time of 10 hr. Based on changes in the concentrations of nitrogen and phosphorus with depth, Damman (1978) suggests that a significant portion of the phosphorus content of peat is mineralized before burial.

In addition to phosphorus, many peatland systems also show shortages of nitrogen. Nitrogen limits the growth of tundra vegetation, and in a multiple fertilization experiment Shaver et al. (1986) found that the response of *Eriophorum vaginatum* in tussock tundra was greater for N than for P. Low temperature limits nitrogen mineralization in the tundra (Marion and Black 1987). Many bog ecosystems show significant amounts of nitrogen fixation (Waughman and Bellamy 1980, Schwintzer 1983, Barsdate and Alexander 1975), which is likely to be in excess of denitrification in field conditions (Bowden 1986, Koerselman et al. 1989). Thus, many bogs show a net accumulation of nitrogen in peat (Hemond 1983, Damman 1988, Urban and Eisenreich 1988).

Peatland ecosystems that receive drainage from the surrounding uplands (fens) and forests that receive seasonal floodwaters often show relatively high concentrations of phosphorus and other elements derived from rock weathering (Mitsch et al. 1979, Waughman 1980, Frangi and Lugo 1985). In these ecosystems, phosphorus and sulfur are retained on iron and aluminum minerals that are mixed with the soil organic matter in peat (Richardson 1985, Mowbray and Schlesinger 1988). A significant amount of the SO_4 entering such systems is also immobilized in organic matter (Brown and MacQueen 1985, Wieder and Lang 1988). With greater inputs of nutrient elements from land, net primary production in these systems is likely to be limited by nitrogen (e.g., Tilton 1978). Wetlands in low topographic positions are likely to function as effective nutrient sinks (e.g., Verry and Timmons 1982, Urban et al. 1989). Peatlands exposed to decomposition and erosion can also be sources of nutrients to aquatic ecosystems receiving their runoff (e.g., Crisp 1966).

Primary Production and Biogeochemical Cycling in Lakes

The physical properties of water exert a significant control on net primary productivity and nutrient cycling in lake ecosystems. The input of sunlight energy warms the surface waters, but light energy is rapidly attenuated by depth. Since water shows its greatest density (g cm^{-3}) at 4°C, a stratification of water layers develops in deep lakes, with warmer surface waters known as the *epilimnion* overlying cooler, deep waters known as the *hypolimnion*. The zone of rapid temperature change is known as the thermocline or metalimnion. Many tropical lakes show permanent stratification (e.g., Kling 1988). In temperate regions, the temperature stratification breaks down, and lake waters may circulate freely from top to bottom at the end of the growing season. During summer stratification, phytoplankton, the free-floating algae that contribute the majority of net production are confined to the surface layers that contain only a small portion of the total nutrient content of a lake. When stratification develops, the epilimnetic waters show high redox potential and depletion of nutrients by plant uptake. Dead organic materials sink to the hypolimnion, where their decay leads to the depletion of oxygen, low redox potentials, and greater nutrient availability.

Unlike terrestrial plants, phytoplankton are not bathed in an atmosphere with CO_2. Carbon dioxide dissolves in lake waters according to equilibrium conditions that depend on pH:

$$CO_2 + H_2O \rightleftarrows H^+ + HCO_3^- \rightleftarrows 2H^+ + CO_3^{2-} \qquad (7.20)$$

At pH < 4.3, most carbon dioxide is found as a dissolved gas, between 4.3 and 8.3 as bicarbonate, and >8.3 as carbonate. These forms comprise dissolved inorganic carbon or DIC. The rate of dissolution of CO_2 in water and the subsequent availability of CO_2 or other forms of DIC are potential constraints on primary production in lakes.

Net Primary Production

Methods for assessing the net primary production of phytoplankton necessarily must differ from the harvest methods that are used in studies of land vegetation (Chapter 5). Two approaches are common. In the first method, small samples of lake water are confined in glass bottles, clear and opaque, that are resuspended in the water column. After a period of incubation, the O_2 content of the water is measured. An increase in O_2 in the clear bottle is taken as the equivalent of net primary production—that is, photosynthesis in excess of respiration by the plankton. Net primary production is calculated by assuming a molar equivalent between O_2 production and carbon fixation [equation (5.1)]. Over the same period of incubation, a decrease in O_2 in the dark bottle is taken to be the result of plant respiration. The sum of changes in the light and dark bottles allows a calculation of gross primary production.

Many recent studies use variations of the ^{14}C method to measure primary production in freshwaters. This method also uses clear bottles, which are innoculated with DIC, containing ^{14}C in a form that is available for phytoplankton. Since the pH of most surface waters lies in the range of 4.3 to 8.3, $NaH^{14}CO_3$ is a frequent choice. The bottles are resuspended in the water column, and during the incubation period, photosynthesis is assumed to convert the inorganic ^{14}C to organic forms that accumulate in phytoplankton cells. The bottles are then retrieved and the water is filtered. Radiocarbon that is retained on the filter is counted using a scintillation counter and assumed to represent net primary production by the phytoplankton community.

The O_2 and ^{14}C methods have been reviewed exhaustively by Peterson (1980), who examines a number of sources of error in both methods. The oxygen method is relatively easy and inexpensive to apply to many situations, but it suffers from a number of problems that are enforced by the artificial environment in the bottles. The bottles contain planktonic bacteria and zooplankton that add to the respiration contributed by the phytoplankton. The sensitivity of most O_2 measurements is relatively low, so long incubations are necessary so that some change in O_2 concentration can be measured. During the incubation, nutrients may be depleted in the bottle, lowering the apparent rate of photosynthesis. Also, during long incubations, increasing O_2 concentrations may increase the rate of photorespiration by phytoplankton.

The artificial nature of the environment confined in bottles also affects the ^{14}C method, but since the technique is more sensitive, the incubations are shorter. A more serious problem stems from the loss of soluble products of photosynthesis (e.g., sugars and amino acids) that are excreted from phytoplankters and lost during filtration. These comprise dissolved organic carbon (DOC), which ought to be included as a product of net primary production. Recently, very small phytoplankton, known as picoplankton, have been found in ocean and some freshwaters (Stockner and Antia 1986). These may also pass through the filters, escaping detection by the ^{14}C method. In practice, the O_2 method often gives higher values for production, particularly when the methods are compared in relatively unproductive environments. Neither method is without potential error, and the best studies often use both approaches.

A compilation of the data from many studies, which showed greater lake productivity in tropical than in temperate or boreal regions, led Brylinsky and Mann (1973) to suggest that available sunlight might control the level of net primary production in lakes. In a similar analysis of a larger data set, Schindler (1978) found no correlation with annual irradiance, but a strong relationship between lake production and the total input of nutrients, especially phosphorus (Fig. 7.8). Alternatively, in an evaluation of the pollution impact of phosphorus detergents, a number of workers suggested that lake productivity was limited by the rate at which atmospheric CO_2 could dissolve in surface waters. Subsequent field studies failed to confirm a CO_2 limitation, except under unusual circumstances of high nutrient availability (Schindler et al. 1972). Søballe and Kimmel (1987) confirmed the importance of phosphorus in a comparison of productivity among 345 rivers and 812 lakes and reservoirs of the United States, and noted that algal cell density was directly correlated to the availability of phosphorus and the residence or turnover time of the waters. The evidence for a phosphorus limitation of net primary productivity in lake waters now appears overwhelming (Vollenweider et al. 1974, Dillon and Rigler 1974, Oglesby 1977).

Phosphorus and Nitrogen in Lake Waters

In natural conditions phosphorus inputs to lake ecosystems are relatively small (Ahl 1988). There is little phosphorus in precipitation, and phos-

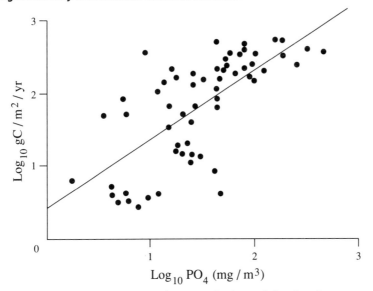

Figure 7.8 Relationship between net primary production and the phosphorus concentration in lakes of the world. From Schindler (1978).

phorus is retained in surrounding terrestrial watersheds by vegetation (Chapter 6) and by chemical interactions with soil minerals (Chapter 4). Analysis of lake water typically shows that a large proportion of the phosphorus is contained in the plankton biomass and only a small portion is found in available form (Fig. 7.9). During a period of stratification, the phosphorus pool in the surface waters is depleted as phytoplankton, and other organisms die and sink to the hypolimnion (Levine et al. 1986).

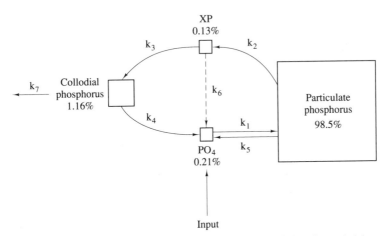

Figure 7.9 Relationships and interactions between the forms of phosphorus in lake waters. From Lean (1973), copyright 1973 by the AAAS.

Available phosphorus may also precipitate with Fe and Mn minerals that are insoluble at high redox potential (Figs. 4.3 and 7.2). Thus, phytoplankton production depends on the rapid cycling of phosphorus between available and organic forms.

With limited supplies of phosphorus available, each atom of phosphorus may cycle through the biotic community several times annually. Studies of phosphorus cycling have shown that the turnover of phosphorus in the epilimnion is dominated by bacterial decomposition of organic material in the water column (Bloesch et al. 1977, Whalen and Cornwell 1985, Levine et al. 1986, Conley et al. 1988). However, when fecal pellets and dead organisms sink through the thermocline, phosphorus remineralization continues in the lower water column and sediments (Lehman 1988, Gächter et al. 1988). Hypolimnetic waters often show high concentrations of P, which is returned to the surface during periods of seasonal mixing. However, during seasonal stratification, the net primary production is largely determined by phosphorus inputs to surface waters (Schindler 1978) and the rate of recycling in the water column. Phosphorus turnover in the epilimnion is mediated by planktonic bacteria and enhanced by the activities of larger grazing organisms (Porter 1976, Lehman 1980, Carpenter et al. 1987).

Of course, turnover of phosphorus through the biotic community is incomplete, and some phosphorus is lost to the sediments. Mortimer (1941, 1942) suggested that the coprecipitation of phosphorus with Fe and Mn minerals, which are insoluble in the high redox conditions of the epilimnion, also carries phosphorus to the sediments. As long as the hypolimnetic waters are aerobic, a microlayer of these minerals remains near the sediment–water interface, which retains phosphorus that may be released from bacterial decomposition or from the dissolution of minerals at low redox potential in the underlying sediments. However, when the hypolimnetic waters are anoxic, the Fe-oxyhydroxide minerals may dissolve, and P is released to the overlying waters (Bostrom et al. 1982, 1988). Recent studies have generally found that the dissolution of Fe is limited (Davison et al. 1982), so that the regeneration of phosphorus from the sediments is usually minor (Levine et al. 1986; Caraco et al. 1990). When phosphorus concentrations in the overlying waters are high, Fe, Mn, and Ca minerals in the sediments can adsorb phosphorus even in anaerobic conditions (Patrick and Khalid 1974). Thus, sedimentary accumulations of undecomposed organic matter and Fe minerals carry P that is permanently lost from the ecosystem.

Interactions between biogeochemical elements may be important in determining the release of P from sediments. In most freshwaters, the concentration of SO_4 is low, and P is strongly absorbed by Fe minerals in the sediment. In the sea, concentrations of SO_4 are higher, and P limitations are less apparent (Chapter 9). Increasing concentrations of SO_4 in lakes affected by acid rain may act through the anion exchange reactions [equation (4.13)] to drive P into solution, enhancing lake productivity (Caraco et al. 1989).

Despite the limited availability of phosphorus in surface waters, we might expect that, as for land vegetation, processes such as denitrification might make nitrogen the nutrient that is in shortest supply in lakes. Although Goldman (1988) has shown that the primary productivity of some lakes is limited by nitrogen, Schindler (1977) suggests why this is not generally the case. When phytoplankton communities grow in limited supplies of nitrogen, there is a shift in algal dominance from green algae to blue-green algae—which fix nitrogen, add to its availability, and raise the N/P ratio. Smith (1983) found that blue-green algae were common only at N/P ratios <29, and Howarth et al. (1988b) found that significant nitrogen fixation by phytoplankton (blue-green algae) occurred only when the N/P ratio was <16. When phosphorus is added as a pollutant and the algal community shifts to species of blue-green algae, total net primary productivity increases. In such conditions, nitrogen fixation can supply up to 82% of the nitrogen input to the phytoplankton community (Howarth et al. 1988a). When the input of phosphorus ceases, blue-green algae decrease in importance (Edmondson and Lehman 1981). Although these shifts in community dominance are not found in all situations (see Canfield et al. 1989), the inputs of nitrogen by blue-green algae tend to maintain a phosphorus shortage for the growth and photosynthesis of phytoplankton (Smith 1982). There is no equivalent biogeochemical process that can increase the supply of phosphorus when it is in short supply.

Other Nutrients

Changes in the dominance of various species of algae are also seen in response to differing availabilities of other nutrient elements. When phosphorus is added to nutrient-poor lakes, the growth of diatoms, which require silicon, may reduce the supply of silicon to low levels, favoring the dominance of other species, such as green algae (Kilham 1971, Schelske et al. 1983, Schelske 1988). Unlike P, much of the organic Si is regenerated from the sediments (Schelske 1985, Conley et al. 1988). Additional inputs of Si in groundwater link the biogeochemistry of lake ecosystems to processes in the surrounding watershed (Hurley et al. 1985). Titman (1976) showed that subtle differences in the ratio of silicon to phosphorus controlled the dominance shared by two species of diatoms, *Asterionella* and *Cyclotella*. Other studies have shown that net primary production is affected by changes in trace micronutrients, such as B (Subba Rao 1981), Fe (Allen 1972) and Cu (Horne and Goldman 1974). Such changes in the phytoplankton community of lakes are perhaps the best examples of how subtle shifts in the biogeochemistry of the environment can alter the distribution and abundance of species and the productivity of a natural ecosystem.

Lake Budgets

Carbon Since phytoplankton show relatively constant ratios between carbon and the uptake of important nutrient elements, such as N and P, studies of the production and fate of organic carbon in lake ecosystems are useful in understanding their biogeochemistry. Rich and Wetzel (1978) present a carbon budget for Lawrence Lake, located in southern Michigan (Table 7.3). Net primary production within the lake ecosystem is known as *autochthonous* production. In this shallow lake, rooted plants contribute 51.3% of the autochthonous net primary productivity, while phytoplankton account for 25.4%. In contrast, Jordan and Likens (1975) report that phytoplankton account for nearly 90% of the production in Mirror Lake, a relatively unproductive lake in New Hampshire with a limited area of shallow water. For Lawrence Lake, DOC that is lost from plants accounts for an additional 10.6% of the annual input of organic carbon to the lake. DOC is also derived from streams entering the lake. The inputs of organic carbon from land are known as *allochthonous* production. Stable isotope ratios, that is, $\delta^{13}C$, in sediments have been used to estimate the comparative contribution of organic carbon from autochthonous phytoplankton production compared to allochthonous inputs from land vegetation (LaZerte 1983).

Nearly three-fourths of the organic carbon entering Lawrence Lake is respired in the lake, with benthic respiration comprising 73.6% of total respiration. In other lakes as much as half of the respiration is due to bacterial decomposition in the water column (e.g., Lehman 1988). Bacterial growth and respiration has proven difficult to study, but a new technique involving the incorporation of [^{3}H]thymidine into bacterial DNA seems to offer an accurate measurement of heterotrophic bacterial growth in fresh water and seawater (Fuhrman and Azam 1982, Fuhrman et al. 1986, Bell et al. 1983).

Only 7.8% of the organic carbon in Lawrence Lake is permanently stored in the sediments, comprising the net ecosystem production of this ecosystem. The sediment storage of 16.8 g C m^{-2} yr^{-1} is similar to the rate of soil organic matter accumulation in many land ecosystems (Table 5.3), but it is derived from a much lower primary production than is typical on land. The greater percentage of net primary production that is permanently stored in aquatic ecosystems speaks for the relative inefficiency of bacterial respiration, often in anaerobic conditions, compared to the importance of aerobic, eukaryotic decomposers (fungi) on land (Benner et al. 1986).

Table 7.3 Origins and Fates of Organic Carbon in Lawrence Lake, Michigan[a]

	g C m^{-2} yr^{-1}	%	%
Net primary productivity (NPP)			
POC			
Phytoplankton	43.3	25.4%	
Epiphytic algae	37.9	22.1%	
Epipelic algae	2.0	1.2%	
Macrophytes	87.9	51.3%	
Total	171.2	100.0%	
DOC			
Littoral	5.5		
Pelagic	14.7		
Total	20.2		
Total NPP	191.4		88.4%
Imports			
POC	4.1	16.3%	
DOC	21.0	83.7%	
Total imports	25.1	100%	11.6%
Total available organic inputs	216.5		100.0%
Respiration			
Benthic	117.5	73.6%	
Water column	42.2	26.4%	
Total respiration	159.7	100.0%	74.2%
Sedimentation	16.8		7.8%
Exports			
POC	2.8	7.3%	
DOC	35.8	92.7%	
Total exports	38.6	100.0%	18.0%
Total removal of carbon	215.1		100.0%

[a] From Rich and Wetzel (1978).

Examining several lakes, Hutchinson (1938) suggested that the rate of depletion of O_2 in the hypolimnion during seasonal stratification was related to the productivity of the overlying waters. Highly productive waters should contribute large quantities of organic carbon for respiration in the hypolimnion, which is seasonally isolated from sources of oxygen. He expressed the consumption of oxygen on an area basis to account for the much greater volume of hypolimnetic water in deep lakes. Although the relationship seems logical, the search for its widespread application has been fraught with controversy. Lasenby (1975) found little evidence for the relationship in 14 lakes of southern Ontario, and Stauffer (1987) shows that a significant amount of oxygen diffuses across the thermocline during periods of stratification. Cornett and Rigler (1979) conclude that "a simple proportionality between biomass in the epilimnion and area hypolimnetic oxygen deficit (AHOD) does not appear to exist." However, these latter workers attempted to refine the relationship by examining the role of hypolimnetic volume and water temperature in a multiple linear regression (Cornett and Rigler 1979, 1980). They found that the greatest O_2 consumption occurred in deep lakes with high retention of phosphorus (i.e., high production), higher water temperatures, and a thick hypolimnion. Presumably water temperature controls the rate of bacterial respiration in the water column and sediments. The relationship to hypolimnion thickness was unexpected, since it suggests that the greatest deficits are found in deep lakes with large hypolimnetic volume. Their findings, while not without criticism (Chang and Moll 1980), suggest that the consumption of oxygen in the hypolimnion may be largely the result of respiration in the water column, which is greatest in deep lakes where the transit time for sinking detritus is long. Evidently sediment respiration plays a lesser role. Despite a history of difficulty with the hypolimnetic oxygen deficit theory, it remains as a useful basis for evaluating the mass balance of organic carbon and the linkage of the carbon and oxygen cycles in fresh waters.

Nutrients   Except for nitrogen fixation, nutrient inputs to lakes are from allochthonous sources. Nutrient budgets are constructed by assessing the inputs of nutrients in precipitation, runoff, and N fixation and the losses of nutrients from lakes due to sedimentation, outflow, and the release of reduced gases. In many cases human impacts dominate the nutrient budget (Edmondson and Lehman 1981). Successful attempts to construct nutrient budgets demand an accurate lake water budget. Losses of nutrients in deep seepage are particularly difficult to estimate (Coleman and Deevey 1987, Deevey 1988). The relative turnover or mean residence time of nutrients compared to water indicates the role of biota in geochemical movements.

Nutrient budgets for most lakes consistently show net retention of N and P in the ecosystem (Table 7.4), although in lakes where outlet streams discharge a large portion of the annual water input, the retention of N and P is relatively small (e.g., Whalen and Cornwell 1985). Losses in discharge tend to yield a balanced budget for Mg, Na, and Cl (Cole and Fisher 1979, Canfield et al. 1984, Jeffries et al. 1988). In highly productive alkaline lakes, calcite ($CaCO_3$) may precipitate during periods when

Table 7.4 Input–Output Balance (tonnes/yr) for Cayuga Lake, New York, 1970–1971, and Rawson Lake, Ontario, 1970–1973[a]

Element	Precipitation Input	Runoff Input	Total Input	Discharge Output	Percent Retained
		Cayuga Lake			
Phosphorus	3	167	170	61	64
Nitrogen	179	2,565	2,744	513	81
Potassium	19	3,480	3,499	3,969	−12
Sulfur	313	24,671	24,984	31,983	−22
		Rawson Lake			
Phosphorus	0.018	0.017	0.035	0.010	71
Nitrogen	0.339	0.346	0.686	0.275	60
Carbon	2.435	19.005	21.440	10.074	53
Potassium	0.059	0.442	0.501	0.434	13
Sulfur	0.055	0.362	0.416	0.331	20

[a] From Likens (1975a).

high photosynthetic rates remove CO_2 from the water column (Brunskill 1969):

$$Ca^{2+} + 2HCO_3^- \rightarrow CaCO_3 \downarrow + H_2O + CO_2 \qquad (7.21)$$

These lakes will show a net retention of Ca, and a relatively short mean residence time for Ca in the water column (Canfield et al. 1984). Schelske (1985) found net retention of Si in nutrient budgets for several of the Great Lakes. In all cases in which biota yield net retention in the ecosystem, biogeochemical control is exerted on the movement of elements at the surface of the Earth. Lake sediments retain a record of the change in biogeochemical function through time (Whitehead et al. 1973, Pennington 1981, Brugam 1978, Schelske et al. 1988).

Many lake nutrient budgets show high retention of P, which is presumably accumulated in the sediments (e.g., Cross and Rigler 1983). Although this would seem to contradict our earlier statements regarding the importance of biological turnover of phosphorus in lakes, much of the P entering lakes is carried with soil minerals (Sonzogni et al. 1982, Froelich 1988, see also Table 4.9), rapidly sedimented, and not recycled. When these forms of phosphorus are subtracted from the total P entering a lake, the budget for organic P typically shows much lower relative retention in the sediments (Lehman 1988).

Nitrogen fixation rates in lakes range from 0.1 kg N ha^{-1} yr^{-1} to >90 kg N ha^{-1} yr^{-1} (Howarth et al. 1988a), roughly spanning the range of nitrogen fixation reported for terrestrial ecosystems (Chapter 6). Lakes with high rates of nitrogen fixation show large apparent accumulations of N (Horne and Galat 1985). Few lake studies have assessed denitrification and other processes of gaseous loss. The total loss of nitrogen by denitrification exceeds the input of nitrogen by fixation in almost all lakes where both processes have been measured (Seitzinger 1988). Gardner et al. (1987) found signficant losses of N from lake sediments due to denitrification, but the process was of limited importance in removing NO_3^- that diffuses into the sediment from overlying hypolimnetic waters. Yoh et al. (1988) found that both nitrification and denitrification were responsible for the production of N_2O in the water column of several lakes in Japan, but the loss of nitrogen from lakes as N_2 greatly exceeds the loss of N_2O (Seitzinger 1988). Ammonia volatilization may occur in alkaline lakes; Murphy and Brownlee (1981) found that the loss of NH_3 exceeded inputs by nitrogen fixation in a highly productive prairie lake.

Few studies have examined whether volatile losses of sulfur can account for the apparent net accumulation of S in lake nutrient budgets. Although volatile losses of sulfur occur (Brinkmann and de Santos 1974), most H_2S appears to be consumed or oxidized in the water column (Mazumder and Dickman 1989). Nriagu and Holdway (1989) found that the loss of dimethylsulfide was a minor component of the S budget of the Great Lakes. When reduced forms of sulfur are cycled through reoxidation pathways, high rates of SO_4 reduction can occur in lake sediments, leading to the mineralization of up to 30% of the particulate organic matter entering the sediment in some highly productive lakes (Smith and Klug 1981).

The input of nutrients relative to lake volume is useful in distinguishing low-productivity *oligotrophic* lakes from high-productivity, *eutrophic* lakes. The nutrient input to oligotrophic lakes is dominated by precipitation (Table 7.5). These lakes are nutrient-poor and seldom have productivity >300 mg C/m^2/day (Likens 1975b). Oligotrophic lakes are often of relatively recent geologic origin (i.e., postglacial), and deep with cold hypolimnetic waters. Such lakes often show a relatively large ratio between lake area and drainage area, and a long mean residence time for water (Dingman and Johnson 1971). In contrast, eutrophic lakes are dominated by nutrient inputs from the surrounding watershed. These

Table 7.5 Sources of Nitrogen and Phosphorus as Percentages of the Total Annual Input to Lake Ecosystems[a]

	Precipitation		Runoff	
	N	P	N	P
Oligotrophic lakes	56	50	44	50
Eutrophic lakes	12	7	88	93

[a] From Likens (1975a).

nutrient-rich lakes are often shallow, with warm, highly productive waters. Of course, sedimentation will eventually convert the physical state of many oligotrophic lakes to shallow, eutrophic conditions, so these concepts have also been used to describe a sequence of lake aging. However, in most cases, nutrient status remains the best contrast between oligotrophic and eutrophic conditions (Fig. 7.10). Humans may cause rapid "cultural eutrophication" by large nutrient additions in pollutants (Schindler 1974, Vallentyne 1974, Goldman 1988), and the eutrophic conditions can be reversed when lake management is applied (Edmondson and Lehman 1981, Levine and Schindler 1989).

Alkalinity   Alkalinity is defined as

$$\text{Alkalinity} = 2CO_3^- + HCO_3^- + OH^- - H^+ \qquad (7.22)$$

It is roughly equivalent to the balance of cations and anions in lake waters, where

$$\text{Alkalinity} = [2Ca^{2+} + 2Mg^{2+} + Na^+ + K^+ + NH_4^+] \\ - [2SO_4^{2-} + NO_3^- + Cl^-] \qquad (7.23)$$

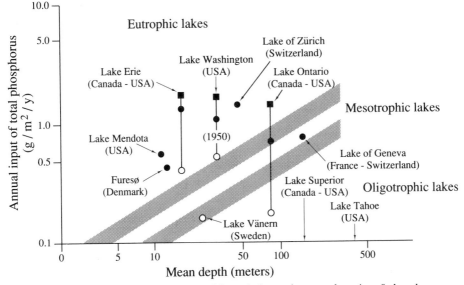

Figure 7.10   The position of important lakes relative to the annual receipt of phosphorus and their mean depth, differentiating oligotrophic and eutrophic lakes. For lakes that have undergone significant pollution, the change from previous conditions (○) to present conditions (●) is shown. From Vollenweider (1968).

Generally, alkalinity is measured in milliequivalents per liter by titration of a water sample to a pH of 4.3. When present, organic anions, such as humic acids, contribute to the apparent alkalinity of lake waters. Thus, the titration of a water sample to a pH of 4.3 is often said to represent acid-neutralizing capacity (ANC), since the protonation of organic anions is not separated from H^+ neutralized by HCO_3^- and other inorganic anions. In acid waters, Al^{3+} reduces alkalinity, since it acts as an "acid" cation (Chapter 4).

Calculation of the alkalinity budget for lakes allows us to link biogeochemical processes to the potential impacts of acid rain. For instance, alkalinity is increased by processes that consume SO_4^{2-} or NO_3^- from the water column, including sulfate reduction, sulfate adsorption on minerals, and dentrification (Rudd et al. 1986b, Baker et al. 1988). Production of organic carbon and the deposition of calcite by phytoplankton reduce alkalinity by consuming HCO_3^- [equations (7.21) and (7.22)]. In most lakes, the drainage basin contributes a large amount of alkalinity, since the runoff of cations is usually balanced by HCO_3^- (Chapter 4). It is not surprising that highly alkaline waters are found in areas underlaid by limestone.

In many regions terrestrial sources dominate the alkalinity budget of lakes, particularly when the mean residence time for lake water is <1 yr (Shaffer and Church 1989). However, in an oligotrophic lake in Canada where the mean residence time for water is 7–9 yr, D. W. Schindler et al. (1986) found that the generation of alkalinity by biogeochemical processes within the lake was greater than the receipt of alkalinity from the surrounding watershed. Acid rain impacts may be minimized when sulfate reduction increases in response to higher sulfate inputs in rainfall (Kilham 1982, D. W. Schindler et al. 1986). We can use changes in the alkalinity status of lakes as a diagnostic tool for the effects of acid rain, analogous to the use of H^+ budgets in terrestrial watersheds (Chapter 6).

Summary

Nutrient cycling in wetland ecosystems and lakes is controlled by redox potential and by the microbial transformations of nutrient elements that occur in conditions in which O_2 is not always abundant. Under these conditions, decomposition is inhibited, and organic carbon accumulates in peat and sediments. Wetland ecosystems are significant to the global cycle of sulfur through the emission of reduced sulfur gases. Wetland ecosystems are also the major source of methane, which is rapidly increasing in the atmosphere (Chapter 11). Depending upon their position at the interface between terrestrial and aquatic ecosystems, wetlands receive varying amounts of runoff from land, which affects their net primary productivity and specific nutrient limitations. Wetland ecosystems are the site of nutrient transformations, and the runoff from wetlands often controls the form of nutrient movement in rivers (Chapter 8).

The physical properties of water control many aspects of nutrient cycling in lakes. Generally most lake ecosystems are stratified into an upper zone where photosynthesis occurs and high redox potentials prevail and a lower zone where oxygen is depleted due to bacterial respiration. The circulation of lake waters, microbial turnover, and redox conditions control the turnover of nutrients in lakes. Net primary productivity in lakes is limited by phosphorus in almost all conditions. Most wetland and lake ecosystems are net sinks for nitrogen and phosphorus that enter from upland ecosystems.

Recommended Reading

Mitsch, W.J. and J.G. Gosselink. 1986. Wetlands. Van Nostrand Reinhold, New York.

Wetzel, R.G. 1983. Limnology, 2nd ed. W.B. Saunders Company, Philadelphia.

Zehnder, A.J.B. (ed.). Biology of Anaerobic Microorganisms. Wiley, New York.

8

Rivers and Estuaries

Introduction

Traditionally geochemists have regarded rivers as simple conduits linking land to sea, but ecologists now recognize that this view is too simple. Important biogeochemical reactions occur in rivers, transforming chemical elements during downstream transport. We have seen how measurements of stream-water chemistry are useful in calculating weathering rates (Chapter 4) and nutrient losses from terrestrial ecosystems (Chapter 6). In this chapter we will focus on the biogeochemical processes that occur *within* rivers, including transformations of organic carbon, phosphorus and nitrogen. We will also examine in more detail the factors that control the flow of stream waters and the origin and concentration of stream-water constituents. We will conclude the chapter with a consideration of the biogeochemistry of salt marshes and estuaries, places where rivers empty into the ocean.

Soil Hydraulics and Stream Hydrology

Vegetation and soil characteristics control the genesis of stream waters. On barren land, little precipitation infiltrates into the soil, and large amounts of surface runoff are generated even when the rainfall is not intense. Vegetation lowers the impact energy of raindrops, allowing greater rates of infiltration into the soil profile (Bach et al. 1986). In addition, plant roots, earthworms, termites, and other soil organisms promote the downward percolation of moisture through pores in the soil (Beven and Germann 1982). There is little surface runoff in most forest ecosystems, but overland flow increases strongly when vegetation is removed (Lull and Sopper 1969).

In addition to its effects on infiltration, vegetation exerts a major control on soil moisture content, since large quantities of soil water are taken up by roots to support transpiration (Chapter 5) (Table 8.1). Much of the rooting zone is below the depth from which water might otherwise evaporate from the surface. When vegetation is removed, soil water contents increase (Ting and Chang 1985, Schlesinger et al. 1987), yielding a greater volume of stream flow (Bormann and Likens 1979).

Infiltration rates and soil water contents are also affected by soil texture, especially soil porosity. Soil pore volume is related to bulk density:

$$\text{Porosity} = 1.00 - \left(\frac{\text{Bulk density}}{2.66}\right) \times 100\% \qquad (8.1)$$

so pores comprise about 50% of the volume of a soil with a bulk density of 1.33 g/cm^3, which is not unusual for many soils. Under moist conditions, water enters soils with a high proportion of sands (i.e., particles >2 mm) much faster than those dominated by clays (particles <0.002 mm), which

Table 8.1 Relative Importance of Pathways Leading to the Loss of Water from Terrestrial Ecosystems

Vegetation	Evaporation	Transpiration	Runoff and Recharge	Reference
Tropical rain forest	25.6%	48.5%	25.9%	Salati and Vose (1984)
Tropical rain forest	10	40	50	Shuttleworth (1988)
Temperate forest	13	32	53	Waring et al. (1981)
Temperate grassland	35	65	0	Trlica and Biondini (1990)
Steppe	55	45	0	Floret et al. (1982)
Desert	29	51	20	Schlesinger et al. (1987); Tromble (1988)

have a lower porosity (Saxton et al. 1986). However, as soils dry, clays retain a greater water content at any soil water potential than soils dominated by coarser fractions (Fig. 8.1). This effect is due to the high matric potential of clays, which tend to retain water on their surface.

The flow of water through terrestrial ecosystems to stream waters is often modeled using simplified assumptions about the rate of plant uptake and the downward flow of water through the soil profile (Waring et al. 1981, Knight et al. 1985). Downward movement is assumed during any interval of time in which the percolation of water to a particular depth is in excess of the water-holding capacity of that depth and the rate of plant uptake during the interval. Water-holding capacity is commonly called field capacity, which is the water content that a soil can retain against the force of gravity. When excess water drains to the bottom of the profile, it is assumed to be delivered to the stream channel.

In some models, the flow of soil water is calculated by the application of Darcy's law:

$$\text{Flux} = kIA \tag{8.2}$$

where k is the hydraulic conductivity, I is the hydraulic gradient, and A is the cross-sectional area under consideration. The constant for hydraulic conductivity must be determined empirically, usually by observations of

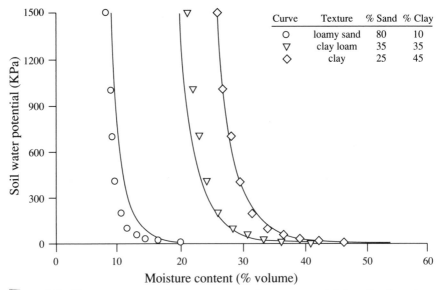

Figure 8.1 Water content remaining at various soil water potentials in soils of varying texture. Modified from Saxton et al. (1986).

the rate of disappearance of water maintained in a ponded cylinder at the surface (Rycroft et al. 1975), where

$$\text{Conductance} = \text{infiltration rate/gradient}$$
$$\text{(cm/s)} \qquad \text{(g cm}^{-2}\text{ s}^{-1}) \quad \text{(g/cm}^3) \qquad \qquad (8.3)$$

The gradient is the difference in the content of water (g/cm^3) between the surface and some known depth of interest. Although the large pores in coarse-textured soils conduct water freely when these soils are wet, drainage of these pores causes a rapid decline in hydraulic conductivity with soil drying. Thus, hydraulic conductivity in dry soils is often greater in clays, on which the adsorbed films of water maintain a continuous path for water movement through the soil. Darcy's law was originally formulated for use in saturated soils and groundwater, but it is often used sucessfully in unsaturated soils (Ward 1967). The method is limited because of the large spatial variation in soil properties (Topp et al. 1980, G. V. Wilson et al. 1989) and the effect of channels caused by roots and soil animals (Beven and Germann 1982). More elaborate treatments of flow in unsaturated soils, known as the vadose zone, are available (e.g., Nielsen et al. 1986), but they are difficult to apply in most field situations.

When precipitation is not occurring, stream flow is largely maintained by the slow drainage of water from the soil profile and from groundwater. This base flow declines slowly as the drought period continues. With rainfall, a number of changes are seen in a stream hydrograph, which relates stream flow to time (Fig. 8.2). An immediate increase in flow, known as quick flow, may result from surface runoff that enters the stream channel during the storm. At the end of rainfall, the effect of surface runoff disappears rapidly, but base flow is reestablished at a new, higher level, which resumes a slow decline as the soil dries. The increased base flow is derived from rainwater that infiltrated the soil profile raising the soil moisture content and the amount of water available for drainage. Long-term observations of streams show that hydrographs are affected by topography, vegetation, and soil characteristics, as well as the pattern and intensity of rainfall in individual storms (Ward 1967). Stream hydrographs show what fraction of the flow is derived from surface runoff, which may carry organic debris and soil particles, and what fraction is derived from the drainage of soil water.

Stream Load

Nutrient transport in streams is often divided into two fractions: that carried in the form of dissolved ions, and that carried as particulates. The dissolved load is largely derived from rainfall and from soil processes, including leaching of plant litter and chemical weathering. The particu-

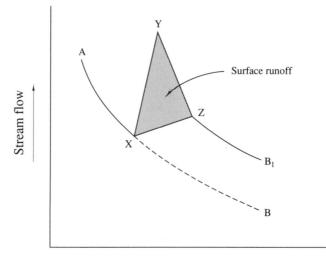

Time (days)

Figure 8.2   A stream hydrograph, showing the effect of a rainstorm at time X on stream runoff, which increases to a peak (Y) during the rainstorm. Streamflow declines rapidly to baseflow (Z), which is reestablished at a higher level (B^1) than without the storm (B). From Ward (1967).

late load, dominated by the products of mechanical weathering, represents erosion and sediment transport from the surface of the soil. Particulate contents include materials ranging in size from colloidal clays to large boulders and from leaves to floating logs. The particulate load includes material suspended in the water—the suspended load—and material that moves long the bottom of the stream channel—the bed load.

Biogeochemical Transformations of C, N, and P

The carbon budget of most small streams is dominated by allochthonous materials, that is, organic carbon that enters from the surrounding terrestrial ecosystem (Chapter 7). Some of these materials are easily observed. When leaves fall into a stream, they are shredded and decomposed during downstream movement. In other cases, dissolved organic compounds from the soil solution account for the major proportion of the allochthonous inputs. Dissolved organic carbon (DOC) compounds include soluble carbohydrates and amino acids, which are leached from decomposing leaves and plant roots (Suberkropp et al. 1976), and humic and fulvic acids from soil organic matter (McDowell and Likens 1988; Chapter 5). Humic acids comprise about 60% of the DOC in the Amazon River (Ertel et al. 1986). The movement of dissolved organic carbon into stream waters is strongly controlled by interactions with clay minerals in the soil

Table 8.2 Yearly Fluxes of Organic Carbon, Nitrogen, and Phosphorus in Bear Brook, New Hampshire[a]

	Organic Carbon (g/m^2)	Nitrogen (g/m^2)	Phosphorus (g/m^2)	Atomic Ratio C : N : P
Inputs				
Total dissolved	260	56	0.39	1700 : 320 : 1
Total fine particulate	12	0.27	0.55	54 : 1 : 1
Total coarse particulate	340	8.2	0.70	1300 : 26 : 1
Total gaseous	1	<0.1	0	—
Total inputs	620	64	1.6	990 : 89 : 1
Outputs				
Total dissolved	260	57	0.29	2300 : 440 : 1
Total fine particulate	25	0.43	1.1	59 : 0.9 : 1
Total coarse particulate	100	1.8	0.38	720 : 10 : 1
Total gaseous	230	?	0	—
Total outputs	620	59	1.8	890 : 72 : 1

[a] From Meyer et al. (1981).

(McDowell and Wood 1984, Fahey and Yavitt 1988). When clay minerals are absent in the soil profile, organic compounds move freely to stream waters, producing deeply stained, blackwater rivers (Beck et al. 1974). Dissolved organic compounds account for 96% of the organic matter in the Qgeechee, a blackwater river draining swamps in southern Georgia (Benke and Meyer 1988).

Bear Brook in New Hampshire is typical of small streams draining temperate forests. Allochthonous inputs of organic carbon are roughly equally divided between dissolved and particulate sources, and net primary production by phytoplankton and mosses supplies only 0.2% of the organic carbon available for respiration in the stream (Fisher and Likens 1973). As these materials pass through Bear Brook, some carbon is respired and lost to the atmosphere as CO_2, and dissolved organic compounds increase at the expense of particulate organic matter (Table 8.2). In small streams, the ratio of dissolved to particulate organic carbon generally increases during downstream transport (S. G. Fisher 1977).

Larger rivers with more slowly moving waters show an increasing importance of net primary production by phytoplankton and rooted plants (Naiman and Sedell 1981, Lewis 1988). This production is of special significance, since it is more easily assimilated by higher trophic levels than is the resistant plant debris derived from land. Despite the large amount of humic material in the Amazon, much of the fish community is supported by the growth of phytoplankton, rather than by organic carbon derived from the rain forest (Araujo-lima et al. 1986). Nevertheless, the rate of plant production in large rivers is often limited by turbidity, so these systems usually retain an overall dominance of allochthonous materials—that is, respiration exceeds net primary production (Edwards and Meyer 1987, Cai et al. 1988).

Large rivers often derive a significant fraction of their organic carbon from their floodplain, especially during seasonal flooding (Cuffney 1988, Grusbaugh and Anderson 1989). Transport of organic materials in the lower reaches of rivers often fuels a large heterotrophic, bacterial population (Edwards and Meyer 1986), which consumes DOC and provides food resources for higher trophic levels (Edwards 1987, Meyer et al. 1987). Despite significant internal production in the Amazon, heterotrophic processes dominate total metabolism (Wissmar et al. 1981, Richey et al. 1990). At least a portion of the heterotrophic activity is anaerobic, probably in floodplain waters, releasing CH_4 to river waters and to the atmosphere (Richey et al. 1988, Tyler et al. 1987, Devol et al. 1988).

Concentrations of DOC and POC increase with increasing stream flow, as a greater portion of the water is derived from overland flow that carries organic material to the stream channel (Meyer et al. 1988). Measurements of small forest streams indicate that stream flow removes about $1–5$ g C m^{-2} yr^{-1} from the watershed (Moeller et al. 1979, Schlesinger and Melack 1981, Naiman 1982). Thus, the metabolism of stream ecosystems is fueled by the transfer of <1% of the net primary production of forests to their runoff waters (Mantoura and Woodward 1983). Somewhat higher values are characteristic of streams draining peatlands and swamps (Mulholland and Kuenzler 1979), while lower values characterize grasslands and deserts (Gifford and Busby 1973, Fisher and Grimm 1985, Mulholland and Watts 1982).

We can approach the question of the total riverine transport of organic carbon using several approaches, which provide a good example of the problems of scaling in biogeochemistry (Chapter 1). For example, we might compile studies of small watersheds in which the transfer of organic carbon from land to stream waters has been carefully measured. These measurements could be multiplied by the area of the globe that is covered by such vegetation, and a global estimate calculated. Alternatively, we might compile studies of large rivers, for which there are measurements of annual riverflow and the load of organic carbon near the mouth. The transport of organic carbon in rivers that have not been studied can be calculated from the regression, allowing us to calculate the total transport for all the major rivers of the world (Fig. 8.3). The large rivers are critical to the second approach, since the 50 largest rivers carry 43% of the total fresh water moving from land to sea. The Amazon alone carries 20%.

The first approach emphasizes the losses of organic carbon from land, while the second approach emphasizes the transfer of organic carbon to the ocean. They will differ by the amount of organic carbon that is metabolized during transport or deposited in floodplains and behind

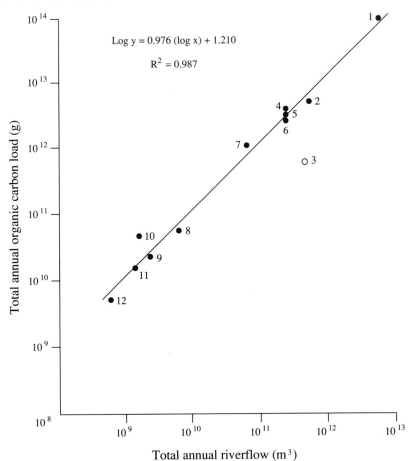

Figure 8.3 Total annual load of organic carbon shown as a logarithmic function of total annual river flow for major rivers of the world. Rivers 1–7 are among the 50 largest, as follows: 1 = Amazon, 2 = Mississippi, 3 = St. Lawrence, 4 = MacKenzie, 5 = Danube, 6 = Volga, and 7 = Rhine. Most of the organic carbon in the St. Lawrence is deposited in the Great Lakes, accounting for its reduced delivery of TOC to the ocean. From Schlesinger and Melack (1981).

dams (Mulholland 1981). Regardless of the approach, most estimates of the riverine transport of organic carbon fall in the range of 0.4–0.5×10^{15} g C/yr (Schlesinger and Melack 1981, Meybeck 1982, Michaelis et al. 1986). The average concentration of total organic carbon (9.5 mg C/l) is derived by dividing the estimate for global transport by that for river volume (4.2×10^{13} m^3/yr; Chapter 10).

Most rivers carry low concentrations of dissolved inorganic nitrogen (NH_4^+, NO_3^-) and phosphorus (HPO_4^{2-}), which are actively taken up by plants and soil microbes and retained on land (Meybeck 1982). Decomposition of coarse particulate matter (leaves) in streams is accompanied by decreases in the C/N and C/P ratios (Table 8.2), just as we see during the decomposition of terrestrial litter (Chapter 6). The decline is due to the immobilization of these essential elements by the microbes involved in litter decay (Meyer 1980, Triska et al. 1984, Qualls 1984). Decomposition rates are limited by substrate quality, and experimental additions of P to streams increase the rate of litter decay (Elwood et al. 1981). Nitrogen additions appear to have less effect (Triska and Sedell 1976). During river transport, phosphorus is also adsorbed on sediments and suspended minerals (Table 4.9) (Meyer 1979). Thus, most upland streams export lower concentrations of dissolved N and P, and a greater fraction of fine particulate N and P, to downstream regions (Meyer and Likens 1979, Triska et al. 1984). Floodplain forests, subject to seasonal innundation, receive a significant input of nutrients in deposited sediments that stimulate net primary production (Mitsch et al. 1979, Brown 1981).

An effective theory for nutrient cycling in stream ecosystems is the concept of nutrient spiraling (Newbold et al. 1983). During downstream transport, dissolved ions are accumulated by bacteria and other stream organisms and converted to organic form. When these organisms die, they are degraded to inorganic forms that are returned to the water, only to be taken up again by organisms that are involved in the further degradation of organic materials. The cycle between inorganic and organic forms may be completed many times while a nutrient atom moves downstream to the ocean. Since the cycle will occur most rapidly when biotic activity is highest, the spiral length or turnover time is an inverse index of ecosystem metabolism (Fig. 8.4). Comparative estimates of spiral length are determined by following the disappearance of isotopic tracers (^{15}N

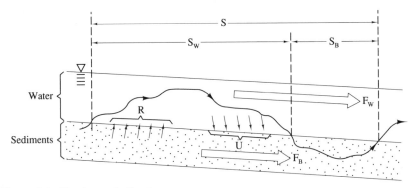

Figure 8.4 Nutrient spiraling in a two-compartment stream. U represents uptake and R represents release by the sediments. Spiraling length S is the sum of the uptake length S_W, and the turnover length S_B; F_W is the downstream flux of dissolved nutrients in the water compartment and F_B is the downward flux in the particulate compartment. Modified from Newbold et al. (1982).

and ^{32}P) from stream waters. In a small stream in Tennessee, Newbold et al. (1983) found that phosphorus moved downstream at an average velocity of 10.4 m/day, cycling once every 18.4 days, so the average spiral length was about 190 m. Through most of that distance, phosphorus traveled as a dissolved ion.

The Apure River, a tributary of the Orinoco River of Venezuela, transports 0.068 g m^{-2} yr^{-1} of P and 0.345 g m^{-2} yr^{-1} of N to the sea (Saunders and Lewis 1988). Dissolved inorganic forms account for only 24% of P and 19% of N. Similarly, on a global basis, rivers transport 21×10^{12} g/yr of phosphorus, nearly all in particulates, and 36×10^{12} g/yr of nitrogen, with about 87% in organic forms (Meybeck 1982, Ittekkot and Zhang 1989). The overall C/P and C/N ratios in global riverflow are 19.0 and 11.1, respectively. Recall that the C/N ratio of plant litter is >100 (Tables 6.5 and 6.7). The lower C/N ratio of particulates in rivers reflects the retention of nitrogen and the respiration of carbon during downstream transport (Meybeck 1982).

The global transport of N and P in rivers has been increased greatly by human activities, such as the widespread use of nitrogen fertilizers and phosphorus detergents. Meybeck (1982) estimates that the total riverload of N has doubled and that for P is tripled over preindustrial levels. These increases are not evenly distributed around the world; they are strongly correlated with human population and energy consumption in the drainage basin. As a result of the ban on the use of phosphorus-based detergents in the United States, total P has declined in rivers, while total nitrate has continued to increase (Table 8.3).

Table 8.3 Recent Changes in the Delivery of Nutrients to Coastal Areas of the United States Show Major Differences in the Transport of Nitrogen and Phosphorus in Rivers[a]

Region	Change in Load, 1974–1981	
	Total Nitrate (%)	Total Phosphorus (%)
Northeast Atlantic Coast	32	−20
Long Island Sound/New York Bight	26	−1
Chesapeake Bay	29	−0.5
Southeast Atlantic Coast	20	12
Albemarle/Pamlico Sound	28	0
Gulf Coast	46	55
Great Lakes	36	−7
Pacific Northwest	6	34
California	−5	−5

[a] From Smith et al. (1987). Copyright 1987 by the AAAS.

Dissolved Constituents

Variations in the concentration of dissolved ions can be linked to changes in discharge and to the origin of waters that contribute to the stream hydrograph (Johnson et al. 1969). In a simple geochemical system, we might expect that stream water concentrations would be highest at periods of low flow, since the water would be derived by drainage from the soil profile where it would be in equilibrium with weathering and ion-exchange reactions (Chapter 4). As stream flow increased, we might expect concentrations to decline as an increasing proportion of the flow is derived from precipitation, surface runoff, and drainage from large soil pores, with little or no equilibration with the soil mineral phases. This simple geochemical model often explains the behavior of major ions in stream water (Ca, Mg, Na, Si, Cl, and HCO_3^-), although there are frequent exceptions (Meyer et al. 1988). These ions are easily soluble, usually nonlimiting to biota, and little affected by *bio*geochemistry. Jennings (1983) shows dilution of Ca and Mg with increasing discharge in areas of limestone in New Zealand. The slope of the relationship changes slightly from summer to winter, reflecting a greater weathering of limestone by the more active respiration of roots in the summer.

Potassium usually shows little relation to stream-water discharge, presumably because it is actively taken up by plant roots (Lewis and Grant 1979, Feller and Kimmins 1979). In the Hubbard Brook Experimental Forest of New Hampshire, the lowest potassium and nitrate concentrations are found during the low flow periods of the summer, when biotic demands are greatest (Johnson et al. 1969, Likens et al. 1977).

During rainfall or seasonal flooding, concentrations are often higher on the rising limb of the stream hydrograph than at equivalent flows on the declining limb (Whitfield and Schreier 1981, McDiffett et al. 1989). The effect, known as hysteresis, is thought to result from an initial flushing of highly concentrated waters that have accumulated in the soil pores during low flow periods. Not all ions show consistent hysteresis patterns, so to calculate the total annual loss of dissolved ions from a watershed, the streamflow discharge for each day must be multiplied by the concentration measured at that discharge and the products must be summed for all 365 days. Even for elements that show lower concentrations at greater discharge, the total removal is greatest during years of high stream flow (Fig. 8.5). Thus, the effects of increasing flow predominate over changes in concentration.

The relationship between the concentration of dissolved ions and river-flow has been compiled for many of the major rivers of the world by Holland (1978). Figure 8.6 shows the relation of the *total* concentration of dissolved substances to stream flow, and serves to summarize similar relationships that are seen for individual ions. Concentrations are great-

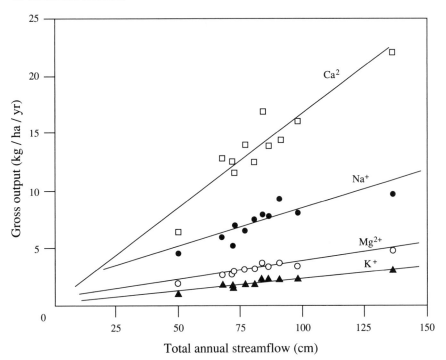

Figure 8.5 Annual stream-water loss of major cations as a function of annual stream discharge in the Hubbard Brook Forest, New Hampshire. From Likens et al. (1977).

est in rivers that drain regions with limited runoff, and they decline with increasing runoff. At relatively high runoff, the relationship between concentration and discharge is nearly inverse, so that the total removal of dissolved materials per unit area (i.e., concentration × discharge) increases only slightly with increasing riverflow among rivers that drain regions with >10 cm/yr of runoff. The apparent contradiction between this observation and the strong relationship of Fig. 8.5 is a matter of scale. The load of small streams is directly affected by year-to-year variations in stream flow from small watersheds, whereas the load of a large river integrates the transport of its tributaries, which are likely to have a wide range in runoff and transport at any given time. Thus, the concentrations and the denudation rate of major rivers are rather constant from year to year.

Gibbs (1970) used the concentrations of ions in major world rivers to suggest the origins of their dissolved constituents and their waters (Fig. 8.7). Rivers dominated by precipitation show low concentrations of dissolved substances, and a high ratio of Cl to the total of Cl + HCO$_3$, reflecting the importance of Cl from rainfall. Rivers in which the dis-

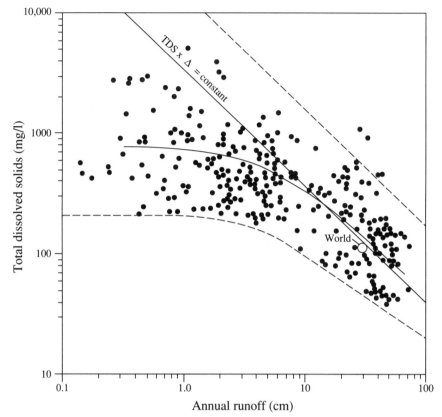

Figure 8.6 Total dissolved solids in river waters of the United States as a function of annual runoff. From Holland (1978).

solved load is largely derived from chemical weathering show higher concentrations of dissolved substances, and HCO_3 is the predominant anion (Chapter 4). Some rivers that pass through arid regions lose a significant amount of water to evaporation before reaching the ocean. These rivers show the greatest concentrations of dissolved ions, and high ratios of $Cl/(Cl+HCO_3)$, since HCO_3 has been removed by the chemical precipitation of minerals such as $CaCO_3$ as the river flows to the ocean (Holland 1978). Seawater represents the end point of the evaporative concentration of river waters. Gibbs's (1970) relationship is similar when Na and Ca are used to scale the x axis, with the relative concentration of Na as an index of rainfall and Ca as an index of chemical weathering.

   The composition of "average" river water was calculated by Livingstone (1963) from measurements on a large number of rivers (Table 8.4). His estimate of total dissolved transport, 37.6×10^{14} g/yr, is confirmed by more recent work (e.g., Meybeck 1976, 1979, see also Tables 4.8 and 9.1). Not all of this transport is derived from rock weathering. A significant

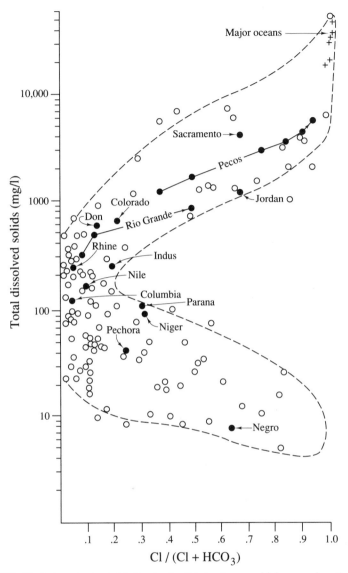

Figure 8.7 Variations in the total dissolved solids in rivers and lakes as a function of the ratio $Cl/(Cl + HCO_3)$ in their waters. From Gibbs (1970), copyright 1970 by the AAAS.

fraction of the Na, Cl, and SO_4 is derived from marine aerosols (cyclic salts) that are deposited on land (Chapter 3 and Table 4.6), and humans have enhanced the atmospheric deposition of NO_3 and SO_4, accounting for the relatively high concentrations of these ions in the runoff from industrialized continents (Table 8.4). Most of the HCO_3^- is also derived directly from the atmosphere, or indirectly, via organic decomposition and root respiration in the soil profile (Holland 1978).

Table 8.4 Mean Composition of River Waters of the World[a]

Continent	HCO_3^-	SO_4^{2-}	Cl^-	NO_3^-	Ca^{2+}	Mg^{2+}	Na^+	K^+	Fe	SiO_2	Sum
North America	68	20	8	1	21	5	9	1.4	0.16	9	142
South America	31	4.8	4.9	0.7	7.2	1.5	4	2	1.4	11.9	69
Europe	95	24	6.9	3.7	31.1	5.6	5.4	1.7	0.8	7.5	182
Asia	79	8.4	8.7	0.7	18.4	5.6	9.3		0.01	11.7	142
Africa	43	13.5	12.1	0.8	12.5	3.8	11		1.3	23.2	121
Australia	31.6	2.6	10	0.05	3.9	2.7	2.9	1.4	0.3	3.9	59
World	58.4	11.2	7.8	1	15	4.1	6.3	2.3	0.67	13.1	120
Anions[b]	0.958	0.233	0.220	0.017							1.428
Cations[b]					0.750	0.342	0.274	0.059			1.425

[a] Livingstone (1963); concentrations in mg/l.
[b] Millequivalents of strongly ionized components.

Nearly all the Ca, Mg, and K in river water is derived from rock weathering (Table 8.5). Weathering of carbonates is the dominant source for Ca, while silicates are the dominant source for Mg and K (Holland 1978). At least some Na is also derived from weathering, since its content in river water is in excess of the molar equivalent of Cl, which would be expected if seasalt were the sole source. The composition of individual streams may differ strongly from these averages, depending upon local conditions. For instance, streams draining areas of carbonate terrain are dominated by Ca and HCO_3, and stream waters may contain high concentrations of Na, Cl, and SO_4 where evaporite minerals are exposed (Stallard and Edmond 1983).

The organic compounds in river water, especially the fulvic acids, are important in the dissolved transport of Fe and Al. These metals form complexes with organic acids (Chapter 4), and are carried at concentrations well in excess of the solubility of Fe and Al hydroxides in river water (Perdue et al. 1976). The importance of dissolved organic acids in the transport of metals to the sea is a good example of the influence of terrestrial biota over simple geochemical processes that might otherwise determine the movement of materials on the surface of the Earth. The river transport of some dissolved ions is also enhanced by human activities, such as mining, that accelerate the natural rate of crustal exposure and rock weathering (Bertine and Goldberg 1971).

Suspended Load

The products of mechanical weathering and erosion are found in the suspended sediments of river water. The concentration of suspended sediment often shows a curvilinear relationship with stream flow, increas-

Table 8.5 Sources of Major Elements in World River Water (in percent of actual concentrations)[a]

Element	Atmospheric Cyclic Salt	Weathering			Pollution
		Carbonates	Silicates	Evaporites	
Ca^{2+}	0.1	65	18	8	9
HCO_3^-	<<1	61	37	0	2
Na^+	8	0	22	42	28
Cl^-	13	0	0	57	30
SO_4^{2-}	2	0	0	22	43
Mg^{2+}	2	36	54	<<1	8
K^+	1	0	87	5	7
H_4SiO_4	<<1	0	99+	0	0

[a] From Berner and Berner (1987).

ing exponentially at high flows (Parker and Troutman 1989) (Fig. 8.8). At low flows, suspended sediments are dominated by organic materials, but the contribution of POC to suspended sediments declines as the amount of suspended sediment increases during high flows, when soil erosion is greatest (Meybeck 1982, Ittekkot and Arain 1986). Long-term records show that the sediment transport during occasional extreme events often exceeds the total transport during long periods of normal conditions (Van Sickle 1981, Swanson et al. 1982). Transport increases when vegetation is removed (Bormann et al. 1974), and large concentrations of suspended sediments are found during flash floods in deserts (Baker 1977, Fisher and Minckley 1978). In many cases the sediment transported during extreme events is deposited in stream channels and floodplains in the lower reaches of the river system (Longmore et al. 1983). Thus, sediment yield per unit area of watershed declines with increasing watershed area (Milliman and Meade 1983). Despite large seasonal variations in volume, the daily sediment transport of the Amazon is rather constant as a result of storage of sediment in the floodplain during periods of rising waters and remobilization during falling waters (Meade et al. 1985).

Transport of suspended sediments in world rivers is affected by many factors, including elevation, topographic relief, and runoff from the watershed. While the rivers draining arid regions show high concentrations of suspended sediments, their total flow is limited, so the loss of soil materials per unit area is rather low (Milliman and Meade 1983). Rivers draining southern Asia carry 70% of the global transport of suspended sediments, 13.5×10^{15} g/yr (Milliman and Meade 1983). Large sediment loads in the rivers of China are derived from erosion of massive deposits of wind-derived soils, loess, in their drainage basin. In contrast, the Amazon River carries only about 1.2×10^{15} g of suspended sediment each year, about 9% of the world's total (Meade et al. 1985). Most of the

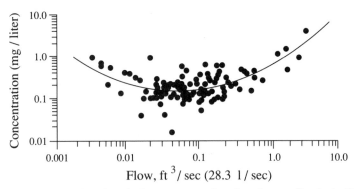

Figure 8.8 Concentration of particulate matter as a function of stream flow in the Hubbard Brook Experimental Forest of New Hampshire. From Bormann et al. (1974).

Amazon Basin occurs at low elevations with limited topographic relief, which accounts for its relatively low yield of suspended sediments (Meybeck 1977).

For elements that are only slightly soluble in water, the majority of the global transport from land to sea is found in the suspended load. Suspended sediments are enriched in phosphorus content as a result of chemical reactions between dissolved P and various soil particles (Avnimelech and McHenry 1984, Sharpley 1985; Chapter 4). When river water mixes with seawater, P may desorb from suspended particles and add to the pool of nutrients available for uptake by marine phytoplankton in coastal waters (Chase and Sayles 1980). Many metals that are micronutrients for biota (e.g., Fe, Cu, Mn) are relatively insoluble. Despite significant transport in combination with dissolved humic materials, most of the river transport of these elements occurs as a component of soil materials in the suspended load (Tables 4.9 and 8.6). The river transport of many metals and P is now greater than under preindustrial conditions (Table 8.6), but it is interesting to note that concentrations of lead have declined recently, presumably as a result of the decreased use of leaded gasoline in automobiles (Smith et al. 1987, Trefry et al. 1985).

Salt Marshes and Estuaries

When large rivers reach sea level, their rate of flow slows, drastically reducing their ability to carry sediment. The load of suspended materials is deposited in the river channel and on the continental shelf. Rivers carrying large sediment loads, such as the Mississippi and the Nile, may form obvious deltas. The river channel is progressively confined and divided by deposited sediments, which may support broad, flat areas of salt-marsh vegetation (Fig. 8.9). The lower reaches of rivers and their salt marshes are subject to daily tidal innudation. An estuarine ecosystem consists of the river channel to the maximum upstream extent of tidal influence, the surrounding salt-marsh vegetation, and the ocean waters to

Table 8.6 Estimates of Some Elemental Fluxes to the Ocean in Rivers (10^{12} g yr^{-1})[a]

	Ca	Na	Mg	Si	Fe	Cu	Pb	Zn
River particulate load	345	110	209	4430	733	1.55	2.3	5.4
River dissolved load	495	131	129	203	1.5	0.37	0.04	1.1
Total river load	840	241	338	4630	734	1.9	2.3	6.5
Theoretical load[b]	946	298	345	5780	754	0.67	0.33	2.6
Discrepancy	N.S.	N.S.	N.S.	N.S.	N.S.	+1.2	+2.0	+3.9
World mining production	—	—	—	—	—	4.4	3.0	3.9

[a] From Martin and Meyback (1979).
[b] Based on weathering of average rock.

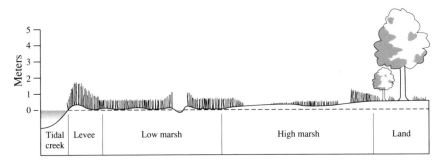

Figure 8.9 Schematic cross section through a salt marsh, showing the relationship between various components of the salt-marsh ecosystem and the open waters of the estuary. From Wiegert et al. (1981).

the maximum seaward extent that they are affected by the addition of fresh water. Estuaries are zones of mixing; within an estuary there is a strong gradient in salinity from land to sea. Estuaries are complicated and dynamic ecosystems that are difficult to model for studies of biogeochemistry (Kempe 1988).

Within estuaries, salt-marsh vegetation exists in a dynamic equilibrium between the rate of sediment accumulation and the rate of coastal subsidence or change in sea level (Frey and Basan 1985). As deposits accumulate, the rate of erosion and the oxidation of organic materials increase, slowing the rate of further accumulation. Conversely, as sea level rises, deposits are innundated more frequently, leading to greater rates of sediment deposition and peat accumulation. Along the Gulf Coast of the United States, the rate of sedimentation has not kept pace with coastal subsidence, and substantial areas of marshland have been lost (DeLaune et al. 1983). Further degradation of marshland areas is expected if sea level rises due to global climatic warming (Gornitz et al. 1982).

Salt-marsh soils are subject to a daily cycle of changing aeration. When the tide is in, the soils are innundated, and anaerobic conditions may develop throughout the profile. When the tide is out, the soils drain, allowing high redox potentials in the surface layers (Chapter 7). The transpirational loss of water from marsh plants may aid in rapid drying of the soil profile and the return of oxidizing conditions (Dacey and Howes 1984). Tidal fluctuations also affect salinity, which is lowest at low tide, when freshwater runoff from land may flush the soil profile. Tides confer spatial variability on estuaries, since the marsh areas that are closest to the sea are innundated more frequently than those in higher topographic positions.

The salinity gradient that develops in the main river channel and in the bordering salt marshes is often used to scale comparative studies of biogeochemical processes in estuaries (e.g., Fisher et al. 1988). Since chloride is much more abundant in seawater than in fresh water, it is used as an index of salinity. Any position in the estuary or salt marsh can be described by a chloride concentration between that of seawater (19,350 mg/l) and that of fresh water (~8 mg/l; Table 8.4). Chloride is particularly useful as an index, since the ion is very soluble, and essentially uninvolved in reactions, such as cation exchange, chelation, and precipitation, that would change its concentration in excess of the changes that would be expected from a simple proportional mixing of fresh waters and seawater. As such, chloride is known as a *conservative* ion. When river waters mix with seawater, other ions may change in concentration due to biotic uptake or exchange with sediments. Changes in the concentration of these ions in excess of changes that would be expected from simple dilution with seawater, shown by the concentration of Cl^-, are used to infer biogeochemical processes in the estuary.

Biogeochemistry in Salt Marshes

Due to the complexity of the estuarine ecosystem, many investigators have examined the salt marshes as a separate subsystem. These areas are often covered with dense vegetation, of which salt-marsh cordgrass, *Spartina alterniflora*, is the best known. Net primary production in salt marshes is high, ranging from 133 to 1153 g C m^{-2} yr^{-1} in estuaries of the eastern United States, with a tendency for greater production in warmer climates (Hatcher and Mann 1975). Root growth accounts for a large portion of the net primary production (Gallagher and Plumley 1979, Pomeroy et al. 1981, Howes et al. 1985), and roots show special adaptations for growth in anaerobic sediments in which high concentrations of potentially toxic substances, such as sulfide, are present (Mendelssohn et al. 1981, Carlson and Forrest 1982).

Salt marshes are effective filters and transformers of nutrients (Correll 1981). Dissolved inorganic nitrogen (NH_4 and NO_3) in runoff from surrounding land is converted to organic forms that are delivered to the open waters of the estuary (Nixon 1980). The flooded, anaerobic sediments allow significant rates of denitrification that also remove NO_3 from the system (Valiela and Teal 1979, Smith et al. 1983). The rate of denitrification varies seasonally depending upon temperature (Kaplan et al. 1979, Jørgensen 1989).

Salt-marsh vegetation is nitrogen-limited and shows increased growth with nitrogen additions. In most cases, N fixation by blue-green algae makes only a small contribution to the nitrogen budget of salt marshes, but in a boreal salt marsh, Bazely and Jefferies (1989) found that the growth of blue-green algae was stimulated when the marsh vegetation was grazed by snow geese. In this case, N fixation by the algae restored most of the nitrogen that was removed as a result of the seasonal migration of the geese from these marshes.

In most salt marshes the dominant form of available nitrogen is NH_4, since nitrification rates are low and denitrifiers remove NO_3. Tidal floods deliver small amounts of NH_4 and NO_3 to salt marshes, but the net effect of tidal flushing is to remove nitrogen from the ecosystem (Nixon 1980, Dankers et al. 1984). Despite long-term storage of organic matter in sediments, most salt marshes are a net source of nitrogen and phosphorus for the open waters of their estuaries (Table 8.7).

Salt-marsh sediments show high rates of sulfate reduction (Chapter 7), since they are rich in organic matter, flushed with high concentrations of SO_4 from seawater, and frequently anaerobic. Although the exact magnitude of sulfate reduction is the subject of some controversy (Howes et al. 1984), various investigators have suggested that more than half of the CO_2 released during decomposition of organic matter in salt marshes is associated with sulfate reduction (Howarth 1984, King 1988). Sulfate-reducing bacteria extract only a portion of the energy from the organic carbon compounds they degrade. The remaining energy is transferred to various sulfide compounds that can be further metabolized [equation (7.13)].

The initial product of sulfate reduction, H_2S, may participate in various reactions in the sediments that produce other reduced sulfur compounds,

Table 8.7 Annual Flux of Carbon and Nutrients from Salt Marshes to Coastal Waters[a]

Marsh	Carbon ($g\ C\ m^{-2}\ yr^{-1}$)			Nitrogen ($g\ N\ m^{-2}\ yr^{-1}$)			Phosphorus ($gP\ m^{-2}\ yr^{-1}$)	
	DOC	POC	TOC	NH_4^+	NO_3^-	Total	PO_4	TP
Great Sippewissett, Massachusetts		−76		−4.2	−3.8	−24.6	−0.6	
Flax Pond, Long Island, New York	−8.4	+61	+53.	−2.0	+1.0		−1.4	−0.3
Canary Creek, Delaware	−38	−62	−100	+0.7	+1.9	−1.2	<−0.1	
Gott's Marsh, Patuxent River, Maryland		−7.3		−0.4	−0.9	−3.7		−0.3
Ware Creek, York River, Virginia	−80	−35	−115	−2.9	+2.3	−2.8	−0.1	+0.7
Carter Creek, York River, Virginia	−25	−116	−142	−0.3	+0.3	−4.0	−0.6	0
Dill Creek, South Carolina		−303					−6.4	
North Inlet, South Carolina			−431					
Barataria Bay, Louisiana	−140	−25	−165					

[a] From Nixon (1980). POC, particulate organic carbon.

of which thiosulfate ($S_2O_3^{2-}$) and organic thiols are most important (Luther et al. 1986). When these reduced molecules diffuse upward in the soil profile or to the open waters of the estuary, they can be metabolized in oxidizing conditions to support bacterial growth. Hydrogen sulfide may diffuse upward to support a large population of photosynthetic sulfur bacteria that reoxidize H_2S and fix CO_2 at the surface [see equation (2.12)]. H_2S and dimethylsulfide [$(CH_3)_2S$] may also escape to the atmosphere at rates of about 5 g S m^{-2} yr^{-1} with much seasonal variation (Goldberg et al. 1981, Steudler and Peterson 1985). Despite these high rates of loss, salt marshes make a small contribution to the global flux of reduced sulfur gases, since the total area of salt marshes is not large (Carroll et al. 1986).

The various pathways for the movement of H_2S and reduced sulfur compounds in salt marshes explain why pyrite accumulations represent only a small fraction of the total sulfate reduction that has occurred in sediments (Fig. 8.10). Nevertheless, in some sediments, the rate of pyrite formation is extremely rapid (Howarth 1979, Giblin 1988), and the long-term accumulation of pyrite accounts for a significant fraction of the total

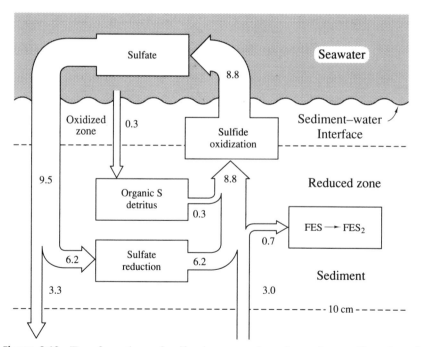

Figure 8.10 Transformations of sulfur in a coastal marine sediment. Note that of 6.2 g S/m^2/yr undergoing sulfate reduction, only 0.7 g S/m^2/yr is permanently stored in the form of pyrite and other reduced minerals. From Jørgensen (1977).

S in the sediment (Haering et al. 1989). Formation of sulfide minerals is important in the retention of other metals, including some metallic pollutants, in salt marshes (Griffin et al. 1989).

In the presence of abundant SO_4 from tidal waters, it is not surprising that the rate of methanogenesis in salt marshes is low, since sulfate-reducing bacteria are more effective competitors when SO_4 is abundant (Chapter 7). At a series of sites along the York River in the Chesapeake Bay estuary, Bartlett et al. (1987) show a gradient of decreasing methanogenesis with increasing salinity, as the SO_4 from seawater progressively inhibits methanogenesis (Fig. 8.11). Howes et al. (1985) found that only about 0.3% of total carbon input to the sediments of Sippewissett marsh in Massachusetts was lost through methanogenesis. Slightly higher rates have been reported for the Sapelo Island estuary in Georgia (King and Wiebe 1978), but globally the methane emissions from saltwater marshes contribute little to the flux of CH_4 to the atmosphere (Chapter 11).

For many years the high fisheries and shellfish productivity of estuaries was attributed to an abundance of organic carbon flushing from salt marshes to the open water. Indeed, the losses of organic carbon from salt marshes are usually >100 g C m^{-2} yr^{-1}, compared to values of $1-5$ g C m^{-2} yr^{-1} from uplands (Nixon 1980, Schlesinger and Melack 1981). Haines (1977), however, suggested that this paradigm was questionable, since the isotopic ratio of carbon in these consumers did not

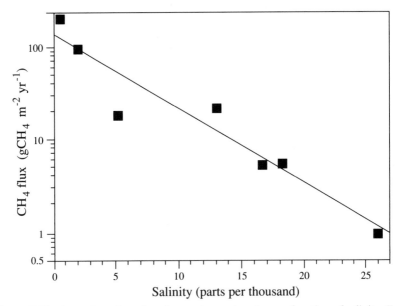

Figure 8.11 Annual methane lost from salt marsh soils as a function of salinity. From Bartlett et al. (1987).

match that of *Spartina*. Using the natural abundance of stable isotopes of both sulfur and carbon, Peterson et al. (1986) have shown that the organic carbon in shellfish of Great Sippewissett Marsh is about equally derived from *Spartina* and from phytoplankton production in the open water (Fig. 8.12). The shellfish show isotopic ratios for C and S that are midway

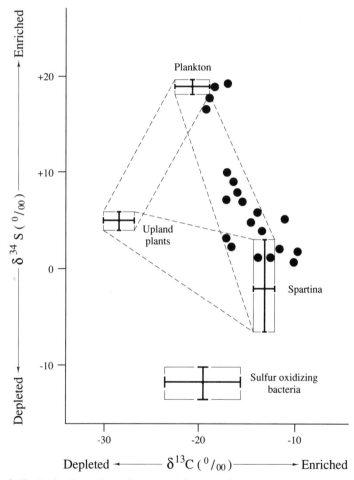

Figure 8.12 In the Great Sippewissett marsh in Massachusetts, the isotope ratio for C and S in estuarine shellfish is shown in relation to the ratios in upland plants, phytoplankton, salt-marsh *Spartina*, and sulfur-oxidizing bacteria. The sulfur in sulfur-oxidizing bacteria has an isotope ratio that is very different from that in any of the shellfish, implying that these bacteria are not a major source of food for the higher trophic levels in the estuary. Similarly, the isotope ratio for carbon in terrestrial plants is much more negative that that in the consumers. Most of the consumers in the estuary fall between the *Spartina* and phytoplankton, implying that these plant materials are the major sources of food. From Peterson et al. (1986).

between these sources. Similar results were found in the Sapelo Island marsh (Peterson and Howarth 1987). Carbon from upland, terrestrial vegetation and carbon fixed by sulfur-oxidizing bacteria in salt-marsh soils play a minor role in supporting abundant marine life in estuaries.

Open Water Habitats

The mixing of fresh water from rivers and saltwater from the ocean occurs in the central channel of an estuary. If the estuary is well mixed, the transition from fresh water to seawater is gradual and progressive as one moves downstream. In other cases, inflowing fresh water extends over a "wedge" of denser saltwater, creating a sharp vertical gradient in salinity throughout much of the estuary (Fig. 8.13). In either case the zone of mixing is an arena of rapid biogeochemical transformation and high productivity.

Seawater is high in pH (~8.3), redox potential, and ionic strength (total dissolved ions), compared to most fresh waters. The mixing of fresh water with seawater causes a rapid precipitation of dissolved humic compounds. The cations in seawater replace H^+ on the exchange sites of the humic materials (Chapter 4), causing these materials to flocculate and sink to the bottom (Boyle et al. 1977). A similar reaction between salts, usually $Al_2(SO_4)_3$, and organic matter is frequently used to cleanse sewage waters of dissolved organic compounds. Most dissolved humic compounds and metals, such as Fe, that are carried with humus subtances are precipitated

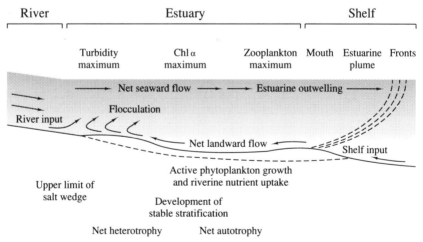

Figure 8.13 Conceptual model of the chemical and biological structure in estuaries. As the suspended load settles from the entering river waters and nutrients are made available, phytoplankton production increases, fueling an increase in zooplankton production and higher trophic levels. From Fisher et al. (1988).

in the estuary or within a short distance of the mouth of the river (Boyle et al. 1974, Sholkovitz 1976, Fox 1983). The flocculation of dissolved organic compounds and the deposition of larger plant debris account for a major portion of the organic carbon in estuarine sediments (Hedges et al. 1988), and there is little evidence that organic matter from land contributes much to marine sediments beyond the continental shelf (Gardner and Menzel 1974, Hedges and Parker 1976, Shultz and Calder 1976). As a result of the removal of terrestrial organic matter, the majority of the organic carbon in estuarine waters is composed of nonhumus substances, presumably from net primary production in the estuary and its salt marshes (Fox 1983).

Most river waters are supersaturated with dissolved CO_2, which is derived from the degradation of organic materials during downstream transport. High concentrations of dissolved CO_2 and humic materials cause river waters to be slightly acid. Under these conditions, phosphorus binds to Fe-hydroxide minerals and is transported in the load of suspended sediment (Fig. 4.3; Table 4.9). Upon mixing in the higher pH of seawater, phosphorus desorbs from these minerals and contributes to dissolved phosphorus in the estuary (Chase and Sayles 1980). Similarly, de Jonge and Villerius (1989) suggest that phosphorus bound to carbonate minerals in seawater is released as these minerals dissolve in the relatively acidic conditions of freshwater. Thus, seawater also contributes to the pool of phosphorus available for uptake in the estuary. This view suggests that the waters of estuaries provide a "window of availability" between fresh water and seawater, where phosphorus is largely found in unavailable forms bound to minerals.

A large amount of effort has been directed toward understanding the nitrogen budget of estuaries, since most river waters do not contain large concentrations of available nitrogen (NO_3 and NH_4), and these forms are removed when the waters pass over coastal salt marshes. Indeed, the filtering action of land and marsh vegetation is so effective that inputs of nitrogen in rain can make a substantial contribution to the nitrogen budget of estuaries (Correll and Ford 1982). As seen for terrestrial ecosystems (Chapter 6), most of the nitrogen that supports estuarine productivity is derived from mineralization and recycling of organic nitrogen within the estuary and in its sediments (Stanley and Hobbie 1981). When storms and tidal currents stir up the sediments in an estuary, large quantities of NH_4 are released to the water column (Simon 1989).

At the pH and redox potential that is maintained by seawater, nitrification occurs rapidly in estuarine waters (Billen 1975, Horrigan et al. 1990). Nitrification also occurs in the upper layers of sediment (Admiraal and Botermans 1989). Denitrification in the lower, anaerobic layers of sediment is primarily supported by nitrate diffusing down from the upper sediment (Seitzinger 1988), although nitrate in the water column may also diffuse back into the sediments where it is reduced (Simon 1988). In Narragansett Bay, Rhode Island, Seitzinger et al. (1980, 1984) found that denitrification removed about 50% of the available NO_3 entering in riverflow and about 35% of that derived from mineralization within the estuary. The major product of denitrification was N_2. In Chesapeake Bay, denitrification leaves the nitrate in the lower water column enriched in $\delta^{15}N$ (Horrigan et al. 1990).

As a result of human inputs of nitrogen, in sewage, agricultural runoff, and acid rain, many estuaries show excessive levels of productivity and conditions that resemble the eutrophication of fresh waters (Officer et al. 1984). The management of polluted estuaries is the subject of much controversy. Some workers argue that an improvement in estuarine conditions will be directly related to efforts to reduce nutrients in inflowing waters (Nixon 1987), while others suggest that the retention of prior inputs and recirculation of nitrogen within the system mean that efforts to reduce human inputs will not necessarily produce immediate improvements in water quality (Kunishi 1988).

Many estuaries show a peak in net primary productivity by phytoplankton at intermediate salinities, reflecting the zone of maximum nutrient availability (Edmond et al. 1981, Fisher et al. 1988) (Fig. 8.14). In other cases, mixing prevents any obvious relationships between net primary production and conservative properties, such as salinity, in the estuary (Powell et al. 1989). Phytoplankton productivity and organic matter derived from the surrounding salt marshes fuel the high productivity of fish and shellfish in estuarine waters. The disruption of estuaries by direct pollution, global sea level rise, and other human perturbations

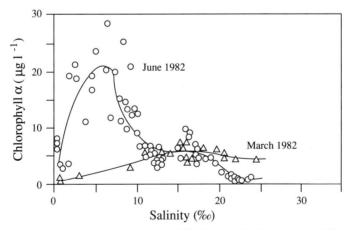

Figure 8.14 Data from Chesapeake Bay in 1982, showing the distribution of chlorophyll a, as an index of phytoplankton production, relative to salinity in the estuary. From Fisher et al. (1988).

may substantially reduce their potential to contribute to future human food supplies.

Summary

Stream ecosystems are directly linked to the surrounding land, since the rate of flow and the properties of stream water are determined by the soil properties and vegetation in the watershed. A number of biogeochemical transformations involving C, N, and P occur during stream flow. Stream ecosystems are heterotrophic—showing an excess of respiration over net primary production. During stream transport, available forms of N and P are removed from water and sequestered in organic and inorganic forms. Streams carry a variety of the other ions of biochemistry in dissolved and suspended forms, which are largely the products of rock weathering in the watershed.

The mixing of fresh water and seawater occurs in estuaries, located at the mouth of major rivers. In response to the change in pH, redox potential and salinity, the river waters feed estuaries with a rich solution of available N and P, and high rates of net primary production fuel a productive coastal marine ecosystem. Despite temporary storage of nutrients in salt marshes and estuarine sediments, river waters are always a net source of nutrients to their estuary and the coastal ocean. As we shall see, rivers are the major input in the global budget of biogeochemical elements in the ocean.

Recommended Reading

Degens, E.T., S. Kempe and J.E. Richey (eds.). 1990. Biogeochemistry of Major World Rivers, Wiley, New York.

Knox, G.A. 1986. Estuarine Ecosystems: A Systems Approach, 2 vols. CRC Press, Boca Raton, Florida.

Ward, R.C. 1967. Principles of Hydrology. McGraw-Hill, London.

9

The Sea

Introduction

The Earth's waters constitute the hydrosphere. Only small quantities of fresh water contribute to the total; most water resides in the ocean. In this chapter we will examine the biogeochemistry of seawater and the contributions that oceans make to global biogeochemical cycles. We will begin with a brief overview of the circulation of the oceans and the mass balance of major elements that contribute to the salinity of seawater. Then we will examine net primary productivity in the surface waters and the fate of organic carbon in the sea. Net primary productivity in the sea is related to the availability of several essential nutrient elements, particularly nitrogen and phosphorus. Thus, patterns of net primary production control the biogeochemistry of these elements. We will examine the biogeochemical cycles of elements essential to marine biota and the processes that lead to the exchange of gaseous components between the oceans and the atmosphere.

Ocean Circulation

Global Patterns

In Chapter 3 we saw that the circulation of the atmosphere was driven by the receipt of solar energy, which heated the atmosphere from the bottom creating instability in the air column. Unlike the atmosphere, the oceans are heated from the top. Since warm water is less dense than cooler water, the receipt of solar energy conveys stability to the water column, preventing exchange between warm surface waters and deep, cold waters over much of the ocean. The surface waters are relatively well mixed internally by the wind (Thorpe 1985). Depending upon the incident radiation, the surface waters range from 75 to 200 m in depth with a mean temperature of about 18°C. The ocean deep waters contain about 95% of the volume with a mean temperature of 3°C.

The atmospheric circulation pattern (Chapter 3) leads to the formation of surface currents in the ocean, such as the well-known Gulf Stream in the Atlantic Ocean. In each ocean the trade winds drive equatorial currents from east to west (Fig. 9.1). When these currents encounter land, the waters divide to form currents moving north and south along the eastern borders of the continents. As they move toward the poles, the currents are deflected to the right by the Coriolis force (Fig. 3.3), so the Gulf Stream crosses the North Atlantic and delivers warm waters to northern Europe. Water returns to the tropical latitudes in cold, surface currents that flow along the west side of continents. The cyclic pattern of surface currents in each of the major oceans is called a gyre. The global circulation of surface currents transfers heat from the tropics to the polar regions of the Earth. More than half of the net excess of solar energy received in the tropics is transfered to the poles by ocean circulation; the remainder is transferred through the atmosphere (Von der Haar and Oort 1973).

The loss of heat from seawater at the poles leads to downwelling, which delivers surface water to the deep ocean. Exchange between the surface ocean and the deep waters is possible when the surface waters cool and their density increases to that of the underlying water. During the winter

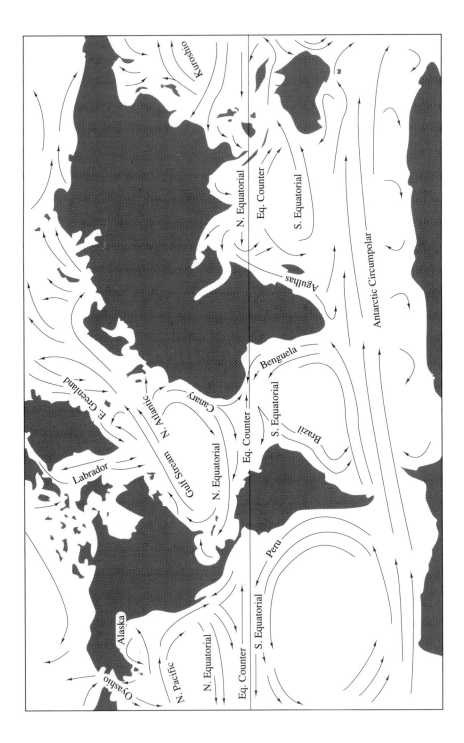

in the Arctic and Antarctic oceans, the density of some polar waters increases as fresh water is "frozen out" of seawater and added to the polar ice caps, leaving behind waters of greater salinity that sink to the deep ocean. During the summer, these oceans have lower surface salinity due to melting from the polar icecaps. Since the seasonal downwelling of cold polar waters is driven by both temperature and salinity, it is known as thermohaline circulation.

Penetration of cold waters to the deep ocean at the poles creates currents in the deep ocean that return waters to the equatorial latitudes. Thus, cold North Atlantic deep water (NADW) moves southward in the deep ocean, and upwells at tropical latitudes. A similar, northward-flowing current returns Antarctic bottom waters (AABW) to the tropics. Deep waters are nutrient-rich, so high levels of oceanic productivity are found in zones of upwelling. Upwelling along the western coast of South America yields high levels of net primary production that support the anchovy fishery of Peru.

These patterns of ocean circulation have important implications for biogeochemistry. One might calculate an overall mean residence time of 34,000 yr for ocean water with respect to stream flow (i.e., total ocean volume/annual river flow). In fact, most river waters enter the smaller volume of the surface ocean, yielding a mean residence time with respect to river waters of about 1800 yr for the 0–200 m layer (Speidel and Agnew 1982). The actual turnover time is much less due to the addition of rain waters and upwelling waters to the surface. For example, turnover of surface waters in the northern Pacific ocean is about 9–15 yr (Michel and Suess 1975). The surface water is also in rapid gaseous equilibrium with the atmosphere. Mean residence time for CO_2 in the surface ocean is about 6 yr (Stuiver 1980).

Renewal of the bottom waters is confined to the polar regions. Downward mixing of ^3H_2O produced from the testing of atomic bombs in the 1950s shows the addition of recent surface waters and the movement of deep waters towards the equator in the Atlantic Ocean (Fig. 9.2). Estimates of the mean age of bottom waters using ^{14}C dating of dissolved CO_2 range from 275 yr for the Atlantic to 510 yr for the Pacific (Stuiver et al. 1983). Since the age of bottom waters is much less than the calculated mean residence time of 34,000 yr, the volume of water entering the deep ocean each year must be much greater than the total rate of river flow to the oceans. Dickson et al. (1990) have recently estimated the downward transport in the North Atlantic (10.7×10^6 m^3/s), roughly 10 times the rate of river flow to the oceans. Additional downwelling, of similar magnitude, occurs in the cold waters of Antarctica. Despite this flux, the deep waters are much less dynamic than the surface ocean; in a very real sense,

Figure 9.1   Major currents in the surface waters of the world's oceans. From Knauss (1978).

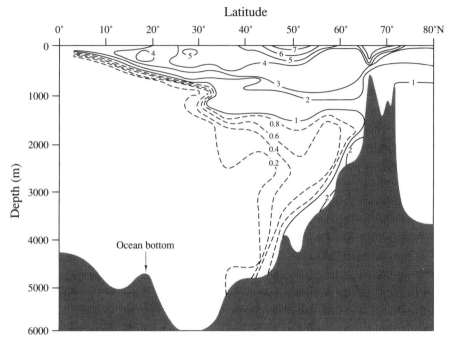

Figure 9.2 Penetration of bomb-derived tritium (^3H_2O) into the North Atlantic Ocean. Data are expressed as a ratio of $H/^3H \times 10^{18}$ in seawater samples. From Ostlund (1983).

the deep waters maintain a historical record of the conditions of the surface ocean several centuries ago.

Deep-water currents also transfer seawater between the major ocean basins as a result of the Antarctic circumpolar current. In the Atlantic, evaporation exceeds the sum of river flow and precipitation, yielding a higher seawater salinity than in the Pacific Ocean (Fig. 9.3). The Atlantic receives a net inflow of less saline waters from the Pacific to restore the water balance. At the same time, dense, saline water flows out of the deep Atlantic to enter the Indian and Pacific Oceans.

Changes in ocean currents, particularly in the formation of deep waters, are associated with changes in global climate. For example, an increase in the rate of formation of cold, deep waters may have led to a decline in atmospheric CO_2 during glacial periods, since CO_2 is more soluble in cold, saline waters (Broecker and Peng 1987). Lower atmospheric CO_2 is consistent with a lower global temperature due to a lesser greenhouse effect, and during the last glacial epoch the concentration of atmospheric CO_2 was about 200 ppm, compared to the 350 ppm of today (Chapter 11). What remains unknown is whether changes in ocean currents led to lower atmospheric CO_2 or vice versa.

The production of deep water is dependent upon the density differ-

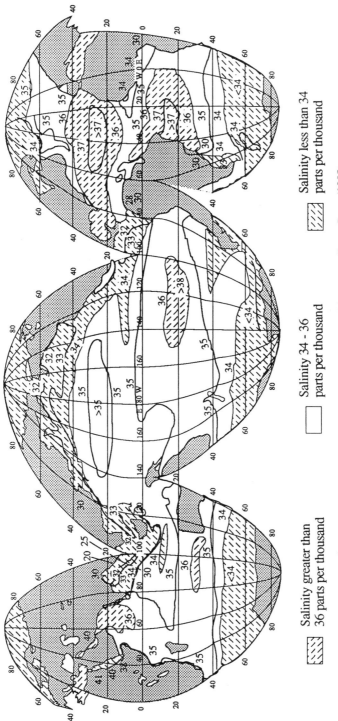

Figure 9.3 Salinity of the surface waters of the world's oceans. From Gross (1982).

Salinity greater than 36 parts per thousand

Salinity 34 - 36 parts per thousand

Salinity less than 34 parts per thousand

ence between a warm surface layer and cold waters that sink due to greater density. As continental glaciers grew and the surface waters cooled, the production of Atlantic deep water is likely to have declined, reducing the transport of atmospheric CO_2 to the deep ocean and allowing warmer conditions to return. These changes during glacial periods also affected the pattern of surface currents; the Gulf Stream is likely to have shifted southward, producing a humid climate in southern Europe during the last glacial period (Keffer et al. 1988).

El Niño

Ocean currents also show year-to-year variations that affect biogeochemistry and global climate. One of the best known variations in current occurs in the central Pacific Ocean. Under normal conditions, the trade winds drive currents that carry warm surface waters to the western Pacific, allowing cold bottom waters to upwell along the coast of Peru. Periodically, the surface transport breaks down in an event known as the El Niño–Southern Oscillation (ENSO). During El Niño years, the warm surface waters remain along the coast of Peru, preventing the upwelling of nutrient-rich water. Phytoplankton growth is limited and the fisheries industry collapses (Glynn 1988).

Associated with the warm surface waters in the eastern Pacific are changes in global climate, for example, exceptionally warm winters and greater rainfall in western North America. At the same time the absence of warm surface waters in the western Pacific reduces the intensity of the monsoon rainfalls in southeast Asia and India. Working with atmospheric scientists, oceanographers now recognize that El Niño events are part of a cycle that yields opposite, but equally extreme, conditions during non-El Niño years. These are known as La Niña conditions (Philander 1989). Although the switch from El Niño to La Niña is poorly understood, it is likely that the conditions at the beginning of each phase reinforce its development, with the cycle averaging between 3 and 5 yr between El Niño events. A similar cyclic pattern of ocean circulation is seen in the Atlantic Ocean (Philander 1989).

The upwelling of cold, deep ocean waters during the La Niña years leads to lower atmospheric temperatures in much of the northern hemisphere. Global cooling during La Niña events adds variation to the global temperature record, complicating efforts to perceive atmospheric warming that may be due to the greenhouse effect. Moreover, the El Niño–La Niña cycle affects the concentrations of atmospheric CO_2, since the release of CO_2 from cold, upwelling waters is lower during years of El Niño (Bacastow 1976). In addition to obvious effects on ocean productivity, El Niña conditions affect other aspects of biogeochemistry in the sea. Codispoti et al. (1986) suggest that greater rates of denitrification in cold La

Niña waters may increase the total marine denitrification rate by as much as 25% over El Niño conditions. Efforts to understand and predict El Niño events are an important component of the current research in global change.

Sea Level

Sea level with respect to the continents has fluctuated widely through geologic time. Sea level is affected by long-term tectonic movements of the continents and shorter-term phenomena, such as the freezing and melting of polar ice caps. When sea level is low, a greater area of the continents is exposed to chemical weathering, and sediment transport to the ocean is greater (Worsley and Davies 1979). Sea level was lower and ocean salinity was higher as a result of the accumulation of polar ice caps during the last glacial maximum. Sea level has risen more than 120 m during the last 10,000 yr (the Holocene) (Fairbanks 1989).

Using a network of 193 stations around the world, Gornitz et al. (1982) found that sea level had increased about 12 cm in the last 100 yr—a rate that is substantially higher than during most of the recent Holocene (Fig. 9.4). Atmospheric and sea surface temperatures have risen over the same period (Jones et al. 1986, Strong 1989), and at least some of the rise in sea

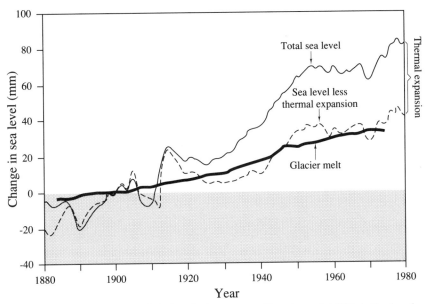

Figure 9.4 Change in sea level during the last century (Gornitz et al. 1982), indicating the proportion due to thermal expansion of the oceans and that due to melting of glaciers. From Jacobs (1986) after Meier (1984). Copyright 1984 by the AAAS.

level is attributed to the thermal expansion of water at warmer temperatures. The recent higher rate of sea-level rise may be the first indication of global warming due to the greenhouse effect (Peltier and Tushingham 1989). The current rise in sea level is likely to be associated with the widespread retreat of continental glaciers in many areas of the world. Although recent measurements of the Greenland ice cap by remote sensing suggest that the southern portion may be increasing in thickness (Zwally 1989), we have no global estimate of the mass balance of continental glaciers. If substantial melting of the Greenland or Antarctic ice begins with global warming, the rate of sea level rise will accelerate dramatically.

Observations of sea level and sea surface temperature (SST) are gathered from a large network of observing stations and ships, and global averages are difficult to calculate. In many areas the observation of sea level rise is complicated by the isostatic rebound of the continents that were covered with a large mass of ice during the last glacial epoch. Fortunately, the sea is easy to observe with remote sensing, and satellite monitoring will improve our estimates of ocean temperature, volume, and circulation. Sea surface temperature has been measured globally by the AVHRR-NOAA satellite (Chapter 5) that records infrared and microwave emission from the ocean surface (Strong 1988). MODIS, an orbiting spectrometer to be included on the Earth Observing System (EOS), will allow continuous monitoring of ocean temperature as part of the NASA program in understanding global change during the next century. A synthetic aperture radar (SAR) has recently been used to estimate the velocity of surface ocean currents near San Diego (Goldstein et al. 1989).

The Composition of Seawater

Major Ions

Table 9.1 gives the concentration of the major ions in seawater of average salinity, 35‰. The mean residence times for these ions are much longer than the mean residence time for water in the oceans, so the elements are distributed uniformly and seawater salinity is not widely variable. Although seawater varies slightly in salinity throughout the oceans (Fig. 9.3), these ions are conservative in the sense that they maintain the same concentrations relative to one another in most ocean waters. Thus, a good estimate of total salinity can be calculated from the concentration of a single ion. Often chloride is used, and the relationship is

$$\text{Salinity} = 1.81(\text{Cl}), \tag{9.1}$$

with both values in ‰. Table 9.1 shows the mean ratio between chloride and other major ions in seawater over a wide range of salinity.

Table 9.1 Major Ion Composition of Seawater, Showing Relationships to Total Salinity and Mean Residence Times for the Elements with Respect to River Water Inputs

Constituent	Concentration in Seawater[a] (mg/kg)	Chlorinity Ratio[a]	Concentration in River Water[b] (mg/kg)	Mean Residence Time (10^6 yr)
Sodium	10,760	0.5561	5.15	75
Magnesium	1294	0.0668	3.35	14
Calcium	412	0.0213	13.4	1.1
Potassium	399	0.0206	1.3	11
Strontium	7.9	0.00041	0.03	12
Chloride	19,350	1.0000	5.75	120
Sulfate	2712	0.1400	8.25	12
Bicarbonate	145	0.0075	52.	0.10
Bromide	67	0.0035	.02	100
Silicate	2.9	0.00015	10.4	0.02
Boron	4.6	0.00024	0.01	10.0
Fluoride	1.3	0.000067	0.10	0.5
Water				0.036

[a] Holland (1978).
[b] Meybeck (1979) and Holland (1978).

Like the atmosphere, the composition of the major elements of seawater has been relatively constant for long periods of time. In the face of continual inputs of new ions in river water, the constant composition of seawater must be maintained by processes that remove ions from the oceans. Table 9.1 shows that the time for rivers to supply the elemental mass in the ocean, that is, mean residence time, varies from 120 million years for Cl to 1.1 million years for Ca. Biological processes, such as the deposition of calcium carbonate in the shells of animals, are responsible for the relatively rapid cycling of Ca. But even for Cl the mean residence time is much shorter than the age of the oceans.

A number of processes act to remove the major elements from seawater. Earlier, we saw that the effect of wind on the ocean surface produces seaspray and marine aerosols that contain the elements of seawater (Chapter 3). From 10 to 50% of the river transport of Cl from land is derived directly from the sea (Table 8.5). When the composition of river water is corrected for these cyclic salts, the mean residence time for Cl in seawater may increase up to 230 million years.

The atmospheric transport of cyclic salts removes ions from the sea roughly in proportion to their concentration in seawater. Other processes must act differentially on the major ions, since their concentrations in seawater are much different from the concentrations in rivers. For example, whatever process removes Na from seawater must not be effective until the concentration of Na has built up to high levels (Drever 1988). On the other hand, Ca is the major cation in river water, but it is maintained at relatively low, stable concentrations in seawater.

Ions are removed from the oceans when the suspended clays carried by rivers undergo ion exchange in equilibrium with seawater. In rivers, most of the cation exchange sites (Chapter 4) are occupied by Ca. When the clays arrive in ocean water, Ca is released and other ions, especially Na and K, are added (Sayles and Mangelsdorf 1977). Chemical weathering that began on land may continue in the ocean, yielding the clay minerals illite and montmorillonite that contain K and Mg as part of their crystal structure (Chapter 4). Most deep sea clays show higher concentrations of Na, K, and Mg than found in the suspended matter of river water (Martin and Meybeck 1979). The clays eventually settle to the ocean floor, causing a net loss of these ions from ocean waters.

A second mechanism of loss occurs in ocean sediments. Sediments are porous and the pores contain seawater. Burial of ocean sediments and their pore waters is significant in the removal of Na and Cl, which are the most concentrated ions in seawater. During some periods of the Earth's history, vast deposits of minerals have formed when seawater evaporated from shallow, closed basins. The extensive salt flats, or sabkhas, in the Persian Gulf region are among the most well known. Although the area of such seas is limited today, the formation of evaporite minerals is an important mechanism for the removal of Na, Cl, and SO_4 from the oceans over geologic time (Holland 1978).

Biological processes are also involved in the burial of elements in sediments. As we will discuss in more detail in a later section, the deposition of $CaCO_3$ by animals is the major process removing Ca from seawater. Biological processes also cause the removal of SO_4, which is consumed in sulfate reduction, and S is deposited as pyrite in ocean sediments (see Chapters 7 and 8). Over long periods of time, ocean sediments are subducted to the Earth's mantle, where they are converted into primary silicate minerals (Fig. 2.8), with volatile components being released in volcanic gases (H_2O, CO_2, Cl_2, SO_2, etc.).

So far, the processes that we have discussed for the removal of elements from seawater cannot explain the removal of much of the annual river flow of Mg or K to the sea. For a time, marine geochemists postulated several reactions of "reverse weathering," whereby silicate minerals were reconstituted in ocean sediments, removing Mg and other cations from the ocean (MacKenzie and Garrels 1966). One type of reverse weathering, which removes K, appears to occur when basaltic rocks are exposed to seawater at ordinary temperatures of the ocean bottom. Aside from this mechanism, little evidence for reverse weathering, or its products, has been found (Drever 1988).

In the late 1970s, Corliss et al. (1979) examined the emissions from subsurface hydrothermal vents in the sea. One of the best-known hydrothermal systems is found near the Galapagos Islands in the eastern Pacific Ocean. Hot fluids emanating from these vents were substantially depleted in Mg and SO_4 and enriched in Ca and other elements compared to the seawaters that feed the hydrothermal system (Fig. 9.5). Globally the annual sink of Mg in hydrothermal vents exceeds the delivery of Mg to the oceans in river water. The flux of Ca to the oceans in rivers, 12×10^{12} moles/yr, is incremented by an additional flux of 2.1–4.3×10^{12} moles/yr from hydrothermal vents (Edmond et al. 1979). What had been a problem for geochemists seeking to explain the loss of Mg from the ocean was converted to a problem in explaining the fate of additional Ca and other elements added to the sea from hydrothermal vents.

In sum, it appears that most Na and Cl are removed from the sea in pore waters, sea spray, and evaporites. Magnesium is largely removed in hydrothermal exchange, and calcium and sulfate by the deposition of biogenic sediments. The mass balance of potassium is not well under-

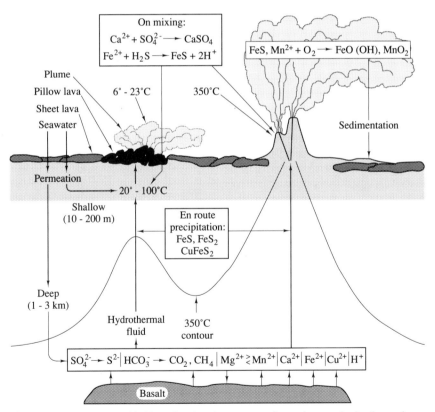

Figure 9.5 Circulation of fluids and major element transformations at a hydrothermal vent system. From Jannasch and Mottl (1985). Copyright 1985 by the AAAS.

stood, but K appears to be removed by exchange with clay minerals, leading to the formation of illite, and by some reactions with basalt. Most of these processes are not completely understood at this time, and the speed with which recent discoveries have changed our perception of ocean chemistry adds excitement to studies of biogeochemistry and global change.

Net Primary Production

Global Patterns

As much as half of the global net primary production (NPP) may occur in the sea (Table 5.2). Compared to massive forests, the organic carbon produced in the ocean is easy to overlook, since it is largely the result of phytoplankton that are small and ephemeral. Phytoplankton production occurs in the surface mixed layer, in which the distribution of dissolved O_2 is an indirect measure of the rate of photosynthesis (Fig. 9.6). Net

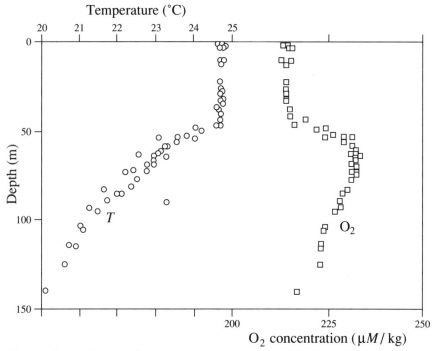

Figure 9.6 Distribution of temperature and O_2 with depth in the North Pacific Ocean. From Craig and Hayward (1987). See also Figure 9.13 for the distribution of O_2 to 1700 m. Copyright 1987 by the AAAS.

primary production in the sea is usually measured with the oxygen bottle and ^{14}C techniques outlined for lake waters in Chapter 7.

Controversy surrounding the exact magnitude of marine production derives from the tendency for O_2-bottle measurements of NPP to exceed those made using ^{14}C in the same waters (Peterson 1980). Part of the problem can be explained by recent observations of a large biomass of picoplankton, which passes through the filtration steps of the ^{14}C procedure. In the waters of the eastern tropical Pacific Ocean, Li et al. (1983) found that 25–90% of the photosynthetic biomass passes a 1-μm filter, and Stockner and Anita (1986) suggest that such picoplankton may regularly account for up to 50% of ocean production.

These methodological problems probably account for much of the variation in estimates of global marine production, ranging from 20 to 44×10^{15} g C/yr (de Vooys 1979). A widely cited estimate of marine NPP, 31×10^{15} g C/yr (Platt and Subba Rao 1975), is likely to be too low. Figure 9.7 shows the global distribution of marine NPP. The highest values are found in coastal regions, where nutrient-rich estuarine waters mix with seawater, and in regions of upwelling, where nutrient-rich water reaches the surface. However, as a result of their large area, the open oceans account for about 80% of the total marine NPP, with continental shelf areas accounting for the remainder (Martin et al. 1987). Although massive beds of kelp are found along some coasts, such as the *Macrocystis* kelps of southern California, seaweed accounts for only approximately 0.1% of marine production globally (Smith 1981, Walsh 1984).

Remote sensing offers significant potential for improving estimates of marine NPP. In 1978 the National Oceanic and Atmospheric Administration (NOAA) launched the Coastal Zone Color Scanner (CZCS) aboard the Nimbus-7 satellite (Hovis et al. 1980). The CZCS recorded the various wavelengths of radiation reflected from the ocean surface. Where ocean waters contain little phytoplankton, there is limited absorption of incident radiation by chlorophyll, and the reflected radiation is blue. Where chlorophyll is abundant, the reflectance contains a greater proportion of green wavelengths (Prezelin and Boczar 1986). The reflected light is indicative of algal biomass in the upper 20–30% of the euphotic zone, where most NPP is found. CZCS images show dramatically the distribution of chlorophyll in the coastal ocean (Plate 2). The reflectance data can be used to calculate the concentration of chlorophyll and hence production (Platt and Lewis 1987, Platt and Sathyendranath 1988). The new spectral radiometer (MODIS) being developed by NASA for the Earth Observing System will allow greater satellite coverage of the world's oceans and the potential to measure long-term trends in oceanic NPP (e.g., Venrick et al. 1987).

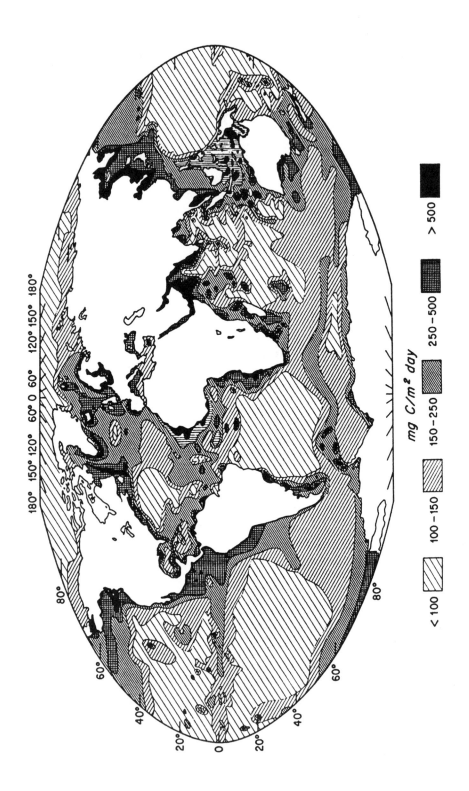

180° 150° 180°
120° 150° 180°
60° 0° 60°
150° 120°

80°

60°

20°

0°

20°

40°

60°

80°

60°

40°

mg C/m² day

< 100

100 – 150

150 – 250

250 – 500

> 500

Fate of Marine Net Primary Production

Most marine NPP is consumed by zooplankton and free-living bacteria, known as bacterioplankton, in the surface waters. Whereas the zooplankton represent the first step in a trophic chain that eventually leads to large consumers such as fish, the bacteria are consumed by a large population of bacteriovores, mineralizing nutrients and releasing CO_2 in the surface waters. Reviewing a large number of studies from marine and freshwater systems, Cole et al. (1988) found that bacterial growth (production) is about twice that of zooplankton and consumed 40% of NPP in the photic zone. Cho and Azam (1988) also concluded that free-living bacteria were more important than zooplankton in the consumption of particulate organic carbon in the ocean. Thus, a large fraction of the carbon fixed in NPP in the sea is not passed to higher trophic levels (Ducklow et al. 1986). When bacterial growth is inhibited by cold waters, more NPP is available to consumers (Pomeroy and Deibel 1986).

There is general agreement among oceanographers that about 90% of the NPP is degraded to inorganic compounds (CO_2, NO_3, PO_4, etc.) in the surface waters, and the remainder sinks below the euphotic zone to the deep ocean. The estimates of sinking are constrained, since greater rates of sinking would remove unreasonbly large quantities of nutrients from the surface ocean (Broecker 1974, Eppley and Peterson 1979). The downward flux of organic matter varies seasonally depending upon productivity in the surface water (Deuser et al. 1981). Bacterial degradation continues as particulate organic material (POM) sinks through the water column of the deep ocean. The mean sinking rate is about 350 m/day, so the average particle spends about 10 days in transit to the bottom (Honjo et al. 1982). Bacterial respiration accounts for the consumption of O_2 and the production of CO_2 in the deep water. Honjo et al. (1982) found that respiration rates averaged 2.2 mg C m^{-2} day^{-1} in the deep ocean, where the rate of bacterial respiration is probably limited by cold temperatures. About 95% of the particulate carbon is degraded within a depth of 3000 m, and only small quantities reach the sediments of the deep ocean (Martin et al. 1987). Significant rates of decomposition also continue in the sediments (Emerson et al. 1985, Cole et al. 1987, Bender et al. 1989).

If the current estimates of marine NPP are correct, then approximately $3.4–4.7 \times 10^{15}$ g C/yr sink to the deep waters of the ocean (Eppley and Peterson 1979). Martin et al. (1987) suggest that this flux may be as large as 6.0×10^{15} g/yr. From a compilation of data from sediment cores taken throughout the oceans, Berner (1982) estimates the rate of incorporation of organic carbon in sediments is 0.157×10^{15} g C/yr. These values

Figure 9.7 Global net primary production in the world's oceans. From Degens and Mopper (1976).

suggest that about 95% of the sinking organic materials are degraded in the deep ocean. Degradation of organic carbon continues in marine sediments, and the ultimate rate of burial of organic carbon in the ocean is about $0.085-0.126 \times 10^{15}$ g C/yr (Lein 1984, Berner 1982, Henrichs and Reeburgh 1987). Even the larger value is much less than 1% of marine NPP.

Maps of the distribution of organic carbon in ocean sediments are similar to maps of the distribution of net primary production in the surface waters (Fig. 9.7), except that a greater fraction of the total burial (83%) occurs on the continental shelf (Premuzic et al. 1982, Berner 1982). Isotopic analyses show that nearly all the sedimentary organic matter is derived from marine production and not from land. Indeed, degradation of river-borne organic materials must continue in the ocean, since the total burial of organic carbon in the ocean is less than the global transport in rivers, 0.4×10^{15} g C/yr (Schlesinger and Melack 1981). This has led to the curious suggestion that the ocean is a net heterotrophic system, since the ratio of total respiration to autochthonous production is >1.0 (Smith and MacKenzie 1987).

Sediment Diagenesis

Changes in the chemical composition of sediments after deposition are known as diagenesis. Organic marine sediments undergo substantial diagensis after burial as a result of sulfate reduction (Berner 1984). In organic-rich sediments sulfate reduction may begin within a few centimeters of the sediment surface as O_2 is depleted by aerobic respiration. In marine environments, sulfate reduction leads to the deposition of pyrite and the release of reduced sulfur compounds, such as H_2S, that are largely oxidized when they diffuse upward to aerobic zones. The escape of reduced gases is greater when the total rate of sulfate reduction is high (Thode-Andersen and Jørgensen 1989). The total net ecosystem production of marine ecosystems is represented by the presence of sedimentary organic matter and by sedimentary pyrite, which results from the transformation of organic carbon to reduced sulfur [equation (7.13)].

The importance of sulfate reduction is much greater in organic-rich, near-shore sediments compared to sediments of the open ocean (Skyring 1987, Canfield 1989). In a coastal marine basin, Martens and Klump (1984) found that 149 moles of carbon $m^{-2} yr^{-1}$ were deposited, of which 35.6 moles were respired annually. The respiratory pathways included 27% in aerobic respiration, 57% in sulfate reduction leading to CO_2, and 16% in methanogenesis. Globally, Lein (1984) suggests that 14% of the sedimentary organic carbon may be oxidized through sulfate reduction. However, only a small fraction of the sulfate reduction is retained as sedimentary pyrite, and the remainder of the sulfide is oxidized in aerobic zones (Jørgensen 1977).

In organic-rich sediments, the rate of pyrite formation is is often limited by the amount of available iron, since SO_4^{2-} is abundant in seawater (Boudreau and Westrich 1984). Many of these near-shore areas are characterized by high rates of sedimentation and the rapid development of anaerobic conditions in the sediments (Canfield 1989). Pelagic (open-ocean) sediments are generally aerobic (Murray and Grundmanis 1980, Murray and Kuivila 1990), and in these areas aerobic respiration exceeds sulfate reduction by a large factor (Canfield 1989). As a result, little organic matter remains to support sulfate reduction (Berner 1984). There is a strong correlation between the content of organic carbon and pyrite sulfur in most sediments (Berner 1984), but it is important to remember that the deposition of pyrite occurs at the expense of organic carbon (Fig. 1.1).

Permanent burial of reduced compounds (organic carbon and pyrite) accounts for the release of O_2 to the atmosphere. The molar ratio is 1.0 for organic carbon, but as a result of the partial oxidation of organic matter that occurs during sulfate reduction, the burial of 1 mole of reduced sulfur accounts for only about 0.5 moles of O_2 (Fig. 1.1) (Berner and Berner 1987). The weight ratio of C/S in most marine shales is about 2.8, equivalent to a molar ratio of 7.5 (Raiswell and Berner 1986). Thus, through geologic time the deposition of reduced sulfur in pyrite may account for about 7% of the O_2 in the atmosphere. As discussed in Chapter 3, the burial of these reduced substances is thought to regulate the content of O_2 in the atmosphere. As O_2 increases, the area and depth of anoxic sediments decrease.

In Chapter 7 we saw that redox potential controls the order of anaerobic metabolism by microbes in sediments. The zone of methanogenesis underlies the zone of sulfate reduction, because the sulfate-reducing bacteria are more effective competitors for reduced substrates. As a result of high concentrations of SO_4 in seawater, methanogenesis in the ocean is limited (Lovley and Klug 1986). Nearly all methanogensis is the result of CO_2 reduction, because normally acetate is depleted before SO_4 is fully removed from the sediment (Crill and Martens 1986, Whiticar et al. 1986). There is some seasonal variation in the use of CO_2 and acetate that appears to be due to microbial response to temperature (Martens et al. 1986).

Since methane is not highly soluble in seawater, even modest rates of marine methanogenesis are enough to keep ocean waters supersaturated with methane (Ward et al. 1987). A small amount of CH_4 is also released by hydrothermal vents (Charlou et al. 1988). Some methane is oxidized in the water column, and the global flux of methane from the ocean to the atmosphere, $<10 \times 10^{12}$ g/yr, is limited compared to other sources (Liss and Slater 1974, Conrad and Seiler 1988; see Table 11.1).

Biogenic Carbonates

A large number of marine organisms precipitate carbonate in skeletal and protective tissues. Clams, oysters, and other commercial shellfish are obvious examples, but a vast quantity of $CaCO_3$ is contained in foraminifera, pteropods, and other small zooplankton that are found in the sea (Krumbein 1979, Simkiss and Wilbur 1989). The coccolithophores, a group of marine algae, are responsible for a large amount of $CaCO_3$ deposited on the seafloor of the open ocean. The annual production of $CaCO_3$ by these organisms is much larger than the supply of Ca to the oceans in river flow (Broecker 1974). However, not all of the $CaCO_3$ produced is stored permanently in the sediment.

Recall that CO_2 is produced in the deep ocean by the degradation of organic materials that sink from the surface waters. Deep ocean waters are supersaturated with CO_2 with respect to the atmosphere as a result of their long isolation from the surface and the progressive accumulation of respiratory CO_2. CO_2 is also more soluble at the low temperatures and high pressures that are found in deep ocean water. (Note that CO_2 effervesces when the pressure of a warm soda bottle is released upon opening.) The accumulation of CO_2 makes the deep waters undersaturated with respect to $CaCO_3$, as a result of the formation of carbonic acid:

$$H_2O + CO_2 \rightleftarrows H^+ + HCO_3^- \rightleftarrows H_2CO_3 \qquad (9.2)$$

When the skeletal remains of $CaCO_3$−producing organisms sink to the deep ocean, they dissolve:

$$CaCO_3 + H_2CO_3 \rightleftarrows Ca^{2+} + 2HCO_3^- \qquad (9.3)$$

Their dissolution increases the alkalinity, roughly the concentration of HCO_3^-, in the deep ocean. The depth at which dissolution is complete is called the carbonate compensation depth (CCD), which is found at about 3700 m in the Atlantic and 1000 m in the Pacific, although there is wide variation (Holland 1978). The tendency for a shallower CCD in the Pacific is the result of the longer mean residence time of Pacific deep water, which allows a greater accumulation of respiratory CO_2. Small particles may dissolve totally during transit to the bottom, while large particles may survive the journey, and dissolution occurs as part of sediment diagenesis. Dissolution of $CaCO_3$ means that calcareous sediments are found only in shallow ocean basins. Of about 8.5×10^{15} g/yr of $CaCO_3$ produced in the surface layer, only about 1.5×10^{15} g is preserved in shallow, calcareous sediments (Wollast 1981). This carbonate carries about 15×10^{12} moles/yr of Ca to the sediments, which is just about enough to balance the annual input of Ca to the oceans from river flow and from hydrothermal fluids.

Many studies of carbonate dissolution have employed sediment traps that are anchored at varying depths to capture the sinking particles. In most areas, biogenic particles constitute most of the material caught in sediment traps, and most of the $CaCO_3$ is found in the form of calcite. Pteropods, however, deposit an alternative form of $CaCO_3$ known as aragonite in their skeletal tissues. The downward movement of aragonite has been long overlooked since it is more easily dissolved than calcite and often disappears from sediment traps that are deployed for long periods. As much as 12% of the movement of biogenic carbonate to the deep ocean may occur as aragonite (Berner and Honjo 1981, Betzer et al. 1984).

Geochemists have long puzzled that dolomite [$(Ca,Mg)CO_3$] does not appear to be deposited in the modern ocean, despite the large concentration of Mg in seawater and the occurrence of massive dolomites in the geologic record. There are few organisms that precipitate Mg-calcites in their skeletal carbonates, but thermodynamic considerations would predict that calcite should be converted to dolomite in marine sediments. Baker and Kastner (1981) show that the formation of dolomite is inhibited by SO_4^{2-}, but dolomite can form in organic-rich marine sediments in which SO_4^{2-} is depleted and HCO_3^- is enriched by sulfate reduction. Thus, dolomite is indirectly the result of biotic processes. Burns and Baker (1987) further show that dolomite forming in the zone of sulfate reduction contains lower concentrations of Fe and Mn, which are preferentially precipitated as sulfide minerals at these depths. Although dolomite has been a significant sink for marine Mg in the geologic past, its contribution today is minor.

Models of the Carbon Cycle in the Ocean

The surface ocean is in theoretical equilibrium with atmospheric CO_2 due to the dissolution of CO_2 in seawater to form bicarbonate (HCO_3^-). However, as a result of the uptake of CO_2 (as bicarbonate) in photosynthesis, the surface ocean remains undersaturated with respect to CO_2. Sinking organic materials remove HCO_3 from the surface ocean, and it is replaced by the dissolution of new CO_2 from the atmosphere.

The production and sinking of $CaCO_3$ also delivers calcium to the deep ocean. Most of the Ca^{2+} is derived from carbonation weathering on land and is balanced in riverwater by $2HCO_3^-$. Whether it is preserved in a shallow-water calcareous sediment or sinks to the deep ocean, $CaCO_3$ carries the equivalent of one CO_2 and leaves behind the equivalent of one CO_2 in the surface ocean; that is,

$$Ca^{2+} + 2HCO_3^- \rightarrow CaCO_3 \downarrow + H_2O + CO_2 \tag{9.4}$$

Globally the CO_2 sink in $CaCO_3$ is about four times larger than the sink in organic sediments (Li 1972), but only the organic sediments will increase in response to higher atmospheric CO_2. In the rest of the ocean, $CaCO_3$ will dissolve in the reverse of reaction (9.4), providing a sink for CO_2 in the form of dissolved HCO_3^- in the ocean.

CO_2 dissolves in water as a function of the concentration of CO_2 in the overlying atmosphere. [Recall Henry's Law, equation (2.6).] The solubility of CO_2 in seawater also depends on temperature. CO_2 is about twice as soluble at 0°C as at 20°C (Broecker 1974). Thus, CO_2 also enters the deep oceans in the downward flux of cold water at polar latitudes.

On a time scale of hundreds to thousands of years, most CO_2 in the deep ocean is returned to the atmosphere when cold, deep waters upwell at tropical latitudes. The small amount of carbon that is permanently buried in ocean sediments is released by volcanoes, following the subduction and metamorphism of sedimentary rocks on a time scale of millions of years (Fig. 2.8).

Equilibrium with ocean waters controls the concentration of CO_2 in the atmosphere, but the equilibrium can be upset when the changes in CO_2 in the atmosphere exceed the rate at which the ocean system can buffer the concentration. The seasonal cycle of photosynthesis and the burning of fossil fuels are two processes that affect the concentration of atmospheric CO_2 more rapidly than the ocean can buffer the system. As a result we observe a seasonal oscillation of atmospheric CO_2 and an exponential increase in the mean annual concentration (Fig. 1.3). Given enough time, the oceans could take up all of the CO_2 released from fossil fuels, and the atmosphere would once again show stable concentrations at only slightly higher levels than today (Laurmann 1979). As the ocean takes up additional CO_2, the pH of the ocean water is buffered at about 8.0 by the dissolution of carbonates in the reverse of reaction 9.4. Already, there is some indication that the concentration of CO_2 dissolved in the surface ocean has increased in response to increasing concentrations of atmospheric CO_2 (Fig. 9.8), but there is little evidence that the dissolution of marine carbonates has begun (Broecker et al. 1979).

A large number of models have been developed to explain the response of the ocean to higher concentrations of atmospheric CO_2 (Bacastow and Bjorkstrom 1981, Emanuel et al. 1985b). Most of these models are constructed to follow parcels of water as they circulate in a simplified ocean basin and to calculate the diffusion of CO_2 between layers that do not mix directly. Figure 9.9 shows a multibox model in which the surface ocean is divided into cold polar waters and warmer waters. In this model, cold waters mix downward to eight layers of the deep ocean, while upwelling returns deep water to the surface. The rate of mixing is calculated using oceanographic data for the rate at which ^{14}C and ^3H_2O from atomic bombs has mixed into the ocean (Killough and Emanuel 1981) and

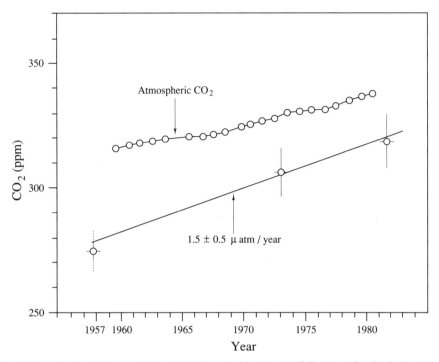

Figure 9.8 Measured changes in CO_2 dissolved in waters of the central Atlantic Ocean during the last two decades. The trend in atmospheric CO_2 is also shown for comparison. From Takahashi et al. (1983).

known constants for the dissolution of CO_2 in water as a function of temperature and pressure (Sundquist et al. 1979). The models then adjust the chemistry of the water in each layer according to the carbonate equilibrium reactions given above.

As atmospheric carbon dioxide increases, we would expect an increased dissolution of CO_2 in the oceans, following Henry's law [equation (2.6)] (Tans et al. 1990). However, the surface ocean provides only a limited volume, and the atmosphere is not in immediate contact with the much larger volume of the deep ocean. It is the rate of formation of bottom waters in polar regions that limits the rate at which the oceans can take up CO_2. One interesting result of such models is that the current rate of CO_2 release to the atmosphere exceeds the rate of buffering by the oceans, even with the most liberal assumptions regarding ocean circulation (Keeling 1983, Tans et al. 1990). These ocean models do not yet incorporate the effects of biotic productivity in the sea, nor do they incorporate the full three-dimensional complexity of ocean basins in both hemispheres of the globe, but they do allow predictions about future global conditions and hypotheses for testing.

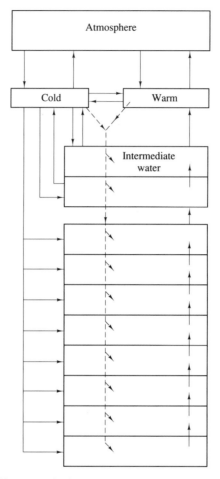

Figure 9.9 A box-diffusion model for the oceans, separating the surface oceans into cold polar waters and warmer waters at other latitudes. Cold polar waters mix with deeper waters as a result of downwelling. Other exchanges are by diffusion. From Emanuel et al. (1985b).

Nutrient Cycling in the Ocean

Net primary productivity in the sea is limited by nutrients. Net primary production is highest in regions of high nutrient availability—the continental shelf and regions of upwelling (Fig. 9.7)—and in the open ocean the concentrations of available N, P, and Si are very low. Nutrients are continuously removed from the surface water by the downward sinking of dead organisms and fecal pellets. Shanks and Trent (1980) found that 4–22% of the nitrogen found as particulates (particulate organic nitrogen, PON) was removed from the surface waters each day. Nutrients are regenerated in the deep ocean, where the concentrations are much higher. The mean residence time of N, P, and Si in the surface ocean is much less than the mean residence time of water, and there are wide differences in the concentration of these elements between the surface and the deep ocean. These are the nonconservative elements of seawater; their behavior is largely controlled by biogeochemistry.

Internal Cycles

In 1958, Albert Redfield published a paper that has served as a focal point in marine biogeochemistry for the last 30 years. Redfield noted that the organic debris falling to the deep ocean contained N and P in a fairly constant atom ratio to the content of carbon: 80C : 15N : 1P (Redfield et al. 1963). Upwelling waters contained available C, N, and P (i.e., HCO_3, NO_3, and PO_4) in a ratio of approximately 800C : 15N : 1P. Even in the face of the high productivity found in upwelling waters, only about 10% of the HCO_3 could be consumed in photosynthesis before the N and P were exhausted. The remaining HCO_3 was lost to the atmosphere as CO_2. Redfield concluded that biota controlled the movement of N and P in the ocean, and he noted that the biotic demand for N and P was closely matched to the availability of these elements in upwelling waters.

Recognizing that the downward flux of biogenic particles carries $CaCO_3$ as well as organic carbon, Broecker (1974) recalculated Redfield's ratios to include $CaCO_3$. His modified Redfield ratio in sinking particles is 120C : 15N : 1P : 40Ca. The ratio in upwelling waters is 800C : 15N : 1P : 3200Ca. Based on these quantities of N and P, net production in the surface water could remove all the N and P but only 1.25% of the Ca in upwelling waters. Although biogenic $CaCO_3$ is the main sink for Ca in the ocean, biota exert only a tiny control on the availability of Ca in surface waters. Thus, calcium is a constant, well-mixed, and conservative element in seawater.

The modified Redfield ratio allows us to calculate the importance of river flow, upward transport, and internal recycling to the net primary production of the surface ocean. Assuming that NPP in the ocean fixes about 45×10^{15} g C/yr, phytoplankton must take up $\sim 6.5 \times 10^{15}$ g N and 1.0×10^{15} g P (Table 9.2). Rivers supply $\sim 0.036 \times 10^{15}$ g N and 0.021×10^{15} g P to the oceans each year (Chapter 8). Estimates of the rate of vertical mixing in the ocean are derived from the depth distribution of radioactive isotopes from atomic bombs. About half of the upward nutrient transport from the deep ocean occurs by diffusion and eddy diffusion and the remainder from upwelling (Wollast 1981). Vertical movement supplies only a small fraction of the nutrient requirement in the surface ocean (Table 9.2); thus, nutrient recycling in the surface waters must supply $\sim 90\%$ of the nutrient requirement each year. Rapid turnover of

Table 9.2 Calculation of the Sources of Nutrients to Sustain a Global Net Primary Production of 45×10^{15} gC/yr in the Surface Waters of the Oceans[a]

Flux	Carbon (10^{12} g)	Nitrogen (10^{12} g)	Phosphorus (10^{12} g)
Net Primary Production[b]	45,000	6500	1000
Amounts Supplied			
By rivers[c]		36	21
By upwelling[d]		542	71
Recycling (by difference)		5922	908

[a] Based on an approach developed by Peterson (1981).
[b] Assuming a Redfield atom ratio of 120 : 15 : 1.
[c] Meybeck (1982).
[d] Assuming 3 m/yr following Wollast (1981), as modified by more recent data.

nutrients in the surface ocean is consistent with the rapid turnover of organic carbon that is also observed there.

Oceanographers differentiate "new production," representing NPP derived from nutrients supplied by rivers, upwelling, and upward diffusion, from "recycled production," as a result of nutrient turnover in the surface waters. How marine phytoplankton maintain high relative growth rates in waters in which the available NO_3 is below analytical sensitivity has puzzled biologists for many years. Much of the nutrient turnover in the surface waters probably occurs very rapidly. McCarthy and Goldman (1979) showed that much of the recycling in the surface waters occurs in a small zone, perhaps in a nanoliter (10^{-9} l) of seawater, which surrounds a dying phytoplankton cell. Growing phytoplankters in the immediate vicinity are able to assimilate the nitrogen that is mineralized, but the process occurs rapidly and modern analytical techniques do not allow us to see nutrient cycling on such a small scale (Shanks and Trent 1979, Goldman and Gilbert 1982).

Nitrogen and Phosphorus: Inputs and Outputs

Redfield ratios suggest that the demand for N and P by phytoplankton is closely matched to their concentrations in upwelling waters. Both elements show low concentrations in surface waters and the concentrations of N and P are correlated with a slope near the Redfield ratio. These observations suggest that both N and P might simultaneously limit marine productivity, in contrast to the widespread limitation by P in freshwaters. In fact, NPP in many ocean waters may show a tendency for limitation by available N (Howarth 1988). What processes lead to a N limitation in the sea?

Despite the importance of internal recycling and upwelling waters, the ultimate source of P to the sea is found in river flow. Much of the river P that is bound to Fe-hydroxide minerals is released from these minerals when river water mixes with seawater, which is higher in pH (Chase and Sayles 1980). An ion exchange with the high concentrations of sulfate in seawater may also play a role (Caraco et al. 1989). Since the N/P ratio in river flow is 4.4, versus the Redfield ratio of 15, we would expect that nitrogen would be limiting in nearshore areas (Ryther and Dunstan 1971), and in the absence of additional sources of N, such as atmospheric deposition or nitrogen fixation, nitrogen would be limiting in the oceans in general.

In contrast to the high rates of nitrogen-fixation by blue-green algae in freshwater habitats, N fixation in the sea is very limited (Capone and Carpenter 1982, Howarth et al. 1988a). Recall that the enzyme of nitrogen fixation requires molybdenum and iron (Chapter 2). Howarth and Cole (1985) showed that the uptake of molybdenum by phytoplankton is inhibited by the high concentrations of SO_4 in seawater. They suggest that the limited assimilation of molybdenum generally limits N fixation by bluegreen algae in the sea. In the lower concentrations of SO_4 in lake waters, blue-green algae dominate at low N/P ratios, adding nitrogen to the ecosystem through nitrogen fixation (Chapter 7).

Paerl et al. (1987) tested this hypothesis in coastal marine waters. They found that additions of Mo, Fe, and P did not stimulate N fixation, but the supply of dissolved organic carbon was critical. Inasmuch as N fixation occurs in blue-green algae, the DOC was probably not as important as an energy source as in the chelation of Mo and Fe that would increase the availability of these elements to N fixers (Howarth et al. 1988b, Paerl and Carlton 1988). Floculation of organic matter in the sea creates small microzones of anaerobic conditions in which the availability of trace micronutrients and low redox potentials could stimulate N fixation (Shanks and Trent 1979, Alldredge and Cohen 1987). Nitrogen fixation also occurs in specialized cells known as heterocysts in blue-green algae, in anaerobic microzones that develop in bundles of filamentous blue-green algae (Paerl and Bebout 1988), and in endosymbiotic bacteria in diatoms (Martinéz et al. 1983). In the latter case, N fixation was estimated to supply 14% of the total nitrogen required to support the observed NPP in the open ocean. Thus, while N fixation in the sea appears to be minor, the process and its local occurrence are deserving of further study.

The anaerobic microzones created by flocculations of organic matter, known as marine snow, also allow significant rates of denitrification in the sea, despite the high redox potential of seawater. Denitrification in a zone of low O_2 concentration in the eastern Pacific Ocean results in the loss of $50-60 \times 10^{12}$ g N/yr from the sea (Lui and Kaplan 1984, Codispoti and Christensen 1985). This denitrification is associated with the accumulation of residual nitrate in seawater that shows a high content of ^{15}N (Lui and Kaplan 1989). As we saw in terrestrial ecosystems (Chapter 6), ^{14}NO$_3$ is used preferentially as a substrate in the production of N$_2$ and N$_2$O during denitrification. The oceans also appear to be a net source of N$_2$O to the atmosphere as a result of nitrification in the water column (Cohen and Gordon 1979, Oudot et al. 1990, Kim and Craig 1990).

Additional denitrification in the oceans is also observed in sediments. Christensen et al. (1987) estimate that over 50×10^{12} g N/yr may be lost from the sea by sedimentary denitrification in coastal regions. Most of the gaseous nitrogen lost from marine sediments is N$_2$ produced by denitrification, and N$_2$O is less important (Seitzinger et al. 1984, Jørgensen et al 1984). The overall gaseous losses of nitrogen from the ocean may exceed the gaseous inputs and atmospheric deposition (Fig. 9.10), so that the oceans are currently declining in nitrogen content (McElroy 1983, Christensen et al., 1987, Smith and Hollibaugh 1989).

Limited inputs of nitrogen in river waters and by nitrogen fixation, and the potential for large losses by denitrification, all reinforce N limitation in the sea. In most areas of the ocean, nitrate is not measurable in surface waters, and phytoplankton respond to nanomolar additions of nitrogen to seawater (Glover et al. 1988). In the open ocean, direct atmospheric deposition of nitrate in rainfall and dryfall may assume special significance, since these areas are distant from rivers and upwelling. Prospero and Savoie (1989) found that 40–70% of the nitrate in the atmosphere over the north Pacific Ocean was derived from soil dusts, presumably from the desert regions of China. Desert dust also contributes P to the central ocean (Graham and Duce 1979, Duce 1983). Deposition of dust links the NPP of the ocean to the biogeochemistry of distant terrestrial ecosystems. Increased deposition of nitrate from air pollution may be responsible for higher marine NPP in some areas (Paerl 1985, Fanning 1989).

Mass-Balance Models for N and P in the Sea

Models for the N and P cycles of the ocean are shown in Figs. 9.10 and 9.11. These models offer a deceptive level of tidiness to our understanding of marine biogeochemistry, and the reader should realize that many fluxes, for example, nitrogen fixation, denitrification, and sedimentary preservation, are not known to better than a factor of 2. Nevertheless, both models show that most NPP is supported by nutrient mineralization in the surface waters and only small quantities of nutrients are lost to the deep ocean. For both elements the mean residence time of the available pool in the surface ocean is $<<1$ yr, while the mean residence time of the *total* pool in the surface ocean is about 10 yr. Thus, each atom of N and P cycles through the biota many times. Upon sinking and mineralization

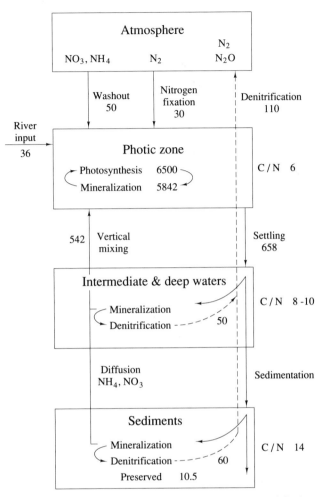

Figure 9.10 A nitrogen budget for the world's oceans. From an original conception by Wollast (1981), with newer data added for some fluxes. All values are 10^{12} g N/yr.

in the deep ocean, N and P enter pools with a mean residence time of ~500 yr. Their movement is largely controlled by the circulation of water through the deep ocean. Mineralization of N and P occurs more rapidly than C as particles settle through the deep waters, so the C/N and C/P ratios of POC increase with depth (Honjo et al. 1982, Copin-Montegut and Copin-Montegut 1983). In both models, vertical mixing includes both upwelling and upward diffusion. Upwelling accounts for about half of the global upward flux, but it is centered in coastal areas where the resulting nutrient-rich waters yield high productivity. In the open ocean, diffusion dominates the upward flux (Table 9.3), but the total supply of

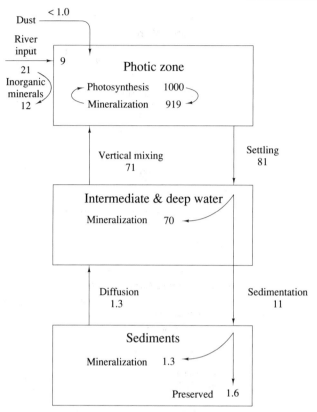

Figure 9.11 A phosphorus budget for the world's oceans. All values given in 10^{12} g P/yr. From an original conception by Wollast (1981), with new data added for dust inputs (Graham and Duce 1979), river flow (Meybeck 1982), and sedimentary preservation (Froelich 1984). Uptake and mineralization in the surface waters are calculated using an NPP of 45×10^{15} g C/yr and the Redfield atom ratio of 120/1. Other fluxes are adjusted for mass balance.

Table 9.3   Sources of Fe, PO_4, and NO_3 in Surface Waters of the North Pacific Ocean[a]

Source	Fe	PO_4	NO_3
Concentration at 150 m (μmol m^{-3})	0.075	330	4300
Upwelling (μmol m^{-2} day^{-1})	0.00090	4.0	52
Net upward diffusion (μmol m^{-2} day^{-1})	0.0034	30	400
Atmospheric flux (μmol m^{-2} day^{-1})	0.16	0.102	26
Total fluxes (μmol m^{-2} day^{-1})	0.164	34	480
Percent from advective input	0.5	12	11
Percent from diffusive input	2	88	83
Percent from atmospheric input	98	0	5

[a]From Martin and Gordon (1988).

nutrients per unit area is limited (Lewis et al. 1986, Martin and Gordon 1988). Diffusive flux is globally significant only as a result of the large area of open ocean by which it is multiplied.

Although the estimates are subject to large uncertainty, nitrogen losses to the atmosphere roughly balance inputs from all sources, and there is little loss of N to sediment. A steady state in the marine biogeochemical cycle of N is maintained by denitrification. In the absence of denitrification higher concentrations of NO_3 would be found in the ocean and lower concentrations of N_2 in the atmosphere (Chapter 12).

In contrast to N, phosphorus has no gaseous losses from the sea. At steady state, the inputs to the sea in river water are balanced by the permanent burial of phosphorus in ocean sediments. The total burial of phosphorus is probably not known within a factor of 10. Note that the model (Fig. 9.11) requires a sedimentary deposit of about 10×10^{12} g/yr in the open ocean to maintain a steady state, while actual measurements of sediment accumulation suggest a flux of about 1.0×10^{12} g/yr (Mach et al 1987). Much of the river input of P is undoubtedly deposited on the continental shelf without ever having much interaction in ocean biogeochemistry. About one-third of the ocean burial occurs as phosphorite (apatite) that is produced during sediment diagenesis (Froelich 1984). Phosphorite is formed when PO_4^{3-} produced from the mineralization of organic P combines with Ca and F to form fluorapatite. The process is apparently limited by the rate of diffusion of F^- into the sediment from the overlying waters (Froelich et al. 1982). In some areas of the ocean, phosphate nodules composed of phosphorite accumulate on the sea floor. These nodules are an enigma; they remain on the surface of the sediment despite growing at rates slower than the rate of sediment accumulation (Burnett et al. 1982). Other phosphorus is buried in organic form or in complexes with biogenic carbonates (Mach et al. 1987). All forms of buried phosphorus complete a global biogeochemical cycle when geologic processes lift the sedimentary rocks above sea level and weathering begins again. Relative to N, the global cycle of P turns very slowly.

Human Perturbations of Marine Nutrient Cycling

Through the direct release of sewage and indirect losses of fertilizers, the river input of N and P to the oceans has increased in recent years (Meybeck 1982). Fossil fuel pollutants have also increased the atmospheric deposition of N and S on the ocean surface. These inputs have enhanced the productivity of coastal and estuarine ecosystems (Chapter 8) and perhaps the productivity of the entire ocean. Greater net primary production in the surface ocean should result in a greater transport of particulate carbon to the deep ocean, potentially serving as a sink for increasing atmospheric CO_2.

Using a Redfield-ratio approach, Peterson (1981) and Peterson and Melillo (1985) have shown that the enhanced biotic sink for CO_2 in the ocean is very small. In the open ocean, NPP of 26×10^{15} g C/yr is supported by nitrogen derived from the atmosphere, from upwelling, and from internal recycling (Fig. 9.12). Peterson and Melillo (1985) estimate that an additional 6×10^{12} g N/yr are deposited in the surface waters from atmospheric pollution. That "excess" nitrogen could result in an increase in the downward flux of organic carbon of about 0.04×10^{15} g/yr, assuming a Redfield atom ratio of 120C/15N. Similar calculations using the "excess" flux of N and P in rivers suggest an increased storage of 0.05×10^{15} g C/yr in coastal zones. In the face of a net release of carbon dioxide to the atmosphere of at least 5×10^{15} g C/yr, these ocean sinks are minimal. The major ocean sink for CO_2 is found as a result of an increased dissolution of CO_2 in cold waters of the polar oceans. As we discussed earlier, this inorganic sink for CO_2 is limited by the area of polar oceans and the amount of downwelling water.

Silicon and Trace Elements

Diatoms constitute a large proportion of the marine phytoplankton, and they require silicon as a constituent of their cell walls, where it is deposited

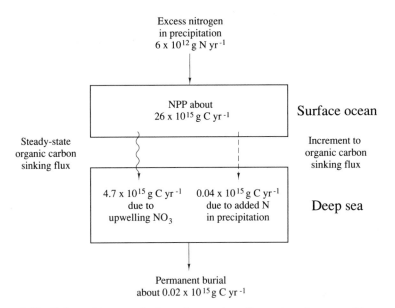

Figure 9.12 Estimated increase in the sedimentation of organic carbon caused by excess nitrogen in precipitation. From Peterson and Melillo (1985).

as opal. As a result of biotic uptake, the concentration of silicon (Si) in the surface waters is very low. Upon the death of diatoms, most of the Si is recycled in the surface waters, since the dissolution of opal is dependent on temperature (Lal 1977, Honjo et al. 1982). Concentrations increase with depth and are fairly constant in the deep ocean. The mean residence time for Si in the oceans is about 20,000 yr (Table 9.1), which is consistent with its nonconservative behavior in seawater.

DeMaster (1981) has developed a mass-balance model for Si in the ocean. Rivers (4.2×10^{14} g/yr) and hydrothermal vents (1.9×10^{14} g/yr) are the main sources, and sedimentation of biogenic opal is the only significant sink. Sedimentation in the cold waters of the Antarctic Ocean accounts for 41% of the global sink. Other cold water areas comprise most of the remaining sinks, and about 13% of the sink is found in estuaries.

Similar to the use of Si by diatoms, marine protists known as acantharians require strontium (Sr). These organisms precipitate celestite ($SrSO_4$) as a skeletal component. Uptake of Sr in surface waters and dissolution of SrO_4 from sinking acantharians confer a slightly nonconservative distribution of Sr in seawater (Bernstein et al. 1987). The mean residence time for Sr is about 12,000,000 yr (Table 9.1), but the Sr/Cl ratio varies from about 392 μg/g in surface waters to >405 with depth (Bernstein et al. 1987).

All phytoplankton require a suite of micronutrients, for example, Fe, Cu, and Zn. These elements are taken up from surface waters and mineralized as dead organisms decay and sink to the deep ocean. Many of these elements are generally insoluble at high redox potentials (Chapter 7). They are normally found at low concentrations in the surface ocean, and concentrations increase with depth (Fig. 9.13). Near the continents, the concentrations of these elements are normally adequate to support phytoplankton growth. In the central Pacific Ocean, however, Martin and Gordon (1988) found that the upward flux of iron from the deep ocean could supply only a small percent of the observed NPP. They suggest that as much as 95% of the new production in this area is supported by Fe derived from dust deposited from the atmosphere. Most of the dust is probably transported from the deserts of central China. Growth of phytoplankton appears to be limited by Fe, accounting for the small, measureable concentrations of NO_3 and PO_4 that remain in these waters during periods of peak production (Martin et al. 1989).

Relative to seawater, phytoplankton show high concentrations of Fe and other trace metals that are taken up as micronutrients (Table 9.4). Uptake and accumulation of trace metals also accounts for the tendency for some nonessential, toxic metals, such as mercury (Hg), to accumulate in phytoplankton and at higher levels of the food chain (e.g., Cross et al. 1973). Despite its toxic properties, cadmium (Cd) is well correlated with available P in waters of the Pacific ocean, implying that it is cycled by biotic

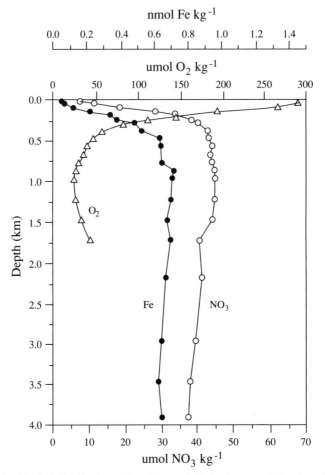

Figure 9.13 Vertical distribution of Fe, NO₃, and O₂ in the central North Pacific Ocean. From Martin et al. (1989).

processes (Boyle et al. 1976). In fact, the concentration of Cd in marine sediments is taken as an indication of the availability of P in seawater of the geologic past (Hester and Boyle 1982). Cadmium appears to substitute for zinc in biochemical molecules, allowing diatoms to maintain growth in zinc-deficient seawater (Price and Morel 1990).

When nonessential elements (e.g., Al, Ba, Hg, and Cd) and essential elements (e.g., Si and P) show similar variations in concentration with depth, it is tempting to suggest that both are affected by biotic processes, but the correlation does not indicate whether the association is active or passive. Organisms actively accumulate essential micronutrients by enzymatic uptake; other elements show passive accumulations, as a result of

Table 9.4 Ratio of the Concentration of Elements in Phytoplankton to the Concentration of Elements in Seawater[a]

Element	Ratio
Al	25,000
Cd	910
Cu	17,000
Fe	87,000
Mg	0.59
Mn	9400
N	19,000
Na	0.14
P	15,000
S	1.7
Zn	65,000

[a]From Bowen (1966).

coprecipitation or adsorption on dead, sinking particles. For instance, widespread observations of nonconservative behavior of barium (Ba) in seawater do not appear to be due to direct biotic uptake. $BaSO_4$ precipitates on dead, sinking phytoplankton, especially diatoms, as a result of high concentrations of SO_4 that surround these organisms during decomposition (Bishop 1988). In the Mediterranean Sea, aluminum shows a concentration minimum at a depth where Si and NO_3 are also depleted and O_2 is high. MacKenzie et al. (1978) suggest that this distribution is the result of biotic activity, and active uptake has been confirmed in laboratory studies (Moran and Moore 1988). Other workers have found that organic particles carry Al to the deep ocean, but the association is passive (Hydes 1979, Deuser et al. 1983). High Al in surface waters is due to atmospheric inputs of dust (Orians and Bruland 1985, 1986). Aluminum declines in concentration with depth as a result of scavenging by organic particles and by sedimentation of mineral particles.

Like Al, manganese (Mn) is found at higher concentrations in the surface waters (0.1 μg/l) than in the deep waters (0.02 μg/l) of the ocean. Based on a Mn budget for the ocean, Bender et al. (1977) attribute the high surface concentrations to the input of dust to the ocean surface. Other sources of Mn are found in river flow and in releases from hydrothermal vents (Edmond et al. 1979). The Mn budget of the ocean has long puzzled oceanographers, who recognized that the Mn concentration in ocean sediments greatly exceeds that found in the average continental rock (Broecker 1974, Martin and Meybeck 1979). Various deep-sea bacteria appear to concentrate Mn by oxidizing Mn^{2+} in seawater to Mn^{4+} that is deposited in sediment (Krumbein 1971, Ehrlich 1975, 1982). The most impressive sedimentary accumulations are seen in Mn nodules that range in diameter from 1 to 15 cm and cover portions of the sea floor (Broecker 1974, McKelvey 1980). As we discussed for phosphorus nodules, the rate of growth of Mn nodules, about 1–100 mm/million years, is slower than the mean rate of sediment accumulation, yet they remain on the surface of the sea floor. Various hypotheses invoking sediment stirring by biota have been have been suggested to explain the enigma, but none is proven. Mn nodules also contain high concentrations of Fe, Ni, Cu, and Co, and are a potential economic mineral resource.

These diverse observations suggest that the geochemistry of many trace elements in seawater is contolled directly and indirectly by biota. Cherry et al. (1978) show that the mean residence time for 14 trace elements in ocean water is inversely related to their concentration in sinking fecal pellets. Some of these elements are mineralized in the deep ocean, but the fate for many trace constituents is downward transport in organic particles to the sediments of the deep sea (Turekian 1977, Lal 1977, Li 1981). Elements with less interaction with biota remain as the major constituents of seawater (Table 9.1).

Biogeochemistry in Hydrothermal Vent Communities

At a depth of 2500 m a remarkable community of organisms is found in association with hydrothermal vents in the east Pacific Ocean. Discovered in 1977, these communities consist of bacteria, tube worms, molluscs, and other organisms, many of which are recognized as new species (Corliss et al. 1979, Grassle 1985). Similar communities are also found at hydrothermal vents in the Gulf of Mexico. In total darkness, these communities are supported by bacterial chemosynthesis, in which hydrogen sulfide (H_2S) from the hydrothermal emissions is metabolized using O_2 and CO_2 from the deep seawaters to produce carbohydrate (Jannasch and Wirsen 1979, Jannasch and Mottl 1985):

$$O_2 + H_2S + CO_2 \rightarrow CH_2O + 4S \downarrow + 3H_2O \tag{9.5}$$

Consumption of H_2S by chemosynthetic bacteria is correlated with declines in O_2 when the hydrothermal waters mix with seawater (K. S. Johnson et al. 1986). At first glance the reaction would appear to result in the production of organic matter without photosynthesis. We must remember, however, that the dependence of this reaction on O_2 links chemosynthesis in the deep sea to photosynthesis occurring in other locations on Earth. Other bacteria at hydrothermal vents employ chemosynthetic reactions based on methane, hydrogen, and reduced metals that are emitted in conjunction with H_2S (Jannasch and Mottl 1985).

On the basis of the chemosynthetic reactions, bacterial growth feeds the higher organisms found in the hydrothermal communities (Grassle

1985). Some of the bacteria are symbiotic in higher organisms. Symbiotic bacteria in the tube worm *Riftia* deposit elemental sulfur, leading to tubular columns of sulfur up to 1.5 m long (Cavanaugh et al. 1981). Filter-feeding clams up to 30 cm in diameter occur in dense mats near the vents. These communities are dynamic; a particular vent may be active for only about 10 yr. Since they are below the carbonate compensation depth, the clam shells slowly dissolve when the vent activity ceases (Grassle 1985). The offspring of these organisms must then colonize new vent systems.

Various metallic elements are soluble in the hot, low redox conditions of hydrothermal vents. Upon mixing with seawater, the precipitation of metallic sulfides removes about 96×10^{12} g S/yr from the ocean (Edmond et al. 1979; Jannasch 1989). Mn and Fe are also deposited as insoluble oxides (MnO_2, FeO) on the sea floor (Fig. 9.5). The iron oxides also act to scavenge vanadium (V) and other elements from seawater and may remove 25% of the annual riverine input of V to the ocean each year (Trefry and Metz 1989).

Hydrothermal vents attain global significance for their effect on the Ca, Mg, and SO_4 budgets of the oceans, but these bizarre chemosynthetic communities speak strongly for the potential for life in unusual locations where oxidized and reduced substances are brought together by global biogeochemical cycles.

The Marine Sulfur Cycle and Global Climate

Sulfur is abundant in the oceans, where it is found as SO_4^{2-}. Sulfate shows highly conservative behavior in seawater and a mean residence time of about 3,000,000 years relative to total inputs (Fig. 9.14; cf. Table 9.1). Except at hydrothermal vents, marine biota do not appear to be limited by available sulfur. Nevertheless, the sulfur cycle of the oceans is dynamic, and our understanding of many of its features has developed only within the last 20 yr (e.g., hydrothermal vents). Of greatest significance, the oceans are a major source of dimethylsulfide [$(CH_3)_2S$] to the atmosphere. Trace quantities of this gas impart the "odor of the sea" to coastal regions (Andreae 1986).

Dimethylsulfide (DMS) is produced during the decomposition of dimethylsulfoniopropionate (DMSP) from dying phytoplankton cells (Andreae and Barnard 1984). In an effort to balance the global sulfur cycle, DMS was first proposed as a gaseous output of the sea by Lovelock et al. (1972). In 1977, Maroulis and Bandy were able to measure DMS as an atmospheric constituent near the eastern coast of the United States. It is now widely recognized as a trace constituent in seawater and in the marine atmosphere, and the diffusion gradient across the sea–air interface indicates a global flux of $15-40 \times 10^{12}$ g S/yr to the atmosphere (Andreae and Raemdonck 1983, Ferek et al. 1986, Toon et al. 1987,

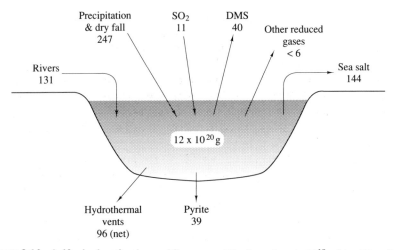

Figure 9.14 Sulfur budget for the world's oceans. All values given in 10^{12} g S/yr. River flux is from Husar and Husar (1985), gaseous outputs from Toon et al. (1987), hydrothermal flux from Jannasch (1989), and pyrite from Berner (1982). All other fluxes are from Brimblecombe et al. (1989). As a result of pollution, the present-day river flux may be $>200 \times 10^{12}$ g S/yr. (See also Fig. 13.1).

Erickson et al. 1990). This accounts for 50% of the natural emission of sulfur gases to the atmosphere globally (Möller 1984). The flux of DMS to the atmosphere would be significantly greater if it were not for microbial degradation of DMS in the surface waters (Kiene and Bates 1990). In the atmosphere, DMS is largely oxidized by OH radicals forming sulfate that is deposited in precipitation (Chapter 3). Nearly 80% of the sulfate in the atmosphere over the North Pacific Ocean appears to be derived from DMS, with the soil dust and pollution contributing the rest (Savoie and Prospero 1989). Marine DMS is estimated to contribute about 16% of the atmospheric sulfur over industrial Europe (Turner et al. 1988).

In contrast to terrestrial and freshwater wetland environments, where H_2S dominates the losses of gaseous sulfur, the oceans emit only small quantities of H_2S. Losses of carbonyl sulfide (COS) are also a small component of the marine sulfur budget (~0.5×10^{14} g S/yr), although the oceans are the major source of COS to the atmosphere (Chapter 13). Dimethylsulfide is the major form of gaseous sulfur lost from the oceans. Iverson et al. (1989) show that DMS increases in relation to increasing salinity as river water mixes with seawater in estuaries of the eastern United States.

In addition to helping balance the marine sulfur budget, dimethylsulfide attains global significance for its potential effects on climate. Charlson et al. (1987) recognized that the oxidation of DMS to sulfate aerosols would increase the cloud condensation nucleii in the atmosphere, leading to greater cloudiness (Bates et al. 1987). Clouds over the sea reflect incoming sunlight, leading to global cooling. The production of DMS is directly related to the growth of marine phytoplankton (Andreae and Barnard 1984, Turner et al. 1988). An increase in marine NPP from additions of nutrients or higher atmospheric CO_2 may increase the production of DMS. Thus, DMS has the potential to act as a negative feedback on global warming that might otherwise occur by the greenhouse effect. This hypothesis for a biotic regulation on global temperature is intriguing, for it may be responsible for the moderation of global climate throughout geologic time.

Given the strong arguments for global warming by increased atmospheric CO_2, the negative feedback mechanism of DMS is the subject of intense scientific scrutiny and debate. Schwartz (1988) argues that anthropogenic emissions of SO_2 to the atmosphere should have the same effect as natural emissions of DMS, since SO_2 is also oxidized to produce condensation nucleii. Yet he finds no evidence for increased cloudiness or cooler temperatures in the Northern Hemisphere, where most SO_2 is emitted. Using a general circulation model for global climate, Wigley (1989) found that climatic cooling by SO_2 may have offset temperature changes owing to the greenhouse effect. These mechanisms may act together. It is possible that an increased flux of both SO_2 and DMS due to human activities will act to dampen the greenhouse effect during the next century.

The Sedimentary Record of Biogeochemistry

Ocean sediments contain a record of the conditions of the oceans through geologic time. Sediments rich in $CaCO_3$ (calcareous ooze) show the location of shallow, productive seas, where foraminifera and coccolithopores were abundant. Sediments rich in opal indicate past environments of diatoms. Sediments of the deep sea are dominated by silicate clay minerals, with high concentrations of Fe and Mn (red clays). Near-shore sediments often contain abundant organic carbon that is isolated from microbial attack by the rapid accumulation of materials carried from land. Direct identification of preserved organisms and changes in their species composition have been used to infer patterns of ocean climate, circulation, and productivity during the geologic past (Weyl 1978, Corliss et al. 1986).

Calcareous sediments contain a record of paleotemperature. When the continental ice caps grew during glacial periods, the water they contained was depleted in ^{18}O, relative to ocean water, since $H_2{}^{16}O$ evaporates more readily from seawater and subsequently contributes to continental rainfall and snowfall. When large quantities of water were lost from the ocean and stored in ice, the waters that remained in the ocean were enriched in

$H_2^{18}O$, compared to today. Thus, analysis of the changes in $\delta^{18}O$ of sedimentary carbonates is an indication of ocean volume and hence of global climate (Fig. 9.15).

The sedimentary record of ^{13}C in organic matter and $CaCO_3$ contains a record of the biotic productivity of Earth. Recall that photosynthesis discriminates against $^{13}CO_2$ relative to $^{12}CO_2$ (Chapter 5), slightly enriching plant materials in ^{12}C compared to the atmosphere. When large amounts of organic matter are stored on land and in ocean sediments, the $^{13}CO_2$ that remains in the atmosphere and the ocean (i.e., $^{13}HCO_3$) is greater. Arthur et al. (1988) suggest that the relatively high ^{13}C content of

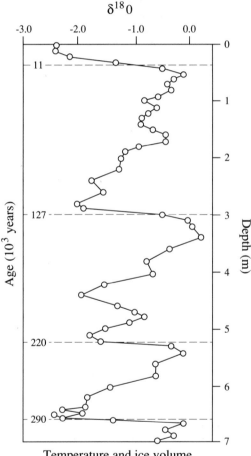

Figure 9.15   Changes in the $\delta^{18}O$ in sedimentary carbonates of the Caribbean Sea during the last 300,000 yr. Enrichment of $\delta^{18}O$ during the last glacial (20,000 ybp) is associated with lower sea level and a greater proportion of $H_2^{18}O$ in seawater. From Broecker (1973).

marine carbonates during the Late Cretaceous reflects a greater storage of organic carbon from photosynthesis. Similar effects are seen in the $\delta^{13}C$ of coal age (Permian) brachiopods (Brand 1989). When the storage of organic carbon is greater, there is the potential for an increase in atmospheric O_2, such as has also been postulated for the Permian (Berner and Landis 1984).

Summary

Biogeochemistry in the sea offers striking contrasts to that on land. The environment on land is spatially heterogeneous; within short distances there are great variations in soil characteristics, including redox potential and nutrient turnover. The sea is relatively well mixed. Large, long-lived plants dominate the primary production on land, versus small, ephemeral phytoplankton in the sea. A fraction of the organic matter in the sea escapes decomposition and accumulates in sediments, whereas soils contain little permanent storage of organic matter.

Through its buffering of atmospheric composition and temperature, the ocean exerts enormous control over the climate of Earth. At a pH of ~8.0 and a redox potential of +200 mV, seawater sets the biogeochemical conditions for 75% of the Earth's surface. Most of the major ions in seawater have long mean residence times and their concentration in seawater has been constant for nearly all of geologic time. All of this reinforces the traditional, and unfortunate, view that the ocean is a constant body that offers nearly infinite dilution potential for the effluents of modern society.

Looking at the sedimentary record, however, we see that the ocean is subject to large changes in volume and productivity, due to changes in global climate and nutrient flux. Already, we have strong reason to suspect that the productivity of coastal waters is affected by human inputs of N and P. Changes in the temperature and productivity of the central ocean basins may well indicate that global climate change is affecting the oceans as a whole (Venrick et al. 1987, Strong 1988).

Recommended Reading

Berger, W. H., V. H. Smetack, and G. Wefer. (eds.) 1989. Productivity of the Ocean: Present and Past. Wiley, New York.

Broecker, W. S. 1974. Chemical Oceanography. Harcourt Brace Jovanovich, New York.

Drever, J. I. 1988. The Geochemistry of Natural Waters. Prentice-Hall, Englewood Cliffs, New Jersey.

Holland, H. D. 1978. The Chemistry of the Atmosphere and Oceans. Wiley, New York.

PART II

Global Cycles

10

The Global Water Cycle

Introduction

The annual circulation of water is the largest movement of a chemical substance at the surface of the Earth. Through evaporation and precipitation, water transfers much of the heat energy received by the Earth from the tropics to the poles, just as a steam heating system transfers heat from the furnace to the rooms of a house. Movements of water determine the climatic patterns of the globe, and the annual availability of water is the single most important factor that determines the growth of land plants (Kramer 1982). Where precipitation exceeds evapotranspiration on land, there is runoff. Runoff carries the products of mechanical and chemical weathering to the sea.

In this chapter we will examine a general outline of the global hydrologic cycle and then look briefly at some indications of past changes in the hydrologic cycle and global water balance. Finally, we will look, speculatively, at some changes in the global water cycle that may accompany future, potential climate changes. These changes may have direct effects on global patterns of vegetation, the rate of rock weathering, and biogeochemical cycles.

The Global Water Cycle

Since the quantities of water in the global water cycle are so large, it is traditional to describe the pools and transfers in units of 1000 km³ (Fig. 10.1). Remember that each cubic meter of water weighs 1 ton, so 1000 km³ weighs 10^{18} g. The flux of water in the water cycle is also expressed in units of depth. For example, if all the rainfall on land were spread evenly over the surface, each weather station would record a depth of about 70 cm/yr. Units of depth can just as easily be used to express evaporation. The annual evaporation from the oceans removes the equivalent of 100 cm of water each year from the surface area of the sea.

Not surprisingly, the oceans are the dominant pool in the global water cycle (Fig. 10.1). Seawater contains about 97% of all the water at the surface of the Earth. The equivalent depth of seawater is ~3500 m, the mean depth of the oceans (Chapter 9). Water held in polar ice caps and continental glaciers is the next largest contributor to the global pool. In contrast, land plants and human society depend on a relatively small pool of liquid freshwater on land. The large pool of freshwater in groundwater is poorly estimated and largely unavailable. As a result of the short mean residence time of water vapor, the pool in the atmosphere is tiny, equiva-

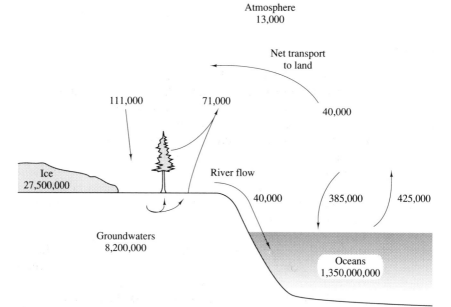

Figure 10.1 The global water cycle. Pools (km³) and flux (km³/yr) are from Spiedel and Agnew (1982).

lent to about 0.3 cm of rainfall at any given time [equation (3.4)]. Nevertheless, enormous quantities of water move through the atmosphere each year.

Evaporation from the world's oceans produces about 425,000 km³ of water vapor each year. Thus, the mean residence time of ocean water with respect to the atmosphere is about 3100 years. Only about 385,000 km³ of this returns to the oceans in rainfall; the rest contributes to precipitation on land. Plant transpiration and evaporation from soil contribute 71,000 km³ to the atmosphere. About 111,000 km³ falls as precipitation on land; the difference is derived from the sea. Since precipitation exceeds evapotranspiration on land, the excess becomes runoff (Table 8.1).

These global average values obscure enormous regional differences in the water cycle. Evaporation from the oceans is not uniform, but ranges from 4 mm/day in tropical latitudes to <1 mm/day at the poles (Mitchell 1983). Although much precipitation falls at tropical latitudes, an excess of evaporation over precipitation in the tropics carries water through the atmosphere to polar regions. Since water vapor carries latent heat, this movement through the atmosphere carries heat to high latitudes (Von der Haar and Oort 1973). Net evaporative loss accounts for the high salinity in tropical oceans (Fig. 9.3).

On land the relative balance of precipitation and evaporation differs strongly between regions. In tropical rain forests, precipitation may greatly exceed evapotranspiration. Shuttleworth (1988) calculates that 50% of the rainfall becomes runoff in the Amazon rain forests. In desert regions, precipitation and evapotranspiration are essentially equal, so there is no runoff and only limited recharge of groundwater (e.g., Phillips et al. 1988). As a global average, rivers carry about one-third of the precipitation from land to the sea.

The concept of potential evapotranspiration (PET), developed by hydrologists, expresses the maximum possible evapotranspiration that would occur under the climatic conditions of a site, if soil water were nonlimiting and plant cover were 100%. Potential evapotranspiration is greater than evaporation from an open pond, as a result of the plant uptake of water from the deep soil and a leaf area index >1.0 in many plant communities (Chapter 5). In tropical rain forests, PET and actual evapotranspiration (AET) are about equal (Vörösmarty et al. 1989). In deserts, PET greatly exceeds actual AET, since the soils are dry for much of the year. In southern New Mexico, precipitation averages about 21 cm/yr, but the receipt of solar energy could potentially evaporate over 200 cm/yr from the soil (Phillips et al. 1988).

The source of water contributing to precipitation also differs greatly in different regions of the Earth. Nearly all the rainfall over the oceans is derived from the oceans. On land, much of the rainfall in maritime and monsoonal climates is also derived from evaporation from ocean regions. In contrast, Salati and Vose (1984) calculate that 50% of the water falling in the Amazon Basin is derived from evapotranspiration in the basin, with the rest derived from long-distance atmospheric transport. The importance of regional evapotranspiration speaks strongly for the long-term implications of forest destruction in the Amazon Basin. Using a general circulation model of the Earth's climate, Lean and Warrilow (1989) show that a replacement of the Amazon rainforest by a savanna would decrease regional evaporation and precipitation and increase surface temperatures. Irreversible declines in precipitation as a result of the removal of vegetation may be responsible for the increasing desertification of semi-arid regions in the Sahel (Nicholson 1988, Schlesinger et al. 1990). Thus, the transpiration of land plants is an important factor determining the Earth's climate (Shukla and Mintz 1982).

Estimates of global river flow range from 33,500 to 47,000 km^3/yr (Speidel and Agnew 1982). Most recent workers assume a value of 42,000 km^3/yr (Lvovitch 1973). The distribution of flow among rivers is highly skewed. The 50 largest rivers carry about 43% of the total river flow, so reasonable estimates of the global transport of organic carbon, inorganic nutrients, and suspended sediments can be based on data from a few large rivers (e.g., Fig. 8.3).

As a result of the position of continents, their surface features, and global climatic patterns, there are large regional differences in the distribution of runoff to the oceans. The average runoff from North America is about 32 cm/yr, whereas the average runoff from Australia, which has a large area of internal drainage and deserts, is only 4 cm/yr (Tamrazyan 1989). Thus, the delivery of dissolved and suspended sediment to the oceans varies greatly between rivers draining the various continents (Table 4.8).

The mean residence time of the oceans with respect to riverflow is about 34,000 yr, which is 10 times less dynamic than the exchange with the atmosphere. Again, mean residence times differ among ocean basins. Mean residence time for the Pacific Ocean, 43,700 yr, is significantly longer than for the Atlantic, 9600 yr, which accounts for the greater accumulation of nutrients and shallower carbonate compensation depth in the Pacific (Chapter 9). Despite the enormous river flow in the Amazon, which carries about 20% of the annual fresh water delivered to the oceans, the continental runoff to the Atlantic ocean is less than the loss of water through evaporation. Thus, the Atlantic Ocean has a net water deficit, which is consistent with its greater salinity (Fig. 9.3). Conversely, the Pacific Ocean receives a greater proportion of the total fresh water returning to the seas each year. Ocean currents carry water from the Pacific and Indian Oceans to the Atlantic Ocean to restore the balance (Chapter 9).

Models of the Hydrologic Cycle

A variety of models have been developed to predict the movement of water through terrestrial ecosystems. Watershed models follow the fate of water received in precipitation and calculate runoff after subtraction of losses due to plant uptake (Waring et al. 1981). In these models, the soil is considered as a collection of small boxes, in which the annual input and output of water must be equal. Water entering the soil in excess of its water-holding capacity is routed to the next lower soil layer, or to the next downslope soil unit on the landscape via subsurface flow (Chapter 8). Models of water movement in the soil can be coupled to models of soil chemistry to predict the loss of elements in runoff (e.g., Nielsen et al. 1986, Knight et al. 1985). The major source of error in these models is the calculation of plant uptake and transpiration loss. This flux is usually computed using a formulation of the basic diffusion law, in which the loss of water is determined by the gradient, or vapor pressure deficit, between plant leaves and the atmosphere. The loss is also mediated by a resistance term, which includes stomatal conductance and wind speed (Waring et al. 1981). In a model of forest hydrology in western Montana, Running et al. (1989) assume that canopy conductance decreases to zero when air temperatures fall below O°C or soil water potential declines below -1.6 MPa. Their model appears to give accurate regional predictions of evapotranspiration and primary productivity for a variety of forest types.

Larger-scale models have been developed to assess the contribution of continental land areas to the global hydrologic cycle. For example, Vörösmarty et al. (1989) divide South America into 5700 boxes, each $\frac{1}{2} \times \frac{1}{2}°$ in size. Large-scale maps of each country are used to characterize the vegetation and soils in each box, and data from local weather stations are used to characterize the climate. A model (Fig. 10.2) is used to calculate the water balance in each unit. During periods of rainfall, soil moisture storage is allowed to increase up to a maximum water-holding capacity determined by soil texture (Fig. 8.1). During dry periods, water is lost to evapotranspiration, with the rate becoming a declining fraction of PET as the soil drys.

This kind of model can be coupled to other models to predict global biogeochemical phenomena. For example, a monthly prediction of the soil moisture content of the South American continent can be used with known relationships between soil denitrification and soil moisture to predict the loss of N_2O and the total loss of nitrogen to the atmosphere. The excess water in the water balance model is routed to stream channels, where it can be used to predict the flow of the major rivers draining the continent. Changes in land use and the destruction of vegetation are easily added to these models, to allow a prediction of future changes in continental-scale biogeochemistry.

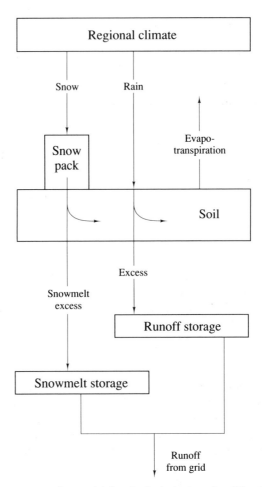

Figure 10.2 Components of a model for the hydrologic cycle of South America. From Vörösmarty et al. (1989).

The History of the Water Cycle

As we learned in Chapter 2, water was delivered to the primitive Earth in planetisimals, meteors, and comets. The accretion of the Earth was largely complete by 3.8 billion years ago (bya). Water vapor was released from the Earth's crust in volcanic eruptions (i.e., degassing). As long as the Earth's temperature was $>100°C$, water vapor was retained in the atmosphere. Water vapor confers a large greenhouse effect on the Earth, and may have slowed the rate of cooling considerably. When the Earth cooled to $<100°C$, nearly nearly all the water condensed to form the oceans. Even so, the small amount of water vapor and CO_2 in the Earth's atmosphere is enough to raise the temperature of the Earth about 33°C; without this greenhouse effect the Earth would be a frozen ball of ice.

There is good evidence of liquid oceans on Earth as early as 3.8 bya, and it is likely that the volume of water in the hydrologic cycle has not changed appreciably since that time. Owing to the low content of water vapor in the atmosphere, only 0.1% of the water on Earth appears to have been lost by the photolysis of H_2O in the upper atmosphere (Walker 1977). Much larger quantities appear to have been lost from Venus, where all water remained as vapor in the atmosphere (Kasting et al. 1988). An estimate of the total degassing of the Earth's crust suggests that 160×10^{22} g of water comprise the hydrologic cycle of today (Table 2.1). The difference between this value and the total of the pools in Fig. 10.1 is largely contained in sedimentary rocks (Table 2.1).

Throughout the Earth's history, changes in relative sea level have accompanied periods of tectonic activity that increase (or decrease) the volume of the mid-ocean ridge system. Changes in sea level also accompany changes in global temperature that lead to glaciation (Degens et al. 1981). The geologic record shows large changes in ocean volume during the 16 continental glaciations that occurred in the Pleistocene Epoch extending to 2 million years ago. During the most recent glaciation, which reached a peak 18,000 yr ago, $42,000 \times 10^3$ km^3 of seawater were sequestered in the polar ice caps (Starkel 1989). This represents 3% of the ocean volume, and it lowered sea level about 120 m from that of the present day. As we saw in Chapter 9, the Pleistocene glaciations are recorded in calcareous marine sediments. During periods of glaciation, the ocean was relatively rich in $H_2^{18}O$, which evaporates more slowly than $H_2^{16}O$. Calcium carbonate precipitated in these oceans shows higher values of $\delta^{18}O$, which can be used as an index of paleotemperature (Fig. 9.15).

Although many causes have been suggested, most workers now believe that ice ages are related to small variations in the Earth's orbit around the sun (Harrington 1987). These variations lead to differences in the receipt of solar energy, particularly at polar regions. Once polar ice begins to accumulate, the cooling accelerates, since snow has a high reflectivity or albedo to incoming solar radiation. Proponents of this theory believe that low concentrations of atmospheric CO_2 and high concentrations of atmospheric dust during the last ice age are probably an effect, rather than a cause, of global cooling, although changes in the atmosphere may have reinforced the rate of cooling (Harvey 1988). At the present time, the Earth is unusually warm; we are about halfway through an interglacial period, which should end about 12,000 A.D.

Continental glaciations represent a major disruption—a loss of steady-state conditions—in the Earth's water cycle. These changes in global climate appear to have affected the circulation of the oceans and the interaction of oceans with the atmosphere (Chapter 9). Global cooling yields lower rates of evaporation, reducing the circulation of moisture through the atmosphere and reducing precipitation. One model of global climate suggests that 18,000 yr ago, total precipitation was 14% lower than today (Gates 1976). Throughout most of the world, the area of deserts expanded (Thompson et al. 1989, Petit et al. 1990). Total net primary productivity on land may have been much lower (Shackleton 1977), and greater wind erosion of desert soils contributed to the accumulation of dust in ocean sediments and loess deposits (Chapter 3). The southwestern United States appears to have been an exception. Over most of this desert area, the climate of 18,000 yr ago was wetter than today (Van Devender and Spaulding 1979, Wells 1983, Marion et al. 1985).

Changes in the rate of global river flow produce changes in the delivery of dissolved and suspended matter to the sea. Broecker (1982) suggests that erosion of exposed continental shelf sediments during the glacial sea-level minimum may have led to a greater nutrient content of seawater and higher marine net primary productivity in glacial times. Worsley and Davies (1979) show that deep sea sedimentation rates throughout geologic time have been greatest during periods of relatively low sea level, when a greater area of continents is displayed.

The Water Cycle under Scenarios of Future Climate

Models of the radiation balance of the atmosphere, known as general circulation models (Chapter 3), predict an increase in global temperature as a result of increasing concentrations of the "greenhouse" gases—CO_2, CH_4, and chlorofluorocarbons. The exact magnitude and distribution of the climate change is controversial, since the interactive effects of clouds and atmospheric aerosols are poorly known. In response to global warming, however, climate models generally predict a more humid world, in which the movement of water in the hydrologic cycle through evaporation and precipitation is enhanced. Increased cloudiness may moderate the degree of warming, but a new steady state would be found at a higher mean global temperature than today (Raval and Ramanathan 1989). Not all areas of the land will be affected equally. Most of the temperature change is confined to high latitudes, and Manabe and Wetherald (1986) show that large areas of the central United States and Asia will experience a reduction in soil moisture, leading to more arid conditions. Such changes in precipitation and temperature will lead to large-scale adjustments in the distribution of vegetation and global net primary production (Emanuel et al. 1985a).

Are changes in the hydrologic cycle consistent with an indication of global warming over the last century? Analyzing the rainfall records of 1487 weather stations, Bradley et al. (1987) find an increase in precipi-

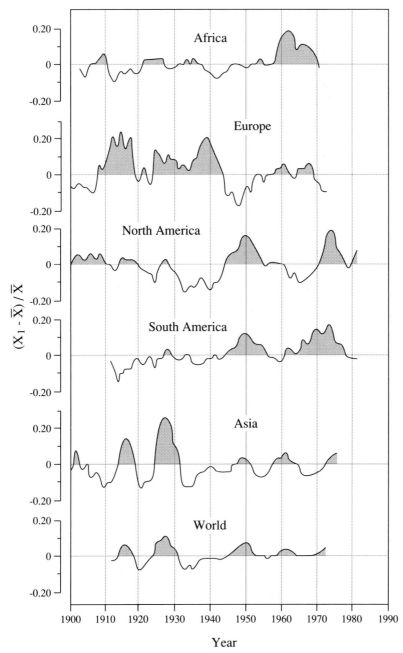

Figure 10.3 A comparison of fluctuations in riverflow draining various continents and averaged for the world. Variation is expressed as the difference between an annual value and the long-term mean, as a fraction of the long-term mean. From Probst and Tardy (1987).

tation over most of the mid-latitudes in the Northern Hemisphere in the last 30–40 years. Their data also show a decrease in precipitation over North Africa and the Middle East—consistent with the expansion of desert in the Sahel. Over the last 65 yr, Probst and Tardy (1987) show a 3% increase in stream flow in major rivers. Increased stream flow may relate directly to greater precipitation or to the human destruction of vegetation leading to greater runoff (Chapter 8). We might also speculate that greater stream flow is due to greater water-use efficiency by vegetation growing in a high-CO_2 atmosphere (Chapter 5; Idso and Brazel 1984).

The historical pattern of runoff for each continent and for the world as a whole shows a cyclic pattern (Fig. 10.3). The cycles for the continents are not synchronous, so the cycles in the global record are "damped," relative to those of the continents. Recent increases in precipitation and stream flow are consistent with predicted changes in the water cycle with global warming, but such observations must be evaluated in the context of long-term cycles in climate that have occurred through geologic time.

Most measurements of meteorological conditions and runoff are made locally, and global extrapolations are difficult. Likewise, current estimates of the volume of ice in the Greenland and Antarctic ice caps are now derived from local studies that show conflicting evidence of change. Our understanding of the global hydrologic cycle and global climate change will improve as these measurements are made at the global scale. The Earth Observing System (EOS) satellite is scheduled to include several microwave sensors (e.g., AMSU). Since water vapor absorbs microwaves, the relative transmission of microwaves through the atmosphere is related to water vapor content and rainfall. This satellite will allow the first integrated measurement of total annual rainfall and begin a long-term record of global rainfall. At the same time the MODIS sensor (Chapter 9) will monitor the global extent of vegetation types and sea surface temperatures, allowing calculation of evapotranspiration. MODIS will also record the extent of snow and ice to improve our understanding of changes in the polar ice caps that may be related to global warming (Chapter 9). These satellites should allow us to refine our understanding of the global water cycle (e.g., Fig. 10.1) and to perceive the rate of global change.

Summary

Through evaporation and precipitation the hydrologic cycle transfers water and heat throughout the global system. Receipt of water is one of the primary factors controlling net primary production on land. Changes in the hydrologic cycle through geologic time are associated with changes in global temperature. All evidence suggests that movements in the hydrologic cycle were slower in glacial time, but they are likely to increase with climatic warming. Movements of water on the surface of the Earth affect the rate of rock weathering and other biogeochemical phenomena.

Recommended Reading

Baumgartner, A. and E. Reichel. 1975. The World Water Balance. R. Olenburg, Munich.

Berner, E.K. and R.A. Berner. 1988. The Global Water Cycle. Prentice-Hall, Englewood Cliffs, New Jersy.

Sumner, G. 1988. Precipitation: Process and Analysis. John Wiley and Sons, New York.

Ward, R.C. 1970. The Principles of Hydrology. McGraw-Hill, New York.

11

The Global Carbon Cycle

Introduction

The carbon cycle is of central interest to biogeochemistry. First, living tissue is primarily composed of carbon, so studies of the global carbon cycle in the past and present give an index of the health of the biosphere. Second, the fixation of carbon by plants through geologic time accounts for the O_2 in our present atmosphere, which sets the oxidation potential for the entire planet. Through oxidation and reduction reactions, the cycles of other elements are closely tied to the global cycles of carbon and oxygen. Finally, there is good evidence that through the burning of fossil fuels and other activities, humans have altered the global cycle of carbon to produce conditions that have not been seen during the past several million years of Earth history.

In this chapter we will consider a simple model for the carbon cycle of the Earth and the human impacts on that cycle. We will then consider the magnitude of past fluctuations in the carbon cycle to gain some perspective of the current human impact. We will look briefly at the budget of methane (CH_4) and carbon monoxide (CO) in the atmosphere. Since increasing concentrations of carbon dioxide and methane are associated with global warming through the greenhouse effect (Fig. 2.4), the global carbon cycle is directly linked to considerations of global climate change. Finally, we will examine the linkage of the carbon and oxygen cycles on Earth.

The Modern Carbon Cycle

The largest fluxes of the global carbon cycle are those that link atmospheric carbon dioxide to land vegetation and to the oceans (Fig 11.1). Considering the land vegetation alone, we find that each molecule of CO_2 in the atmosphere has the potential to be consumed in gross photosynthesis in about 6 yr. The flux of CO_2 to the oceans is of similar magnitude, so the overall mean residence time of CO_2 in the atmosphere is about 3 yr. This mean residence time is close to the mixing time for the atmosphere, so CO_2 shows regional and seasonal differences in concentration that are superimposed on a global average concentration of about 350 ppm (Chapter 3).

Oscillations in the atmospheric content of CO_2 are the result of the seasonal uptake of CO_2 by photosynthesis and seasonal differences in the use of fossil fuels and in the exchange of CO_2 with the ocean. Globally, about two-thirds of the terrestrial vegetation occurs in regions with seasonal periods of growth, and the remainder occurs in the moist tropics

The Global Carbon Cycle

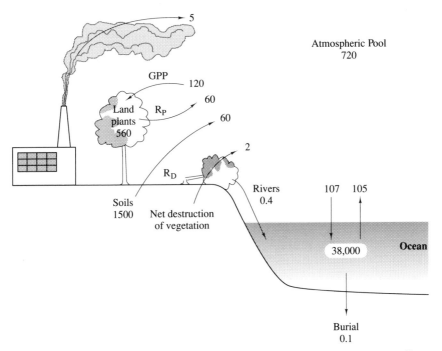

Figure 11.1   The present-day global carbon cycle. All pools are expressed in units of 10^{15} g C and all annual fluxes in units of 10^{15} g C/yr.

(Table 5.2) (Box 1988). The seasonal effect of photosynthesis is most pronounced in the northern hemisphere, which contains most of the world's continental area and temperate vegetation (Fig. 3.6). At high, northern latitudes, vegetation accounts for about 50% of the annual variation in atmospheric CO_2 (D'Arrigo et al. 1987). In the southern hemisphere, the small fluctuations in atmospheric CO_2 appear to be dominated by exchange with ocean waters (Keeling et al. 1984). The oscillation at Mauna Loa, Hawaii (19°N), is about 7 ppm (Fig. 1.3), representing a transfer of about 15×10^{15} g C to and from the atmosphere. We should expect this value to be less than annual net primary productivity (60×10^{15} g C/yr) owing to the asynchrony of terrestrial photosynthesis throughout the globe and buffering of atmospheric CO_2 concentrations by the ocean.

The release of CO_2 in fossil fuels, 5×10^{15} g C yr^{-1}, is one of the best-known values in the global carbon cycle (Rotty and Masters 1985). If all this CO_2 accumulated in the atmosphere, the annual increment would be about 0.7%/yr. In fact, the atmospheric increase is about 0.4%/yr (1.5 ppm), so only 58% of the fossil fuel release accumulates in the atmosphere. This constitutes the "airborne fraction." Where is the remainder?

Using the best models of ocean circulation and CO_2 dissolution in seawater, oceanographers believe that about 40% of the CO_2 released from fossil fuels enters the ocean each year. Thus, the uptake by the oceans (107×10^{15} g/yr) is slightly greater than the return of CO_2 to the atmosphere (105×10^{15} g C/yr). The additional CO_2 dissolves in sea water following Henry's Law [equation (2.6)] and is buffered by the dissolution of marine carbonates (Chapter 9). If the release of CO_2 were curtailed, nearly all the CO_2 that has accumulated in the atmosphere would eventually dissolve in the oceans, and the global carbon cycle would return to a steady state. It is the rate of current release relative to the rate at which the oceans can buffer the global cycle that accounts for the current increase in the atmosphere.

Remembering that the exchange of CO_2 between the atmosphere and the oceans takes place only in the surface waters (Chapter 9), we can calculate the mean residence time of CO_2 in the surface ocean by dividing the pool of carbon in these waters (630×10^{15} g C) by the rate of flux (107×10^{15} g C/yr)—about 6 yr. A similar mixing time is calculated from the distribution of ^{14}C in the surface ocean (Stuiver 1980). Turnover of carbon in the entire ocean is much slower, about 350 yr—consistent with the mixing of deep ocean waters (Chapter 9).

Taken alone, this paints a fairly clear picture of our understanding of the global carbon cycle. Many terrestrial ecologists, however, are not so sanguine. They believe that there have been substantial releases of CO_2 from terrestrial vegetation, caused by the destruction of forest vegetation in favor of agriculture, especially in the tropics (Chapter 5). A net release of carbon from land is suggested by measurements of $\delta^{13}C$ in tree rings and ice cores, which both show a decline in atmospheric $^{13}CO_2$ that is consistent with the destruction of organic carbon on land (Fig. 5.11). Globally the net release from land appears to be about 1.8×10^{15} g C/yr (Houghton et al., 1987). Thus, in the present-day carbon cycle, gross photosynthesis (120×10^{15} g C/yr) is slightly less than total respiration (122×10^{15} g C/yr) on land (Fig. 11.1). If these calculations are accurate, then the atmospheric budget is misbalanced, and a large amount of carbon that ought to be in the atmosphere is missing (Fig. 11.2).

Tans et al. (1990) recently estimated the oceanic uptake of CO_2 as 1.6×10^{15} g C/yr—only about 30% of the annual release from fossil fuels. Their analysis, based on measurements of the partial pressure of CO_2 in the atmosphere and of the alkalinity in seawater, suggests that substantially less CO_2 dissolves in the North Atlantic Ocean than most previous workers have assumed. Downwelling waters north of 50° latitude are estimated to carry 0.23×10^{15} g C/yr to the deep ocean. Most of the rest enters the deep ocean near Antarctica. Independently, Brewer et al. (1989) estimated that 0.26×10^{15} g C/yr moves southward in the deep Atlantic crossing an east-west transect at 25° latitude. Their value substantiates the limited dissolution of CO_2 in the North Atlantic Ocean (Tans et al. 1990). A small sink for CO_2 in the ocean further complicates the balance of the global carbon budget (Fig. 11.2).

Since no current model of the oceans can accomodate an uptake of more than about 6×10^{15} g C/yr (Keeling 1983), oceanographers believe either that the estimates of carbon lost from land are too high or that the rate of photosynthesis by the remaining vegetation is stimulated by higher atmospheric CO_2 concentrations. The extent of stimulation is informally known as the "beta" factor, but there is little evidence to support it (Chapter 5). Thus, our current understanding of the carbon cycle is incomplete, speaking strongly for how poorly we understand the global biogeochemical system. At least 1 billion tons of carbon are lost from our accounting each year!

About half of the carbon fixed by land plants (gross primary production) is respired by the plants themselves, so net primary production is

Net emissions = Net changes in the carbon cyle

$$\underset{\text{fuel}}{\text{Fossil}} + \underset{\substack{\text{of land} \\ \text{vegetation}}}{\text{Destruction}} = \underset{\text{increase}}{\text{Atmospheric}} + \underset{\text{uptake}}{\text{Oceanic}} + \underset{\text{sink?}}{\text{Unknown}}$$

$$5 \quad + \quad 1.8 \quad = \quad 3 \quad + \quad 1.6 \quad + \quad 2.2$$

Figure 11.2 An attempt to balances sources and sinks of CO_2 in the atmosphere shows the misbalance in current budgets for the global carbon cycle. All data are expressed in 10^{15} g C/yr.

only 60×10^{15} g C/yr (Chapter 5). Estimates of current terrestrial bio-mass, 560×10^{15} g C, yield a mean residence time of 9 yr in live biomass. Decomposition of dead plant materials returns CO_2 to the atmosphere. There is little long-term storage in soil organic matter, and transport of organic carbon in rivers is a minor component of the global cycle (Schle-singer and Melack 1981). Assuming a steady state, that is, net primary production equal to decomposition, the mean residence time of dead materials on land is about 25 yr (Schlesinger 1977). The pools of organic matter on land are large, so small changes in their size yield large impacts on the atmosphere. A 2% increase in the rate of net photosynthesis on land would balance the carbon cycle if the organic carbon were not subsequently lost by decomposition.

One approach to estimating net primary production and the potential for a beta factor is through an examination of the amplitude of the annual oscillation in atmospheric CO_2. Since the seasonal decline in atmospheric CO_2 is partially the result of photosynthesis, while the seasonal upswing is partially due to decomposition, an increasing amplitude of the oscillation, after the removal of fossil fuel effects, implies a greater activity of the terrestrial biosphere. Such a trend is evident in an analysis of the Mauna Loa record of CO_2 (Fig. 11.3), in which the *amplitude* has increased by about 0.7%/yr (Bacastow et al. 1985). Variations in the amplitude of the seasonal oscillation of CO_2 appear to result, in part, from the effects of El Niño on terrestrial primary production (Keeling et al. 1989).

Higher productivity of terrestrial vegetation may result from a direct effect of CO_2 on photosynthesis and through a fertilization effect of N, P, and S in the global system. Using a model of the global cycle, Kohlmaier et al. (1989) suggest that as much as 25% of the increase in the amplitude is

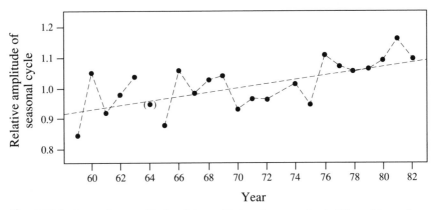

Figure 11.3 Increasing amplitude of the oscillations in atmospheric CO_2 at Mauna Loa, Hawaii. From Bacastow et al. (1985).

due to vegetation response, sequestering as much as 1.3×10^{15} g C/yr on land. If correct, their analysis would help balance current models of the carbon cycle. Other workers disagree, finding no evidence for a sufficient, higher supply of N and P to support increased plant growth on land or in the ocean (Likens et al. 1981, Peterson and Melillo 1985). Moreover, an increasing amplitude does not necessarily imply a greater rate of carbon storage on land. Greater rates of decomposition may balance greater rates of photosynthesis (Houghton 1987).

In our view of the global carbon cycle, it is important to recognize that the annual movements of carbon, rather than the amount stored in various reservoirs, are most important. The ocean contains the largest pool of carbon near the surface of the Earth ($38,000 \times 10^{15}$ g), but most of that pool is not involved with rapid exchange with the atmosphere. Similarly, desert soil carbonates contain more carbon than land vegetation, but the exchange between desert soils and the atmosphere is tiny (0.01×10^{15} g C/yr), yielding a turnover time of 85,000 years (Schlesinger 1985b).

All explanations for increasing concentrations of atmospheric CO_2 must rely on documented, recent changes in the carbon cycle. A flux that has not changed in recent times, no matter how large, is not likely to explain the increase in the atmosphere (Houghton et al. 1983). For example, the release of CO_2 in forest fires is of no consequence to changes in atmospheric CO_2 unless the frequency or area of forest fires has changed in recent times (Adams et al. 1977, Seiler and Crutzen 1980, Kaufman et al. 1990). The carbon flux in rivers or sinking pteropods cannot serve as a net sink for anthropogenic CO_2 in the ocean, unless the flux in these pathways has increased in recent years. The largest global pool of carbon is found in sedimentary rocks, including the fossil fuels. Storage of organic carbon in these deposits accounts for the accumulation of O_2 in the atmosphere through geologic time (Chapter 3). In the absence of human perturbations, the exchange between the fossil pool and the atmosphere could be ignored in global models. Humans affect the global system by creating a large biogeochemical flux where none existed before.

Future changes in the distribution of vegetation as a result of global climate change may reinforce the release of CO_2 that results from fossil fuels and from the destruction of current vegetation. For example, Emanuel et al. (1985a) predict a 6.7-17% increase in the world area of desert land, which presumably will have lower biomass and net primary production than the vegetation it replaces (Schlesinger et al. 1990). Warming of tundra and boreal regions should increase rates of decomposition and CO_2 emission to the atmosphere (Billings et al. 1982). A 1% increase in the rate of decomposition on land would result in the release of nearly 1×10^{15} g C/yr to the atmosphere (Schleser 1982, Schlesinger

1984). Although changes in the distribution of vegetation and in ecosystem function due to climatic change do not appear responsible for much of the increase in atmospheric CO_2 between 1860 and the present, they are potentially important factors to consider for the future (Esser 1987).

Temporal Perspectives of the Carbon Cycle

Studies of the biogeochemistry of carbon must begin with the origin of carbon as an element and with theories that explain its differential abundance on the planets (Chapter 2). These are non-steady-state views; the carbon content of the planet grows with the receipt of planetisimals and meteorites, and the atmospheric content increases as volcanoes release CO_2. The history of atmospheric CO_2 is a good index of the global carbon cycle, since the atmosphere is directly linked to most other compartments, including the biosphere, and rapidly responds to changes in their function (Fig. 11.1). The oldest geologic sediments suggest that atmospheric CO_2 may have been as high as 3% on the primitive Earth, providing a substantial greenhouse effect during a time of low solar output (Walker 1985a). Even today, 350 ppm of atmospheric CO_2 raises the surface temperature of the Earth above freezing and is essential for the persistence of the biosphere (Ramanathan 1988).

A comparison among the planets provides one end to a spectrum of views of the carbon cycle. How did the Earth avoid the "runaway" greenhouse effect that has raised the surface temperature on Venus far above that conducive for life (Chapter 2)? This longest view suggests some stabilizing, or steady-state components, in the global carbon cycle on Earth.

One mechanism for maintaining relatively constant, low concentrations of atmospheric CO_2 is through its interactions with the cycle of carbonate and silicate rocks (Fig. 2.8). Using a long-term model for the Earth, Berner et al. (1983) suggest that this cycle has maintained the concentrations of atmospheric CO_2 between 200 and 6000 ppm for the last 100 million years (Berner and Lasaga 1989). On Mars, where this cycle has slowed or stopped, the atmosphere contains a small amount of CO_2, and the planet is cold (Chapter 2). On Venus, where CO_2 cannot react with crustal minerals, the atmosphere contains a large amount of CO_2, and the planet is very hot (Nozette and Lewis 1982). During periods of extensive volcanism on Earth, the atmospheric concentration of CO_2 on Earth may have been higher, leading to warmer climates (Owen and Rea 1985). Although atmospheric CO_2 on Earth has fluctuated through geologic time, it has remained within limits that produce moderate surface temperatures.

Interactions among the carbon and sulfur cycles provide a further mechanism that buffers atmospheric CO_2 within narrow limits, as illustrated in the model of Garrels and Lerman (1981) (Fig. 1.1). Their model suggests that if atmospheric CO_2 were higher, greater rates of photosynthesis and storage of organic carbon in marine sediments should follow. Although the model does not consider nutrient limitations in the ocean, it shows that fluctuations in atmospheric CO_2 are likely to be small and short-lived, because the atmosphere is in rapid exchange with other compartments of the global carbon cycle.

Holland (1965) points out that during the last several million years, neither gypsum ($CaSO_4 \cdot 2H_2O$) nor dolomite [$(Ca,Mg)CO_3$] has been an important constituent of marine sedimentary rocks. This sets the limits of atmospheric CO_2 between 200 and 1300 ppm, since concentrations of CO_2 greater than 1300 would lead to the precipitation of dolomite, as well as calcite, in the shallow oceans, while concentrations less than 200 ppm would lead to the deposition of gypsum. Holland's view suggests that biogeochemical cycles have buffered atmospheric CO_2 within rather narrow limits during the evolution of most of the species found on the Earth today, including humans.

Collections of gas trapped in ice cores from the Antarctic provide a historical record of atmospheric CO_2 for the last 160,000 yr (Fig. 11.4).

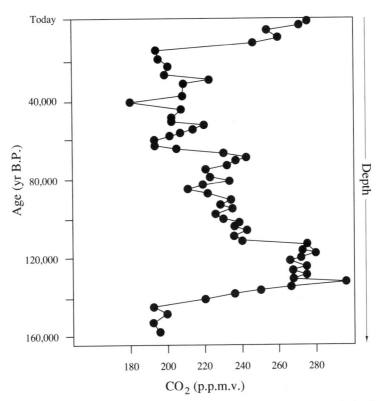

Figure 11.4 Record of atmospheric CO_2 during the last 160,000 yr, as derived from analysis of gas bubbles trapped in the Antarctic ice core. Modified from Barnola et al. (1987).

Concentrations have varied between 200 and 280 ppm, with the lowest values found in layers of ice that were deposited during the most recent ice age. Changes in ocean currents may have been linked to changes in atmospheric CO_2 during the transition from glacial to interglacial conditions (Chapter 9). From the beginning of the industrial age, the atmospheric content of CO_2 has increased from about 270 ppm to about 350 ppm today. This represents a global change of 30% in less than 200 yr! Although the current level of CO_2 is not unprecedented in the geologic record, our concern is the speed at which a basic characteristic of the planet has changed to levels not previously experienced during human history or during the evolution of current ecosystems. Steady-state conditions have been upset.

These perspectives of the global carbon cycle extend from processes that occur on a time scale of 10^9 yr to those that occur annually. Buffering of atmospheric CO_2 over geologic time involves small net changes in carbon storage that occur slowly. For example, the net storage of organic carbon in marine sediments has varied between about 0.03 and 0.08 10^{15} g C/yr during the last 300 million years (Berner and Raiswell 1983); currently the rate is about 0.10×10^{15} g C/yr (Chapter 9). This flux buffers atmospheric CO_2 over geologic time. In contrast, the current flux of CO_2 through the atmosphere is about 200×10^{15} g C/yr, which buffers CO_2 on an annual basis. The global carbon cycle is composed of large, rapid transfers superimposed on an underlying set of smaller, slow transfers. The current change in atmospheric CO_2 results from the ability of humans to change the flux of CO_2 to the atmosphere by an amount that is significant relative to the biogeochemical reactions that buffer the system over short periods of time.

Atmospheric Methane

Fluxes of methane are a minor component in the global carbon cycle, since all are in the range of 10^{13} to 10^{14} g/yr, which is several orders of magnitude less than the values of Fig. 11.1. Globally the atmospheric methane concentration is 1.7 ppm, versus 350 ppm for CO_2; however, methane is currently increasing at about 1%/yr, which is much faster than the rate of CO_2 increase (Fig. 3.7). Each molecule of methane in the atmosphere contributes 20-fold to the Earth's greenhouse warming relative to each molecule of CO_2 (Lacis et al. 1981, Ramanathan et al. 1985, Lashof and Ahuja 1990). Thus, changes in methane have the potential to contribute strongly to global climate change in the future (Dickinson and Cicerone 1986).

The current increase in atmospheric methane adds about 40×10^{12} g/yr to a global pool of about 4.8×10^{15} g. The cause of the current increase in methane is not obvious, since natural sources appear to dominate the annual production of $\sim 500 \times 10^{12}$ g/yr (Table 11.1). This estimate of total flux is fairly robust, for it yields a mean residence time for atmospheric CH_4 of about 10 yr, which is consistent with independent calculations based on methane consumption (Cicerone and Ormeland 1988, Khalil and Rasmussen 1990). The concentration of CH_4 is slightly greater in the northern hemisphere (Steele et al. 1987, Dianov-Klokov et al. 1989), so reactions that affect methane production are likely to be found in that hemisphere.

Despite methane oxidation in surface soils, methanogenesis from wetland habitats is the dominant source of atmospheric methane (Chapter 7). Matthews and Fung (1987) estimate 110×10^{12} g/yr from anaerobic decomposition in natural wetlands, indicating that about 60% of the total methane production appears to derive from peat bogs of 50–70° N latitude (Whalen and Reeburgh 1990a). Although arriving at a similar global flux, Aselmann and Crutzen (1989) found that tropical wetlands comprised a larger fraction of the total. Their estimate of 31×10^{12} g/yr emitted between 10°N latitude and 10°S latitude is in general agreement with an extrapolation of field measurements from the Amazon Basin by Bartlett et al. (1988).

The production of methane in wetland ecosystems shows a seasonal cycle that is correlated with temperature (Aselmann and Crutzen 1989).

Table 11.1 Budget for Atmospheric Methane[a]

Source or Sink	CH_4 (10^{12} g CH_4/yr)
Sources	
Natural wetlands	115
Open freshwaters	5
Rice paddies	110
Animals	80
Termites	40
Oceans	10
Anthropogenic	
Biomass burning	55
Landfills	40
Coal mining	35
Natural gas	45
Methane hydrate	5
Total sources	540
Sinks	
Reactions with OH	490
Soil microbes	10
Atmospheric increase	40
Total sinks	540

[a] From Cicerone and Oremland (1988), Aselmann and Crutzen (1989), and Steudler et al. (1989).

Emissions from seasonal ecosystems are likely to be the major contributor to the annual oscillation observed in methane concentrations in the atmosphere, and thus to total methane flux. Surprisingly, the annual oscillation in the northern hemisphere shows a minimum concentration in midsummer (Steele et al. 1987). J. O. Wilson et al. (1989) show the greatest rates of methane production in the early spring and late summer for a swamp in Virginia. Emissions from wetlands may be lower in midsummer as a result of higher O_2 concentrations from algal photosynthesis and methane oxidation in the surface waters (King 1990).

An increase in methane flux from boreal peatlands may be expected if climatic warming lengthens the season of microbial activity and the zone of saturated soil moisture changes due to permafrost melting. Significantly, Khalil and Rasmussen (1989) observe that methane concentrations in the atmosphere were lower during the last ice age, when most of the current area of boreal peatlands was beneath the continental ice sheet (cf. Raynaud et al. 1988, Chappellaz et al. 1990). Catastrophic release of methane from marine sediments, where it is held as methane hydrate, might yield a large increase in greenhouse warming in the future (Revelle 1983, MacDonald 1990).

Changes in the distribution of wetlands may be related to increases in atmospheric methane over the last century. While many wetlands have been drained, Harriss et al. (1988) found that the current management of wetland areas in southern Florida has potentially enhanced the flux of methane to the atmosphere. A large portion of the current increase in atmospheric methane may derive from an increase in the worldwide area of rice cultivation. Since most rice paddies are found in warm climates, they often yield a large CH_4 flux, which is enhanced by the upward transport through the hollow stems of rice (Seiler et al. 1984a, Schütz et al. 1989a).

Many grazing animals and termites maintain a population of anaerobic microbes that conduct fermentation at low redox potentials in their digestive tract. Digestion in these animals provides the equivalent to a mobile wetland soil! The flatulence of grazing animals makes a significant contribution to the global sources of methane (Table 11.1). Crutzen et al. (1986) estimate 78×10^{12} g/yr from domestic and wild animals. Humans contribute 1×10^{12} g/yr. Early suggestions of a large flux of methane from termites (Zimmerman et al. 1982) have now largely been discounted (Seiler et al. 1984b, Fraser et al. 1986, Khalil et al. 1990), and Crutzen et al. (1986) show that the current increase in atmospheric methane is not related to larger herds of grazing animals.

The ocean is a minor source of atmospheric methane, despite being supersaturated with methane in the surface waters (Chapter 9). In any case, it is unlikely that there has been a change in the flux from the ocean during the last century. Inadvertent releases during the production of natural gas account for about 20% of the annual flux, based on the ^{14}C age of atmospheric methane (Ehhalt 1974, Wahlen et al. 1989). Releases of natural gas and biomass burning appear to have increased the $\delta^{13}C$ of atmospheric methane from a preindustrial value of approximately -50% to the -47% that is observed today (Craig et al. 1988, Quay et al. 1988).

In sum, while the flux from wetlands may have increased in recent years, it is unclear that any source of methane has changed enough to yield an increase in the flux to the atmosphere of at least 40×10^{12} g/yr. Thus, many biogeochemists believe that the current increase in the atmosphere is due to a decrease in the reactions that remove methane. The major process removing methane is through reaction with OH radicals (Table 11.1). A reduction in the strength of this methane sink may be related to releases of carbon monoxide as a pollutant (Chapter 3).

Many aerobic bacteria, methanotrophs, consume methane in soils (Keller et al. 1983, Yavitt et al. 1990b, Whalen and Reeburgh 1990b). Consumption of methane is limited by the rate of diffusion of methane into the soil profile (Born et al. 1990). Globally, these bacteria are thought to consume about 10×10^{12} g of methane annually (Aselmann and Crutzen 1989, Steudler et al. 1989). Recent changes in land use may have reduced the rate of CH_4 consumption in tropical soils, contributing to the increase in the atmosphere (Keller et al., 1990).

Methanotrophic bacteria can outcompete nitrifying bacteria for O_2 in soils where methane is abundant (Megraw and Knowles 1987). Conversely, some nitrifying bacteria can also oxidize CH_4 in the soil atmosphere (Jones and Morita 1983, Hyman and Wood 1983). Steudler et al. (1989) suggest that the consumption of CH_4 by nitrifying bacteria may be lower in forests that currently receive a large atmospheric deposition of NH_4, since the NH_4/CH_4 ratio has greatly increased in these regions. However, even if the consumption of methane by soil bacteria has declined in recent years, it is difficult to imagine that the previous sink was larger by 40×10^{12} g/yr (Table 11.1). Thus, as for the sources of methane, our current understanding of methane sinks sheds little light on the underlying causes of increasing atmospheric concentrations.

Carbon Monoxide

The annual release of CO during the burning of fossil fuels (640×10^{12} g CO/yr) is a significant fraction (5%) of the total carbon released during combustion, but carbon monoxide makes only a minor contribution to other aspects of the global carbon cycle. About half of the global production of carbon monoxide ($\sim 3000 \times 10^{12}$ g/yr, Warneck 1988) is derived from the burning of fossil fuels and biomass (Seiler and Conrad 1987, Kaufman et al. 1990). Thus, the current increase in atmospheric carbon monoxide is easily linked to human activities (Kahlil and Rasmussen 1988). Carbon monoxide is also produced indirectly during the oxidation of natural hydrocarbons and methane (Chapter 3). Concentrations are much lower in the southern hemisphere than in the northern hemisphere, where CO is increasing at about 1.5–2%/yr (Khalil and Rasmussen 1988, Dianov-Klokov et al. 1989). Carbon monoxide has a short atmospheric lifetime (60 days, Table 3.4), and is largely consumed by reaction with OH radicals (Chapter 3).

Carbon monoxide is an insignificant greenhouse gas. The main concern with its increase in the atmosphere is associated with the production of ozone during the reaction of CO with OH radicals (Chapter 3). Increases in atmospheric ozone over the tropical regions of South America appear to be related to the production of CO by forest burning, followed by reaction of CO with OH radicals in the atmosphere (Fishman and Browell 1988).

Synthesis: Linking the Carbon and Oxygen Cycles

During the history of the Earth, atmospheric O_2 first appeared following the advent of autotrophic photosynthesis, and began to accumulate when the annual production of O_2 exceeded its reaction with reduced crustal minerals (Chapter 2). The current atmospheric pool of O_2 is only a small fraction of the total produced over geologic time (Fig. 2.6). All O_2 that has been produced is balanced stoichiometrically by the storage of reduced organic carbon (or sedimentary pyrite) in the Earth's crust [equation (5.1)]. The current atmospheric pool of O_2 is maintained in a dynamic equilibrium between the production of O_2 by photosynthesis and its consumption in respiration (Fig. 11.5). The pool of atmospheric O_2 is well buffered, since increases in O_2 expand the area and depth of aerobic respiration in marine sediments, leading to a greater consumption of O_2 (Chapters 3 and 9). The small amount of organic matter that escapes oxidation and is buried in the sea is balanced over geologic time by the uplift and weathering of organic carbon in sedimentary rocks. Like the carbon cycle, the oxygen cycle is composed of large, annual fluxes superimposed on smaller, slow fluxes that maintain a steady state through long periods of time (Walker 1984). Unlike carbon, human perturbations of the O_2 cycle are obscured by the large size of the atmospheric pool.

The oxygen cycle is also linked to the nitrogen cycle. Globally, about 14% of the annual consumption of O_2 is used to oxidize NH_4 in the nitrification reaction (Walker 1980). Of course, NH_4 exists in soils as a result of the plant uptake and assimilatory reduction of NO_3, which requires energy. So the oxidation of NH_4 consumes O_2 that might otherwise go to the oxidation of organic carbon.

It is interesting to note the role of anaerobic respiration in regulating these cycles. As calculated in Chapter 9, the formation of pyrite through sulfate reduction reduces the storage of organic carbon in marine sediments. Similarly, methanogensis in anaerobic sediments returns CH_4 to the atmosphere, where it is oxidized (Henrichs and Reeburgh 1987). Methane oxidation accounts for about 4% of the total consumption of atmospheric O_2 each year (Walker 1980). In the absence of methanogen-

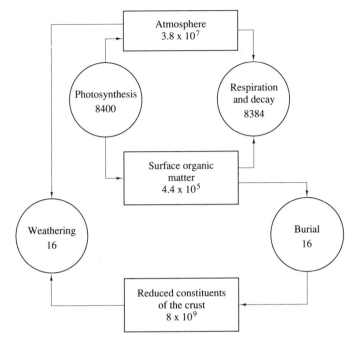

Figure 11.5 A simple model for the global biogeochemical cycle of O_2. Data are expressed in units of 10^{12} moles of O_2 per year or the equivalent amount of reduced compounds. Note that a small misbalance in the ratio of photosynthesis to respiration can result in a net storage of reduced organic materials in the crust and an accumulation of O_2 in the atmosphere. From Walker (1980).

sis, the burial of organic carbon would be greater and the atmospheric content of O_2 might be slightly higher (Watson et al. 1978, Kump and Garrels 1986).

It is also entertaining to speculate whether the carbon cycle on Earth drives the oxygen cycle, or vice versa. Over geologic time, the answer is obvious: the conditions on our neighboring planets are ample evidence that O_2 is derived from life. Now, however, the cycles are inextricably linked, and the discussion seems merely philosophical. The metabolism of eukaryotic organisms, including humans, depends on the flow of electrons from reduced organic molecules to oxygen.

Recommended Reading

Houghton, J.T., G.T. Jenkins, and J.J. Ephraums. (eds.). 1990 Climate Change: The IPCC Scientific Assessment. Cambridge University Press, Cambridge.

Trabalka, J.R. (ed.). 1985. Atmospheric Carbon Dioxide and the Global Carbon Cycle. U.S. Department of Energy, Washington, D.C.

Woodwell, G.M. (ed.). 1984. The Role of Terrestrial Vegetation in the Global Carbon Cycle: Measurement by Remote Sensing. Wiley, New York.

12

The Global Cycles of Nitrogen and Phosphorus

Introduction

The availability of nitrogen and phosphorus controls many aspects of global biogeochemistry. Nitrogen often limits the rate of net primary production on land and in the sea. Nitrogen is an integral part of the enzymes that control the biochemical reactions in which carbon is reduced or oxidized. Phosphorus is an essential component of DNA, ATP, and the phospholipid molecules of cell membranes. Changes in the availability of N and P through geologic time have controlled the size and activity of the biosphere.

A large number of biochemical transformations of nitrogen are possible, since nitrogen is found at valence states ranging from -3 (in NH_3) to $+5$ (in NO_3^-). Various microbes capitalize on the potential for transformations between these states, and use the energy released by the changes in redox potential to maintain their life processes (Rosswall 1982). Collectively these microbial reactions drive the global cycle of nitrogen (Fig. 12.1). The most abundant form of nitrogen at the surface of the Earth, N_2, is the least reactive species. Various processes convert atmospheric N_2

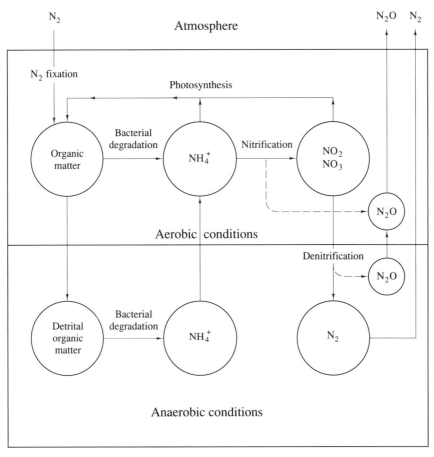

Figure 12.1 Microbial transformations of nitrogen in the global cycle. From Wollast (1981).

to one of the forms of "fixed" nitrogen that can be used by biota. Other bacteria return nitrogen to the atmosphere as N_2.

In contrast, whether it occurs in nature or in biochemistry, phosphorus is almost always found in combinations with oxygen (i.e., as PO_4^{3-}), and the availability of phosphorus is largely controlled by rock weathering and the geochemical reaction of P with soil minerals. These lower the solubility of phosphorus in water, leaving only small quanitites for biota. On land and in the sea, there are few responses of biota to increase the rate of weathering or the availability of P. Transformations of organic phosphorus in soils and ocean waters constitute a biogeochemical cycle that exists on top of the unrelenting flow of phosphorus from weathered rock to ocean sediments. The global P cycle is complete only when sedimentary rocks are lifted above sea level and weathering begins again.

In this chapter we will examine our current understanding of the global cycles of N and P. We will attempt to balance an N and P budget for the world's land area and the sea. For N, the rate of fixation through geologic time determines the nitrogen available to biota and the global biogeochemical cycle. We will review ideas about the rate of nitrogen fixation and denitrification in the geologic past. One of the products of nitrification and denitrification is N_2O (nitrous oxide), which is both a greenhouse gas and a cause of ozone destruction in the stratosphere (Chapter 3). We will formulate a tentative budget for N_2O in the atmosphere, based on our current, limited understanding of the sources of this gas.

The Global Nitrogen Cycle

Land

Figure 12.2 presents the global nitrogen cycle, showing the linkage between the atmosphere, land, and sea. The atmosphere is the largest pool

The Global Nitrogen Cycle

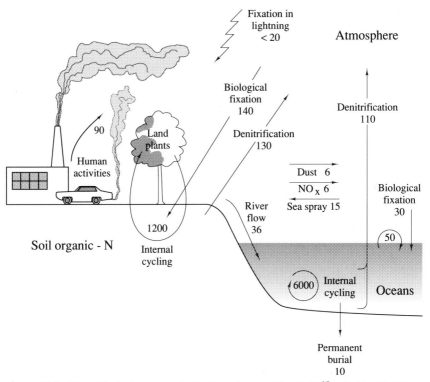

Figure 12.2   The global nitrogen cycle. Pools and annual flux in 10^{12} g N. Modified from Söderlund and Rosswall (1982) based on values derived in the text.

$(3.8 \times 10^{21}$ g N; Table 3.1). Relatively small amounts of N are found in terrestrial biomass $(3.5 \times 10^{15}\text{g})$ and soil organic matter $(95 \times 10^{15}$ g; Post et al. 1985). The mean C/N ratios for terrestrial biomass and soil organic matter are about 160 and 15, respectively. The pool of inorganic nitrogen, NH_4^+ and NO_3^-, on land is very small. The transformations of nitrogen in the soil and the uptake of N by organisms are so rapid that little nitrogen remains in inorganic form, despite a large annual flux through this pool (Chapter 6).

The nitrogen that bathes the terrestrial biosphere is not available to most organisms; the great strength of the triple bond in N_2 makes this molecule practically inert. All nitrogen that is available to the terrestrial biota was originally derived from nitrogen fixation—either by lightning or by free-living and symbiotic microbes (Chapter 6). The rate of nitrogen fixation by lightning, which produces momentary conditions of high pressure and temperature allowing N_2 and O_2 to combine, is poorly known. Most estimates range from 10 to 30×10^{12} g N/yr (Levine et al. 1984), although some recent workers suggest that the rate may be as high as 80 to 100×10^{12} g/yr (Miller et al. 1989, Franzblau and Popp 1989). Only a portion of the nitrogen fixed by lightning is carried to land in precipitation; the remainder is deposited in the sea. Thus, a liberal estimate of abiotic N fixation on land is about 20×10^{12} g N/yr. The annual deposition of fixed nitrogen in precipitation is much larger than 20×10^{12} g/yr; rainfall also contains nitrogen that was fixed in previous years and circulates in the atmosphere from dust, seaspray, volatilized ammonia, etc. (Chapter 3).

Estimates of total biological nitrogen fixation on land range from 44 to 200×10^{12} g N/yr, with a median value of about 140×10^{12} g N/yr—about 10 kg N/yr for each hectare of the Earth's land surface (Burns and Hardy 1975). Most studies of asymbiotic nitrogen fixation on land report values ranging from 1 to 5 kg ha^{-1} yr^{-1} (Boring et al. 1988). A value of 3 kg N ha^{-1} yr^{-1} multiplied by the world's land area suggests that asymbiotic fixation contributes about 44×10^{12} g N/yr to the global total. The remainder is assumed to come from symbiotic fixation in higher plants, and the global estimate may be somewhat too high given the sporadic occurrence of symbiotic nitrogen fixation outside the tropics (Chapter 6). In any case, biotic N fixation exceeds abiotic fixation by a large factor. Taking all forms of N fixation as the only source, the mean residence time of nitrogen in the terrestrial biosphere is about 625 yr (i.e., pool/input).

Assuming that the estimates of terrestrial net primary production, 60×10^{15} g C/yr, are roughly correct and that the mean C/N ratio of net primary production is about 50,[1] the nitrogen requirement of land plants

---

[1] Most primary production consists of short-lived tissues with a C/N ratio that is much lower than that of wood (~160), which comprises most of the terrestrial biomass.

is about 1200×10^{12} g/yr (Chapter 6). Thus, nitrogen fixation supplies about 12% of the nitrogen that is made available for plant use each year. The remaining nitrogen must be derived from internal recycling and the decomposition of dead materials in the soil (Chapter 6). When the turnover in the soil is calculated with respect to the input of dead plant materials, the mean residence time of nitrogen in soil organic matter is about 50 yr.

Human impact on the global nitrogen cycle is significant. The production of nitrogen fertilizers through the Haber process essentially duplicates the fixation of nitrogen in lightning. Fertilizer production supplies about 40×10^{12} g N/yr to terrestrial ecosystems. High-pressure internal combustion in automobile engines also fixes about $40-60 \times 10^{12}$ g N/yr (Rosswall 1981, Warneck 1988). Owing to the short residence time of NO and NO_2 in the atmosphere, most of this nitrogen is deposited by precipitation over land (Chapter 3). Forest ecosystems downwind of major population centers now receive enormous nitrogen inputs that may be related to their decline (Aber et al., 1989, Schulze 1989). Rivers draining industrial regions also show high concentrations of nitrate (Table 8.4).

In total, nitrogen fixation delivers a net increment of about $160-250 \times 10^{12}$ g N/yr to land. In the absence of processes removing nitrogen, a very large pool of nitrogen would be found on land in a relatively short time. Rivers carry about 36×10^{12} g N/yr from land to the sea (Chapter 8). This flux may be larger by about 7×10^{12} g N as a result of human pollution (Meybeck 1982), but in any case, river flow cannot account for a large proportion of the nitrogen that is lost from land. The remaining nitrogen is assumed to be lost by denitrification in terrestrial soils (Chapter 6) and wetlands (Chapter 7) and by N_2 released by forest fires (Lobert et al. 1990).

Estimates of global denitrification from uplands and freshwater wetlands range from 13 to 233×10^{12} g N/yr (Bowden 1986). Our attempt to balance the terrestrial nitrogen budget would suggest that a global rate of $>130 \times 10^{12}$ g N/yr is most likely (cf. Rosswall 1981). At least half of the denitrification on land occurs in wetlands (Bowden 1986). Most of the loss occurs as N_2, but small fractions lost as NO and N_2O during nitrification and denitrification contribute significantly to the global budgets of these gases (Chapter 6).

In balancing the terrestrial N cycle, we concentrate on processes that affect the net movement of fixed nitrogen. Thus, ammonia volatilization (Chapter 6) can be ignored to the extent that the volatilized NH_3 is deposited on land in precipitation. Since NH_3, NO, and NO_2 all have relatively short atmospheric lifetimes, they are usually deposited in precipitation and dryfall near their point of origin (Chapter 3).

The combustion of fossil fuels appears to result in the net transfer of fixed nitrogen (NO_x) from land to sea, where it is deposited in wetfall (Fig. 9.10). Soil dust also carries fixed nitrogen to the sea, whereas seaspray returns an unknown amount of fixed nitrogen to land. All of these transfers appear to be relatively minor components of the global nitrogen cycle (Fig. 12.2).

Sea

The world's oceans receive about 36×10^{12} g N/yr from rivers, about 30×10^{12} g N/yr via biological N fixation, and about 50×10^{12} g N in precipitation (Fig. 9.10). Note that while the flux in rivers is a rather small component of the terrestrial cycle, it contributes about one-third of the total nitrogen delivered annually to the sea. In the surface ocean, the pool of inorganic nitrogen is very small. As we have shown for terrestrial ecosystems, most of the net primary production in the sea is supported by nitrogen recycling in the water column (Table 9.2). The deep ocean contains a large pool of inorganic nitrogen, derived from the decomposition of organic matter. Permanent burial of organic nitrogen in sediments is small, so most of the input to the oceans must be returned to the atmosphere as N_2 by denitrification. Important areas of denitrification are found in the anaerobic deep waters of the eastern Tropical Pacific Ocean and the Arabian Sea (Chapter 9). Losses of N_2O by nitrification and denitrification in ocean waters are also significant (Hahn 1981).

Temporal Variations in the Global Nitrogen Cycle

The earliest atmosphere on Earth is thought to have been dominated by nitrogen, since it is abundant in volcanic emissions and poorly soluble in seawater (Chapter 2). Before the origin of life, nitrogen was fixed by lightning and in the shock waves of meteors, which create local conditions of high temperature and pressure in the atmosphere (Mancinelli and McKay 1988). The rate of N fixation was very low, perhaps about 6% of present-day rates, because fixation in an atmosphere dominated by N_2 and CO_2 is much slower than in an atmosphere of N_2 and O_2 (Kasting and Walker 1981). The best estimates of abiotic fixation suggest that it had a limited effect on the content of atmospheric nitrogen, but it provided a small but important supply of fixed nitrogen, largely $NO_3{}^-$, to the waters of the primitive Earth (Kasting and Walker 1981, Mancinelli and McKay 1988).

The present-day rate of N fixation by lightning could consume the atmospheric content of N_2 in about 100,000,000 yr. The mean residence time of atmospheric nitrogen decreases to about 20,000,000 yr when biological nitrogen fixation is included. In either case, the supply of O_2 would be exhausted long before the process was complete, unless there were compensating changes in the biosphere (Delwiche 1970). The high rate of N fixation in an aerobic atmosphere speaks strongly for the effect of denitrification in returning N_2 to the atmosphere.

The origin of denitrification is uncertain. Mancinelli and McKay (1988) argue for its appearance before the advent of atmospheric O_2, suggesting that a facultative tolerance of O_2 evolved later. Others suggest that denitrification is more recent (Broda 1975, Betlach 1982). These investigators point out that denitrifying bacteria are facultative anaerobes, switching from simple heterotrophic respiration to NO_3 respiration in anaerobic conditions. Even the denitrification reaction itself is somewhat tolerant of O_2 (Bonin et al. 1989).

Requiring nitrogen as a reactant, nitrification clearly arose after photosynthesis and the development of an O_2-rich atmosphere. In any case, the major microbial reactions in the nitrogen cycle (Fig. 12.1) are all likely to have been in place at least 1 billion years ago. Today, the rate of denitrification is controlled by the rate of nitrification, which supplies NO_3 as a substrate (Fig. 6.10).

Because NO_3 is very soluble in seawater, there is little reliable record of changes in the content of NO_3^- in seawater through geologic time. Only changes in the deposition of organic nitrogen are recorded in sediments. Recently, Altabet and Curry (1989) suggested that the $^{15}N/^{14}N$ record in sedimentary foraminifera may be useful in reconstructing the past record of ocean chemistry. The isotope ratio in sedimentary organic matter changes as a result of changes in the global rate of nitrogen fixation and denitrification in the ocean (Chapter 6).

Assuming a steady state in the ocean nitrogen cycle, the mean residence time for an atom of N in the sea is about 8000 yr. During this time, this atom will make several trips through the deep ocean, each lasting 200–500 yr (Chapter 9). Since the turnover of N is much longer than the mixing time for ocean water, NO_3 shows a relatively uniform distribution in deep ocean water. In a provocative paper, McElroy (1983) suggests that the oceans are not presently in steady state; the rate of denitrification exceeds known inputs. He argues that the oceans received a large input of nitrogen during the continental glaciation 10,000 yr ago, and they have been recovering from this input ever since. His suggestion is consistent with sedimentary evidence of greater net primary production in the oceans during the last ice age (Broecker 1982) and with observations of $^{15}N/^{14}N$ in sedimentary foraminifera (Altabet and Curry 1989).

McElroy's paper serves to remind us of several important aspects of biogeochemistry and the global cycle of N. First, while an assumption of a steady state is useful in the construction of global models, such as Fig. 12.2, it is sometimes not realistic. As we saw for the carbon cycle, the current increase in atmospheric CO_2 implies non-steady-state conditions. Recent changes in the global nitrogen cycle are also likely. Humans have greatly accelerated the rate of N fixation, and the atmospheric content of N_2O is increasing rapidly. It is unlikely that the global nitrogen cycle is now in balance (Delwiche 1970). Secondly, without denitrification, the rates of nitrogen fixation would gradually remove nitrogen from the atmosphere and cause nitrate to accumulate in ocean waters. Denitrification closes the global biogeochemical cycle of nitrogen, but it also means that nitrogen remains in short supply for the biosphere.

Nitrous Oxide: An Unbalanced Global Budget

Currently, biogeochemists are devoting a large research effort toward understanding the global budget of nitrous oxide, N_2O. This trace atmospheric constituent has a mean concentration of 300 ppb, which indicates a global pool of 2.3×10^{15} g N_2O or 1.5×10^{15} g N in the atmosphere. The concentration is increasing at an annual rate of 0.3% (Fig. 3.9). Each molecule of N_2O has the potential to contribute 200-fold to the greenhouse effect relative to each molecule of CO_2, so the current increase in the atmosphere has potential consequences for global climate change (Lacis et al. 1981, Ramanathan et al. 1985, Lashof and Ahuja 1990). The only known sink for N_2O, stratospheric destruction (Chapter 3), consumes about 10.5×10^{12} g N as N_2O per year (Logan et al. 1981, Keller et al. 1986). Thus, the mean residence time for N_2O is about 150 yr, consistent with observations of global variability in the concentration of N_2O in the atmosphere (Fig. 3.4). Unfortunately, estimates of sources—particularly sources that have changed greatly in recent years—are poorly constrained.

Based on apparent supersaturation of N_2O in seawater, emissions from the ocean dominated the earliest global estimates of N_2O sources (Liss and Slater 1974, Hahn 1974). When more extensive sampling showed that the area of supersaturation was limited, these workers substantially lowered their estimate of N_2O production in marine ecosystems (Table 12.1) (Hahn 1981, Liss 1983, Bolle et al. 1986, Butler et al. 1989). Similarly, emissions from fossil fuel combustion and forest fires appear to have been overestimated and contribute little to the global budget (Bolle et al. 1986, Muzio and Kramlich 1988, Linak et al. 1990, Hegg et al. 1990).

Soil emissions from nitrification and denitrification are now thought to comprise most of the global source of N_2O (Chapter 6). Keller et al. (1986) suggested that these emissions are greatest in the tropics, where they comprise 6.1×10^{12} g N/yr. Recently, Matson and Vitousek (1990) have lowered the global estimate for tropical soils to 3.7×10^{12} g N/yr, which is still greater than the total estimated flux from temperate forests and other regions (Bowden 1986, Schmidt et al. 1988). Even so, the known sources of N_2O do not balance the stratospheric destruction plus the observed increase in the atmosphere, yielding an unbalanced global budget (Table 12.1).

Table 12.1   Tentative Balance of N_2O in the Atmosphere

Source or Sink	N_2O (10^{12} g N/yr)	Reference
Sources		
Oceans	2.0	Bolle et al. (1986)
Natural soils		
Tropics	3.7	Matson and Vitousek (1990)
Others	2.0	Bolle et al. (1986), Bowden (1986), Schmidt et al. (1988)
Fertilized agriculture	0.7	Eichner (1990)
Change in cultivated land	0.7	Matson and Vitousek (1990)
Biomass burning	2.0	Hegg et al. (1990)
Fossil fuels	0.0	Muzio and Kramlich (1988), Linak et al. (1990)
Total sources	11.1	
Sinks		
Reaction with O_3	10.5	Keller et al. (1986)
Atmospheric increase	3.0	Keller et al. (1986)
Total sinks	13.5	

In a search for additional N_2O sources, especially those that have increased in recent years, Robertson et al. (1988) examined N_2O emission from a range of tropical soils in Costa Rica to see if changes in land use might account for an increase in the loss of N_2O to the atmosphere. Emissions were greatest from undisturbed forests and agricultural fields, but N_2O loss was relatively low in fallow land that was returning to forest. They concluded that the current emissions from this region are probably *lower* than in precolonial times. Looking globally, Matson and Vitousek (1990) estimate an N_2O flux of 0.7×10^{12} g N/yr as a result of land use conversion in other areas of the humid tropics. In agricultural lands, fertilizer application increases the production of N_2O in the soil (Bremner and Blackmer 1978, Conrad et al. 1983, Slemr et al. 1984), perhaps accounting for the increase in global sources of N_2O that lead to increasing concentrations in the atmosphere. Based on the use of fertilizer in worldwide agriculture, Eichner (1990) estimates a release of 0.7×10^{12} g N as N_2O from fertilized land in 1984. Ronen et al. (1988) suggest that groundwater may also be an important source of N_2O to the atmosphere. Denitrifying bacteria were found at a depth of 289 m in coastal South Carolina (Francis et al. 1989). As we saw for methane (Chapter 11), however, it is difficult to see how any of the sources of N_2O have changed to the extent needed to explain an increase in the atmosphere of about 3×10^{12} g N/yr.

Cores extracted from the Antarctic ice cap show that the concentration of N_2O was much lower during the last ice age (Khalil and Rasmussen

1989). At the end of the last ice age concentrations rose to about 270 ppb and remained fairly constant until the Industrial Revolution, when they increased to the present-day value of about 300 ppb (Zardini et al. 1989). The weak seasonal oscillation in atmospheric N_2O suggests that the major source might be associated with the seasonal activity of biota (Fig. 3.9). Khalil and Rasmussen (1989) suggest that the role of wetlands should be further investigated. Relatively high rates of N_2O emission have been observed in a variety of wetlands (Bowden 1986). Deglaciation uncovered large areas of boreal peatland, perhaps accounting for the increase in atmospheric N_2O at the end of the last glaciation. Concentrations of atmospheric N_2O may now be increasing as global warming affects wetlands in tundra and boreal regions.

The Global Phosphorus Cycle

The global cycle of P is unique among the cycles of the major biogeochemical elements in having no significant gaseous component (Fig. 12.3). The redox potential of soils is too high to allow the production of phosphine gas (PH_3; Bartlett 1986), except in very specialized, local conditions (e.g., Dévai et al. 1988). Transfers through the atmosphere in soil dust and seaspray are also several orders of magnitude less important than other transfers in the global P cycle (Graham and Duce 1979).

Unlike transfers in the global nitrogen cycle, the major transfers in the global cycle of P are not driven by microbial reactions. Nearly all the phosphorus on land is originally derived from the weathering of calcium phosphate minerals, especially apatite [$Ca_5(PO_4)_3OH$]. Root exudates and mycorrhizae may increase the rate of rock weathering on land (Chapter 4), but there are no significant responses of biota to limited supplies of P in water (Chapter 7). Although the total P content of soils is large, in most soils only a small fraction is available to biota (Chapter 4). On both land and sea, the biota persist as a result of a well-developed recycling of phosphorus in organic form (Fig. 6.13).

The main flux of P in the global cycle is carried in rivers, which transport about 21×10^{12} g P/yr to the sea (Meybeck 1982). This flux may be slightly higher than in prehistoric time as a result of erosion, pollution and fertilizer runoff. Nearly all of the flux is found in particulate form. Although the concentration of PO_4^{3-} in the surface oceans is low, the large volume of the deep oceans contributes a substantial pool of P to the global cycle. The mean residence time for P in the sea is 4000–80,000 yr, depending on whether dissolved or total river inputs are used as inputs (Fig. 12.3) (Froelich et al. 1982). The turnover through the organic pools in the surface ocean is only a few days (Chapter 9). Eventually, phosphorus is deposited in ocean sediments, which contain the largest pool near the surface of the Earth. On a time scale of hundreds of millions of

The Global Phosphorus Cycle

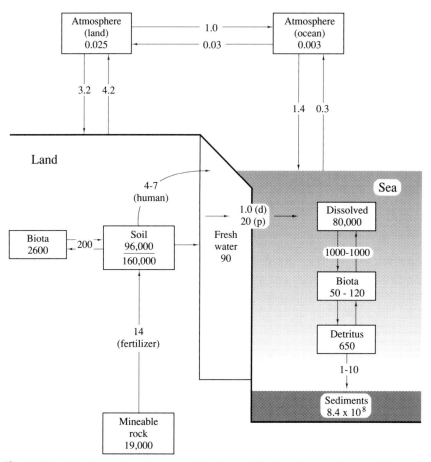

Figure 12.3 The global phosphorus cycle. Pools in 10^{12} g P and annual flux in 10^{12} g P/yr. Modified from Richey (1983), based on data from Meybeck (1982), Graham and Duce (1979), and Fig. 9.11.

years, these sediments are uplifted and subject to rock weathering, completing the global cycle.

In many areas humans have enhanced the availability of P by mining phosphate rocks that can be used as fertilizer. Most of the economic deposits of phosphate are found in sedimentary rocks of marine origin, so the mining activity directly enhances the turnover of the global cycle. In the United States, some of the largest deposits of phosphate rock are found in Florida and North Carolina.

On the primitive Earth, all phosphorus was contained in igneous rocks. Griffith et al. (1977) calculate that it took over 3 billion years for the weathering of igneous rocks to saturate seawater with respect to apatite. The solubility product of apatite is only about 10^{-58} (Lindsay and Vlek 1977), so at pH 8.0, the phosphorus concentration would be about 10^{-8} M (molar) (Fig. 4.3). We would expect that phosphorus has always been in short supply to marine biota. In fact, organic and colloidal forms of P maintain its concentration in excess of that in equilibrium with respect to apatite. The average P content of deep ocean water is about 10^{-2} M (Emsley 1980). Even if all the phosphorus weathered from igneous rocks since the beginning of geologic time still remained in the ocean, the concentration would only be about 5×10^{-2} M (Griffith et al. 1977). At the present day, most of the phosphorus in rivers is derived from the weathering of sedimentary rocks and represents P that has made at least one complete journey through the global cycle.

Linking the Global Cycles of C, N, and P

Cycles of biogeochemical elements are linked at many levels. Stock et al. (1990) describe how P is used to activate a transcriptional protein, stimulating nitrogen fixation in bacteria when nitrogen is in short supply. In this case, the linkage between these elements is seen at the level of cell and molecular biology. In Chapter 5 we saw that the N and P content of plant leaves was strongly correlated to the photosynthetic rate, showing a linkage between these elements and carbon at the level of leaf physiology. In the sea, productivity is easily predicted by the Redfield ratio among these elements (Chapter 9). On land, the amount of available P is correlated to the accumulation of organic carbon in soils (Chapter 6), showing a linkage between C and P at the ecosystem level. Imbalances in the supply of N and P appear to reduce forest growth in industrial regions (Aber et al. 1989). ments (Chapter 9).

Nitrogen fixation by free-living bacteria appears inversely related to the N/P ratio in soil (Fig. 6.3). The rate of accumulation of N is greatest in soils with high P content (Walker and Adams 1958). Similarly, N/P ratios <29 appear to stimulate N fixation in fresh-water ecosystems (Chapter 7). One might speculate that the high demand for P by N-fixing organisms links the global cycles of N and P, with P being the ultimate limit on nitrogen availability and net primary production. Despite these theoretical arguments for phosphorus limitation of the biosphere through geologic time, net primary production in most terrestrial and marine ecosystems shows an immediate response to additions of N (Fig. 12.4). Denitrification appears to maintain small supplies of N in most of the biosphere.

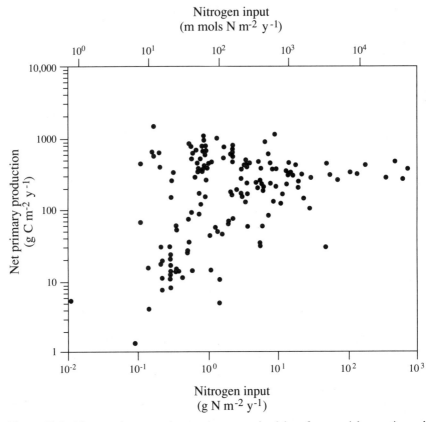

Figure 12.4 Nitrogen inputs and net primary productivity of terrestrial, aquatic, and marine ecosystems. Net primary production increases in direct response to added nitrogen up to inputs of ~ 10 gN m^{-2} yr^{-1}. Inputs in excess of that level are rarely found in natural ecosystems, but are seen in polluted environments and agricultural fields. From Levin (1989).

Summary

For both N and P, a small biogeochemical cycle with relatively rapid turnover is coupled to a large global pool with relatively slow turnover. For P, the large pool is found in unweathered rock and soil. For N, the major pool is found in the atmosphere. The atmospheric pool of N is coupled to the biogeochemical cycle by microbial nitrogen fixation and denitrification. In contrast, there is little evidence for biological processes that link the geochemical pool of P to the biogeochemical cycle.

The large, inorganic pools of N and P have relatively slow turnover. The cycle of N begins with the fixation of atmospheric nitrogen, which transfers a small amount of inert N$_2$ to the biosphere. This transfer is balanced by denitrification, which returns N$_2$ to the atmosphere. The balance of these processes maintains a steady-state concentration of N_2 in the atmosphere with a turnover time of 10^7 yr. While this steady state has been operative for a long period of time, it is unclear how denitrification responds to changes in the rate of nitrogen fixation. In the absence of microbial processes that remove N from the biosphere and return it to the atmosphere as N_2 and N_2O, the N inventory of the Earth would eventually be sequestered in the ocean and in organic sediments. Denitrification closes the global nitrogen cycle, and causes nitrogen to cycle more rapidly than phosphorus, which has no gaseous phase. The mean residence time of phosphorus in sedimentary rocks is measured in 10^8 yr, and the phosphorus cycle is complete only as a result of tectonic movements of the Earth's crust.

The movement of N and P in the biosphere is more rapid than in the global cycle, showing turnover times ranging from hours (for soluble P in the soil) to hundreds of years (for N in biomass). Denitrification maintains relatively small amounts of N in the biosphere. Limited supplies of N have selected for biological nitrogen fixation in certain bacteria, some of which are symbiotic in higher plants. Gutschick (1981) notes that it is surprising, in the face of widespread nitrogen limitation, that only about 2.5% of global net primary production is diverted to nitrogen fixation. Relatively small supplies of P may limit nitrogen fixation on land and in the sea. In response to nutrient limitations, recycling in terrestrial and marine ecosystems allows much greater rates of net primary production than rates of N fixation and rock weathering would otherwise support (Tables 6.1 and 9.2).

Human perturbations of the global nitrogen and phosphorus cycles are widespread. Through the production of fertilizers, we have probably doubled the rate at which nitrogen and phosphorus enter the biogeochemical cycle. River flow of N and P to the oceans has increased, and atmospheric N_2O is increasing at 0.3%/yr. All these changes indicate the effect of the human species in upsetting a steady state in global nutrient cycling.

Recommended Reading

Bolin, B. and R.B. Cook (eds.). 1983. The Major Biogeochemical Cycles and Their Interactions. Wiley, New York.

Porter, R. and D.W. Fitzsimons. 1978. Phosphorus in the Environment: Its Chemistry and Biochemistry. Elsevier, Amsterdam.

Sprent, J.I. 1988. The Ecology of the Nitrogen Cycle. Cambridge University Press, Cambridge.

13

The Global Sulfur Cycle

Introduction

Sulfur is found in valence states ranging from +6 in SO_4^{2-} to −2 in sulfides. The original pool of sulfur was contained in igneous rocks, largely as pyrite (FeS_2). Crustal degassing and, later, weathering under aerobic conditions transferred a large amount of S to the ocean, as SO_4^{2-}. When SO_4 is assimilated by organisms, it is reduced and converted into organic sulfur, which is an essential component of proteins. However, the live biosphere contains relatively little S. Today, the major global pools of S are found in sedimentary pyrite, the ocean, and evaporites derived from ocean water (Table 13.1).

As we saw for nitrogen, microbial transformations between valence states drive the global S cycle. In anaerobic conditions, SO_4 is a substrate for sulfate reduction, which may lead to the release of reduced gases to the atmosphere and the deposition of sedimentary pyrite (Chapters 7 and 9). Anaerobic environments also can support sulfur-based photosynthesis, which is likely to have been the first form of photosynthesis on Earth (Chapter 2). In aerobic conditions, reduced sulfur compounds are oxidized by microbes. In some cases, the oxidation is coupled to the reduction of CO_2, in reactions of S-based chemosynthesis.

Understanding the global biogeochemical cycle of S has enormous economic significance. Many metals are mined from sulfide minerals in

Table 13.1 Reservoirs of Sulfur near the Surface of the Earth[a]

Reservoir	10^{18} g S		
Deep oceanic rocks			
Sediments	75	±	20
Mafic rocks	2300	±	800
Sedimentary rocks			
Sandstone	250	±	60
Shale	2000	±	580
Limestone	380	±	110
Evaporites	5100	±	1600
Volcanics	50	±	18
Connate water	27	±	5
Total sediments	7800	±	1700
Freshwater	0.003	±	0.002
Ice	0.006	±	0.002
Atmosphere	3.6		
Sea	1280	±	55
Organic reservoir			
Land plants	0.6	×	10^{-3}
Marine plants	0.024	×	10^{-3}
Dead organic matter	5.0	×	10^{-3}
Total organic	5.62	×	10^{-3}

[a] From Trudinger (1979).

hydrothermal deposits (Meyer 1985). In other cases, microbial reactions are used to precipitate economic concentrations of metal sulfides from dilute solutions. Sulfur is an important constituent of coal and oil, and SO_2 is emitted to the atmosphere when fossil fuels are burned. A large amount of SO_2 is also emitted during the smelting of copper ores (Cullis and Hirschler 1980, Oppenheimer et al. 1985). An understanding of the relative importance of natural sulfur compounds in the atmosphere compared to anthropogenic SO_2 is important in evaluating the causes of acid rain and the impact of acid rain on natural ecosystems.

In this chapter we will review the global sulfur cycle. As for carbon (Chapter 11), nitrogen, and phosphorus (Chapter 12), we will attempt to establish a budget for S on land and in the atmosphere. We will couple those compartments to the budget for marine S (Fig. 9.14) to form a picture of the global S cycle. The biogeochemical cycle of S has changed through Earth history as a result of the appearance of new metabolic pathways and changes in their importance. We will review the history of the S cycle as it is told by sedimentary rocks. Finally, we will evaluate human impact on the S cycle and the global production of acidic sulfur substances in acid rain.

The Global Sulfur Cycle

All attempts to model the global S cycle must balance the inputs and outputs to the atmosphere, since no sulfur gas is a long-lived or major constituent of the atmosphere. The short mean residence time for atmospheric sulfur compounds, as a result of oxidation to SO_4, allows us to express all the fluxes in the global budget in terms of 10^{12} g S, without regard to the original form of emission. Despite a small atmospheric content of S compounds, the total annual flux of S through the atmosphere rivals that of the nitrogen cycle (compare Fig. 13.1 to Fig. 12.2).

In 1960, Eriksson examined the potential origins of SO_4 in Swedish rainfall, and hence, indirectly, sources of SO_4 in the atmosphere. He reasoned that the Cl^- in rainfall must be derived from the ocean and that the seaspray should also carry SO_4 roughly in the proportion of SO_4^{2-} to

The Global Sulfur Cycle

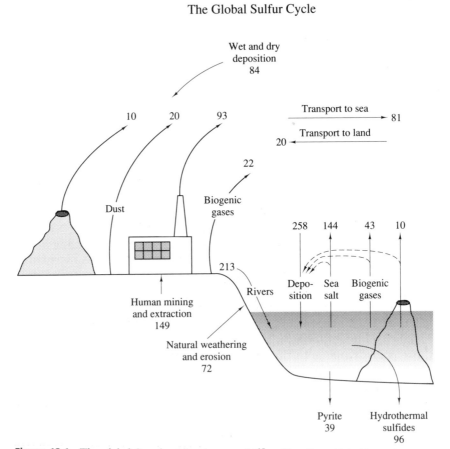

Figure 13.1 The global S cycle. All values are 10^{12} g S/yr. From Brimblecombe et al. (1989). See also Fig. 9.14.

Cl^- in seawater. His calculation suggested that about 4×10^{12} g S yr^{-1} deposited on land must be derived from the sea. At about the same time, however, Junge (1960) was evaluating the deposition of SO_4 in continental rainfall, and he calculated that about 73×10^{12} g S/yr was deposited on land. Clearly, there were other sources of SO_4 in the atmosphere and in rainfall. Junge's maps showed that SO_4 was abundant not only in the rainfall of coastal areas, but also downwind of desert and industrial regions (Fig. 3.10). Deserts are a source of gypsum ($CaSO_4 \cdot 2H_2O$) dust, and the burning of fossil fuels in industrial regions contributes SO_2 to air pollution.

In the intervening years, new sources of S in the atmosphere have been recognized and global flux estimates have been revised repeatedly. Nevertheless, our understanding of the global S cycle is primitive, and most of the estimates illustrated in Fig. 13.1 are subject to considerable uncertainty. Episodic events, including volcanic eruptions and duststorms, contribute to the global biogeochemical cycle of S, and disrupt steady-state conditions in the atmosphere. Many episodic events are difficult to quantify. Legrand and Delmas (1987) used the deposition of SO_4 in the Antarctic ice pack to estimate the contribution of sulfur gases by volcanoes to the global S cycle during the last 220 yr. The Tambora eruption of 1815 was the largest, releasing 150×10^{12} g of H_2SO_4 to the atmosphere. Typically, large eruptions release $17–30 \times 10^{12}$g of H_2SO_4. When volcanic eruptions are averaged over many years, the annual global flux is about $12–20 \times 10^{12}$ g S/yr (Berresheim and Jaeschke 1983, Andreae 1985, Brimblecombe et al. 1989). The movement of S in soil dust is also episodic and poorly understood. Many of the large particles are deposited locally. Ivanov (1983) gives a *net* global flux of 8×10^{12} g S/yr, about 8% of the fossil fuel release.

Estimates of the flux of biogenic gases from land differ by a factor of 10 (cf. Adams et al. 1981 vs. Goldan et al. 1987). The dominant gas emitted from freshwater wetlands and anaerobic soils is H_2S, with dimethylsulfide and carbonyl sulfide (COS) playing a lesser role (Chapters 6 and 7). Emissions from plants are poorly understood and deserving of further study. The total flux of biogenic gas from land carries about 20×10^{12} g S/yr to the atmosphere. The flux of various sulfur gases from other ecosystems is shown in Table 13.2. Although the oxidation of biogenic S contributes to the SO_4 in rainfall in regions downwind of wetlands (e.g., Nriagu et al. 1987), it is certain that emissions derived from human activity are the largest additions of S gases to the atmosphere (Möller 1984). Ice cores from Greenland show large increases in the deposition of SO_4 from the atmosphere in recent years (Herron et al. 1977, Mayewski et al. 1986).

Owing to the reactivity of S gases in the atmosphere, most of the anthropogenic emission of SO_2 is deposited locally in precipitation and dryfall. Deposition in dryfall and the direct absorption of SO_2 are poorly

Table 13.2 Ranges of Estimated Rates of Emission of Volatile Sulfur Compounds to the Atmosphere from Natural Sources[a]

Source	SO$_2$	H$_2$S	DMS	DMDS (and others)	CS$_2$	COS	Total
			Sulfur compound released (10^{12}gS/yr)				
Oceanic		0–15	38–40	0–1	0.3	0.4	38.7–56.7
Salt marsh		0.8–0.9	0.58	0.13	0.07	0.12	1.7–1.8
Inland swamps		11.7	0.84	0.2	2.8	1.85	17.4
Soil and plants		3–41	0.2–4.0	1	0.6–1.5	0.2–1.0	5.0–48.5
Burning of biomass	7	0–1		0–1		0.11	7.1–9.1
Volcanoes and fumaroles	8	1		0–0.02	0.01	0.01	9.0
Total	15	16.5–70.6	39.6–45.4	1.3–3.4	3.8–4.7	2.7–3.5	78.9–142.6

[a] From Kelley and Smith (1990).

understood, and the global estimate may be subject to revision. Total deposition of S on land is now thought to be ~84 × 10^{12} g S/yr (Brimblecombe et al. 1989). This accounts for about half of the total emissions from land. The remainder undergoes long-distance transport in the atmosphere and accounts for the net transfer of S from land to sea (Galloway and Whelpdale 1980, 1987). Without human effects on the global S cycle, net transport through the atmosphere would carry a small amount of S in the reverse direction—from sea to land.

Human activities also affect the transport of S in rivers. Berner (1971) estimates that at least 28% of the current SO$_4$ content of rivers is derived from air pollution, mining, erosion, and other human activities, whereas Ivanov et al. (1983) suggest that the current river transport of about 200 × 10^{12} g S/yr is roughly double that of preindustrial conditions. The natural river load of SO$_4$ is derived from rock weathering and rainfall, which includes cyclic salts that are carried through the atmosphere from the ocean. Weathering of pyrite and gypsum also contributes to the SO$_4$ content of river water (Table 8.5).

The marine portion of the global S cycle is taken from Toon et al. (1987) and Fig. 9.14. The ocean is a large source of aerosols that contain SO$_4$, but most of these are redeposited in the ocean in precipitation and dryfall. Dimethylsulfide [(CH$_3$)$_2$S or DMS] is the major biogenic gas emitted from the sea (Table 13.2). Erickson et al. (1990) suggest that the annual flux of DMS from the sea may be only slightly greater than 15 × 10^{12} g S/yr (cf. Fig. 13.1), but in any case, DMS accounts for about 50% of all biogenic S gases emitted to the atmosphere (Ferek et al. 1986). The mean residence time of DMS is about 1 day (Table 3.4) as a result of oxidation to SO$_4$. Thus, most of the sulfur from DMS is also redeposited in the ocean. The net transport of sulfate from sea to land is about

20×10^{12} g S/yr. The ocean receives a net input of SO_4 in river flow and precipitation.

Although they are subject to great revision, the current estimates of inputs to the ocean are in excess of the estimate of total sinks, implying that the oceans are increasing in SO_4 by over 10^{14} g S/yr. Such an increase will be difficult to document, since the content in the oceans is 12×10^{20} g. As calculated in Chapter 9, the mean residence time for SO_4 in seawater is over 3,000,000 yr.

Temporal Perspectives of the Global Sulfur Cycle

During the accretion of the primordial Earth, sulfur was among the gases that were released during crustal outgassing to form the secondary atmosphere (Chapter 2). Even today, volcanic emissions contain appreciable concentrations of SO_2 and H_2S (Table 2.2). Crustal outgassing on Venus has resulted in large concentrations of SO_2 in its atmosphere (Oyama et al. 1979). When the oceans condensed on Earth, the atmosphere was essentially swept clear of S gases, owing to their high solubility in water. The dominant form of S in the earliest ocean is likely to have been SO_4; high concentrations of Fe^{2+} in the primitive ocean would have precipitated any sulfides, which are insoluble in anaerobic conditions (Walker 1985b). The SO_4 content of the oceans apparently increased until about 400,000,000 yr ago and then decreased slightly to the amount found today (Zehnder and Zinder 1980). The total inventory of S compounds on the surface of the Earth ($\sim 10^{22}$ g S) represents the total crustal outgassing of S through geologic time.

The ratio of ^{32}S to ^{34}S in the total inventory on Earth is thought to be similar to the ratio of 22.22 measured in the Canyon Diablo Triolite (CDT), a meteorite collected in California. The sulfur isotope ratio in this rock is accepted as the international standard, and is assigned a value of 0.00. In other samples, deviations from this ratio are expressed as $\delta^{34}S$, with the units of parts per thousand parts (‰), just as we saw for isotopes of carbon (Chapter 5) and nitrogen (Chapter 6). Presumably the ratio in the earliest oceans was 0.00, since there is no reason to expect any discrimination between the isotopes during crustal degassing.

When evaporite minerals precipitate from seawater, there is little differentiation among the isotopes, so geologic deposits of gypsum and barite ($BaSO_4$) carry a record of the isotopic composition of S in seawater. Sedimentary rocks of 3.8 bya contain gypsum, confirming significant concentrations of SO_4 in the earliest oceans (Walker 1983). In all cases, $\delta^{34}S$ in these deposits is close to 0.00 (Schidlowski et al. 1983).

Dissimilatory sulfate reduction by bacteria strongly differentiates among the isotopes of sulfur, as a result of a more rapid enzymatic reaction with $^{32}SO_4$. Thus, the products of sulfate reduction, including

H_2S and sedimentary pyrite, show $\delta^{34}S$ ranging from $-46‰$ to $-9.0‰$, depending upon the rate of reaction and the concentration of SO_4 as a reactant (Chambers and Trudinger 1979). The evolution of sulfate reduction dates to 2.4–2.7 bya, based on the first occurrence of sedimentary rocks with depletion of ^{34}S (Cameron 1982, Schidlowski et al. 1983). The average $\delta^{34}S$ in sedimentary sulfides is about -10 to $-12‰$ (Holser and Kaplan 1966, Migdisov et al. 1983).

During periods of Earth history when large amounts of sedimentary pyrite were formed from sulfate reduction, seawater SO_4 was enriched in ^{34}S. Since there is little differentiation among isotopes during the precipitation of evaporites, the geologic record of $\delta^{34}S$ in evaporites indicates the relative importance of sulfate reduction leading to pyrite. Figure 13.2 shows a three-box model for the S cycle, in which marine SO_4 and sedimentary sulfides are connected through microbial oxidation and reduction reactions, which discriminate between sulfur isotopes. Shifts be-

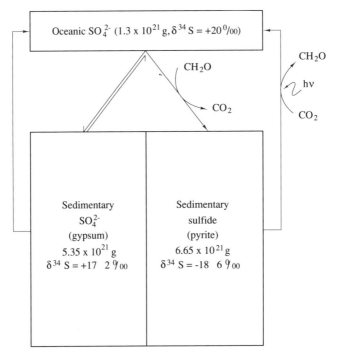

Figure 13.2 A model for the global sulfur cycle showing linkage and partitioning of S between oxidized and reduced forms near the surface of the Earth. Transfers of S from seawater to pyrite involve a major fractionation between ^{34}S and ^{32}S isotopes, whereas exchange between seawater SO_4 and sedimentary SO_4 (largely gypsum) involves only minor fractionation. The sum of all pools, 1.3×10^{22} g, represents the total outgassing of S from the crust (cf. Table 2.1). Only about 10% now resides in the ocean. From Schidlowski et al. (1983).

tween the pool of S in the sea and in sedimentary sulfides are reflected by the isotopic composition of evaporites. Currently, about 50% of the pool of S near the surface of the Earth is found in reduced form (Li 1972, Holser et al. 1989). Changes in the relative size of these reservoirs through geologic time indicate the importance of sulfate reduction or the oxidation of sulfides. By contrast, the uptake of S by plants (assimilatory reduction) and other microbial reactions involving S are insignificant to the global cycles of S and C.

During the last 600,000,000 yr, seawater SO_4 has varied between $+10$ and $+30‰$ in $\delta^{34}S$ (Fig. 13.3), with an average value close to that of today, $+21‰$ (Kaplan 1975, Rees et al. 1978). Seawater sulfate shows a marked positive excursion in $\delta^{34}S$ during the Cambrian ($+32‰$), when the deposition of pyrite must have been greatly in excess of the oxidation of biogenic sulfide minerals exposed on land. Seawater sulfate was less concentrated in ^{34}S, that is, $\delta^{34}S$ of $+10‰$, during the Carboniferous and Permian, when net primary production shifted from the ocean to freshwater swamps in which SO_4, sulfate reduction, and pyrite deposition

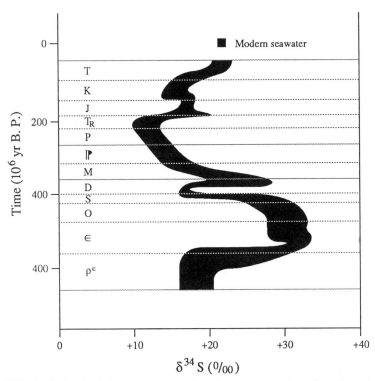

Figure 13.3 Variations in the isotopic composition of seawater SO_4 through geologic time. Adapted from Krouse and McCready (1979).

was limited (Berner 1984). Presumably the concentration of SO_4 in sea-water was also greater during that interval, since the rate of pyrite formation was depressed.

Although the sulfur cycle has shown shifts between net sulfur oxidation and net sulfur reduction in the geologic past, the rate of current human impact is probably unprecedented in the geologic record. As for the carbon cycle, the present-day cycle of S is not in steady state. Human activities have led to a net flux from land to sea through the atmosphere, where a net flux in the reverse direction is likely 100 yr ago. Humans are mining coal and extracting petroleum from the Earth's crust at a rate that mobilizes 150×10^{12} g S/yr, more than double the rate of 100 yr ago (Brimblecombe et al. 1989). The net effect of these processes is to increase the pool of oxidized sulfur (SO_4) in the global cycle, at the expense of the storage of reduced sulfur. The human activities cause only a tiny change in the global pools of S, but they are massive changes in the annual flux of S through the atmosphere.

Various workers have attempted to use measurements of $\delta^{34}S$ to deduce the origin of the SO_4 in rainfall and the extent of human impact on the movement of S in the atmosphere. Unfortunately, the potential sources of SO_4 show a wide range of values for $\delta^{34}S$, making the identification of specific sources equivocal (Nielsen 1974). For example, the sulfur in coal may be depleted in $\delta^{34}S$ if it is found as pyrite or enriched in $\delta^{34}S$ if it is derived from sulfur that was assimilated by the original plant materials forming coal. Thus, coals show a wide range in $\delta^{34}S$. Similarly, petroleum shows a range of -10.0 to $+25‰$ in $\delta^{34}S$ (Krouse and McCready 1979). Desert dusts containing SO_4 range in $\delta^{34}S$ from $-35‰$ to $+17‰$, depending on the parent material of soil formation (Schlesinger and Peterjohn 1988). In the eastern United States, $\delta^{34}S$ of rainfall varies seasonally between $+6.4‰$ in winter and $+2.9‰$ in summer, consistent with any of these sources or a combination of them (Nriagu and Coker 1978). Lower values in the summer are thought to reflect the influence of biogenic sulfur from sulfate reduction in wetlands (Nriagu et al. 1987).

When SO_2 is emitted as an air pollutant, it forms sulfuric acid through heterogeneous reactions with water in the atmosphere (Chapter 3). As a strong acid that is completely dissociated in water, H_2SO_4 suppresses the disassociation of natural, weak acids in rainfall. For example, in the absence of strong acids, the dissolution of CO_2 in water will form a weak solution of carbonic acid, H_2CO_3, and rainfall pH will be about 5.6:

$$CO_2 + H_2O \rightleftarrows H^+ + HCO_3^- \tag{13.1}$$

In the presence of strong acids that lower pH below 4.3, this reaction moves to the left, and carbonic acid makes no contribution to free acidity.

In many industrialized areas, free acidity in precipitation is almost wholly determined by the concentration of the strong acid anions, SO_4 and NO_3 (Table 13.3). Rock weathering that was primarily driven by carbonation weathering in the preindustrial age is now driven by anthropogenic H^+ (Johnson et al. 1972).

It is interesting to estimate the global sources of acidity in the atmosphere. In this analysis, we are interested in reactions that are net sources of H^+ in the atmosphere. We can ignore the movements soil dusts and seaspray, because the strong-acid anions they contain, NO_3 and SO_4, are largely balanced by cations (especially Ca and Na) emitted at the same time. If the pH of all rainfall on Earth was 5.6 as a result of equilibrium with atmospheric CO_2, the total deposition of H^+ ions would be 1.24×10^{12} moles/yr. The production of NO by lightning produces acidity, since NO dissolves in rainwater, forming HNO_3. Globally, N fixation by lightning contributes 1.4×10^{12} moles H^+/yr. Similarly, volcanic emanations of SO_2 contribute 1.3×10^{12} moles H^+/yr, and the oxidation of biogenic S gases produces 4.1×10^{12} moles H^+/yr. In contrast, the anthropogenic production of NO_x and SO_2 produces about 7.4×10^{12} moles H^+/yr; this is nearly as much as all natural sources of acidity combined.

The only net source of alkalinity in the atmosphere comes from the reaction of NH_3 with the strong acids H_2SO_4 and HNO_3 to form aerosols, $(NH_4)_2SO_4$ and NH_4NO_3. However, the global emission of NH_3, $\sim 50 \times 10^{12}$ g/yr (Warneck 1988), reduces the production of H^+ by only about 3×10^{12} moles/yr. Thus, even though the current acidity of the

Table 13.3 Sources of Acidity in Acid Rainfall Collected in Ithaca, New York, on July 11, 1975. Ambient pH was 3.84[a]

Component	Concentration in Precipitation (mg/l)	Contribution to	
		Free Acidity at pH 3.84 (μeq/l)	Total Acidity in a Titration to pH 9.0 (μeq/l)
H_2CO_3	0.62	0	20
Clay	5	0	5
NH_4^+	0.53	0	29
Dissolved Al	0.050	0	5
Dissolved Fe	0.040	0	2
Dissolved Mn	0.005	0	0.1
Total organic acids	0.43	2	5.7
HNO_3	2.80	40	40
H_2SO_4	5.60	102	103
Total		144	210

[a] From Galloway et al. (1976). Copyright 1976 by the AAAS.

atmosphere is much higher as a result of human activities, the atmosphere has acted as an acidic medium throughout geologic time.

The Atmospheric Budget of COS

Showing an average concentration of about 500 parts per trillion, carbonyl sulfide (COS) is the most abundant sulfur gas in the atmosphere (Warneck 1988). The pool in the atmosphere is 4.6×10^{12} g S (Servant 1989). Based on the global budget of Table 13.4, its mean residence time in the atmosphere is about 5 yr. Our understanding of COS is primitive. The apparent minor imbalance in Table 13.4 is not associated with a well-documented increase in COS in the atmosphere (Hofmann 1990).

The major source of COS appears to be the ocean, where it is produced by a photochemical reaction with dissolved organic matter (Ferek and Andreae 1984). Even though the emission of COS from the sea is dwarfed by the emission of dimethylsulfide, the marine source accounts for 42% of the total input of COS to the atmosphere (Toon et al. 1987, Servant 1989). Other sources include biomass burning, fossil fuel combustion, and oxidation of CS_2 by OH radicals in the atmosphere. Early indications of a large source of COS from upland soils (Adams et al. 1981) have been reduced by more recent measurements (Goldan et al. 1987). Emissions of COS from salt marshes are limited by the global extent of salt-marsh vegetation (Steudler and Peterson 1985, Carroll et al. 1986).

Until recently, the global budget of COS was grossly out of balance, for the known sinks could account for only a small portion of the estimated annual production (Khalil and Rasmussen 1984). Some COS is oxidized

Table 13.4 Global Budget for Carbonyl Sulfide (COS) in the Atmosphere[a]

Source or Sink	COS (10^{12} g S/yr)
Sources	
Oceans	0.40
Soils	0.06
Biomass burning	0.175
Fossil fuels	0.10
Oxidation of CS_2	0.205
Total sources	0.94
Sinks	
Oxidation by OH	0.18
Stratospheric photolysis	0.09
Vegetation uptake	0.63
Total sinks	0.90

[a] From Servant (1989).

in the troposphere via OH radicals, but the major tropospheric sink for COS, first reported in 1988 (Goldan et al. 1988), appears to be uptake by vegetation. Servant (1989) now suggests that uptake by vegetation may account for >70% of the total global destruction of COS. Carbonyl sulfide may also be removed from the troposphere when it acts as a corrosive air pollutant (Graedel et al. 1981).

A small amount of COS is mixed into the stratosphere, where it is destroyed by a photochemical reaction involving the OH radical, producing SO_4. Aside from episodic eruptions of volcanoes, COS appears to be the main source of SO_4 aerosols in the stratosphere (Hoffman and Rosen 1983, Servant 1986). There is some evidence that these aerosols have increased in recent years (Hoffman 1990). These aerosols affect the amount of solar radiation entering the troposphere and are an important component of the radiation budget of the Earth (Turco et al. 1980). Although there is no evidence that the atmospheric content of COS is increasing, direct human sources contribute to its budget (Table 13.4), and any increase in the stratosphere has potential consequences for global climate (Hoffman and Rosen 1980).

Summary

The major pool of S in the global cycle is found in the crustal minerals, gypsum and pyrite. Additional S is found dissolved in ocean water. With respect to pools, the global S cycle resembles the global cycle of phosphorus (Chapter 12). In contrast, the largest pool of the global N cycle is found in the atmosphere.

In other respects, however, there are strong similarities between the global cycles of N and S. In both cases, the major annual movement of the element is through the atmosphere, and a large portion of the movement is through the production of reduced gases of N and S by biota. These gases return N and S to the atmosphere, providing a closed global cycle with a relatively rapid turnover. In contrast, the ultimate fate for P is incorporation into ocean sediments; its cycle is complete only as a result of long-term sedimentary uplift.

Biogeochemistry exerts a major influence on the global S cycle. The largest pool of S near the surface of the Earth is found in pyrite, as a result of sulfate reduction. In the absence of sulfate-reducing bacteria, the concentration of SO_4 in seawater and that of O_2 in the atmosphere would be likely to be higher. The sedimentary record shows that the relative extent of sulfate reduction has varied through geologic time.

Current human perturbation of the sulfur cycle is extreme—roughly doubling the annual mobilization of sulfur from the crust of the Earth. As a result of fossil fuel combustion, areas that are downwind of industrial regions now receive massive amounts of acidic deposition from the atmosphere. This excess acidity is likely to lead to changes in rock weathering (Chapter 4), forest growth (Chapter 6), and ocean productivity (Chapter 9).

Recommended Reading

Brimblecombe, P. and A.Y. Lein (eds.). 1989. Evolution of the Global Biogeo-chemical Sulphur Cycle. Wiley, Chichester.

Ivanov, M.V. and J.R. Freney (eds.). 1983. The Global Biogeochemical Sulphur Cycle. Wiley, Chichester.

14

A Perspective

There are few basic axioms of ecology, but one of the most fundamental is that which predicts the ultimate collapse of a population showing exponential growth in a closed environment. Attempts to construct self-perpetuating and stable ecosystems in the laboratory are usually unsuccessful. Often a small initial population grows rapidly—even exponentially—in the closed culture. In a variety of experiments during the 1930s, Gause (1934) showed that stable populations of the freshwater protozoan, *Paramecium*, could be maintained in small aquaria only if supplies of food and water were replenished continuously. Initially the population grew rapidly, but without fresh water, toxic wastes accumulated and the population perished.

With respect to all resources except sunlight energy, the Earth is a "closed" ecological system, and with the application of modern medicine, the human population is now growing exponentially in this closed environment. Certainly, we have the ability to control some of the environmental consequences of our population, just as we developed the modern medical techniques that have produced its rapid growth. Indeed, we have developed plants that are especially efficient in their capture of sunlight energy to provide a greater supply of food. As economic incentives demand it, our ingenuity often yields increasing supplies of fresh water and mineral resources. Most of the metabolic byproducts of human society can be contained and cleansed with the application of appropriate technology and human interest in doing so. Yet as long as the human population is increasing exponentially, a day of reckoning is inevitable. Experience tells us that populations do not grow exponentially in a closed or limited environment.

Already the global effects of the human species are easy to see, and we have emphasized the human effect on global biogeochemical cycles in many chapters of this book. Despite our increasing need for food, most evidence suggests that we have reduced, not increased, the net primary production of the biosphere (Chapter 5). Despite our need for fresh water, we have drastically lowered the quality of most surface waters flowing to the sea (Chapter 8). Despite the enormous size of the atmosphere, our fossil fuel combustion during the last 100 yr has changed its composition and perhaps global climate patterns (Chapter 3).

It is striking that a single species can force a change in the composition of the atmosphere. Perhaps not since the first photosynthetic organisms added oxygen to the environment has a single species had such a dramatic effect on the chemistry of the surface of the Earth and the quality of the environment for life. The problem is global; pollution is no longer a matter for litigation at the end of a local effluent pipe.

Changes in habitat, loss of productivity, and chemical pollution have led to a rapid and continuing reduction in the abundance of other species on this planet. A harbinger of our effects is seen in the declining populations of songbirds that migrate from the tropics to the local city park or to the woodlot near our summer cottage each spring (Holmes et al. 1986, Robbins et al. 1989). "Just aren't as many birds as there used to be," we may say. More likely, we are seeing the effects of expanding human population, agricultural development, and human resource exploitation at the expense of the biosphere.

Many of the species that share this planet are directly responsible for maintaining the stability of the closed biogeochemical system in which we live. For example, a large variety of soil bacteria consume atmospheric methane. Many of these species are less effective at this activity in response to acid rain. Their demise is less obvious than that of songbirds. The Audubon Society does no Christmas count of soil bacteria! Yet their loss may contribute to higher methane concentrations in the atmosphere and the potential for global warming. Certainly, there are other causes of increasing methane in the atmosphere (Chapters 3 and 11), and policy-makers may argue endlessly about appropriate action. All the time, however, the atmospheric methane concentration is increasing as a result of human activities, compounded by the rise in human population.

Thus, I end this book with this perspective: Human population growth is the basis of every major environmental issue facing world nations today. Population growth and its demand for energy in the United States made inevitable the exploration and production of oil in Alaska. Accidents happen and when they do they are most unfortunate, but how many of us have blamed ourselves for what happened in Prince William Sound in 1989? Similarly, if the climate of Brazil—or the globe—changes as the rain forest is harvested to support its expanding population, we may deplore the extinction of rain-forest species. We may gain some comfort from the potential for agricultural research to respond with new crops to feed these peoples from impoverished lands in a changing climate. Of course, less optimistically, it is possible that we will fail in these efforts. When the dust settles, how many of us will ask whether we might have fostered the economic conditions and family planning efforts that would have slowed population growth?

With exponentially increasing population in a closed environment, we will reach the carrying capacity of our planetary aquarium. Whether that is occuring now, with our population of 5.3 billion, or during the next century, when it will reach 14 billion, is not known. But, our focus for maintaining life and quality of life on this planet should be on controlling human population growth.

References

Aber, J.D. and J.M. Melillo. 1980. Litter decomposition: Measuring relative contributions of organic matter and nitrogen to forest soils. Canadian Journal of Botany 58: 416–421.

Aber, J.D., K.J. Nadelhoffer, P. Steudler and J.M. Melillo. 1989. Nitrogen saturation in northern forest ecosystems. BioScience 39:378–386.

Adams, J.A.S., M.S.M. Mantovani and L.L. Lundell. 1977. Wood versus fossil fuel as a source of excess carbon dioxide in the atmosphere: A preliminary report. Science 196: 54–56.

Adams, D.F., S.O. Farwell, E. Robinson, M.R. Pack and W.L. Bamesberger. 1981. Biogenic sulfur source strengths. Environmental Science and Technology 15: 1493–1498.

Adams, M.A. and P.M. Attiwill. 1982. Nitrate reductase activity and growth responses of forest species to ammonium and nitrate sources of nitrogen. Plant and Soil 66: 373–381.

Adams, W.A., M.A. Raza and L. J. Evans. 1980. Relationships between net redistribution of Al and Fe and extractable levels in podzolic soils derived from lower Palaeozoic sedimentary rocks. Journal of Soil Science 31: 533–545.

Admiraal, W. and Y.J.H. Botermans. 1989. Comparison of nitrification rates in three branches of the lower river Rhine. Biogeochemistry 8: 135–151.

Ae, N., J. Arihara, K. Okada, T. Yoshihara and C. Johansen. 1990. Phosphorus uptake by pigeon pea and its role in cropping systems of the Indian subcontinent. Science 248: 477–480.

Aguilar, R. and R.D. Heil. 1988. Soil organic carbon, nitrogen, and phosphorus quantities in northern Great Plains rangeland. Soil Science Society of America Journal 52: 1076–1081.

Ahl, T. 1988. Background yield of phosphorus from drainage area and atmosphere: An empirical approach. Hydrobiologia 170: 35–44.

Alexander, E.B. 1988. Rates of soil formation: Implications for soil-loss tolerance. Soil Science 145: 37–45.

Alldredge, A.L. and Y. Cohen. 1987. Can microscale chemical patches persist in the sea? Microelectrode study of marine snow, fecal pellets. Science 235: 689–691.

Allen, H.L. 1972. Phytoplankton photosynthesis, micronutrient interactions, and inorganic carbon availability in a soft-water Vermont lake. pp. 63–80. In G.E. Likens (ed.), Nutrients and Eutrophication. American Society of Limnology and Oceanography, Lawrence, Kansas.

Altabet, M.A. and W.B. Curry. 1989. Testing models of past ocean chemistry using foraminifera $^{15}N/^{14}N$. Global Biogeochemical Cycles 3: 107–119.

Altschuler, Z.S., M.M. Schnepfe, C.C. Silber and F.O. Simon. 1983. Sulfur diagenesis in Everglades peat and the origin of pyrite in coal. Science 221: 221–227.

Anders, E., R. Hayatsu and M.H. Studier. 1973. Organic compounds in meteorites. Science 182: 781–790.

Anders, E. and T. Owen. 1977. Mars and Earth: Origin and abundance of volatiles. Science 198: 453–465.

Anders, E. 1989. Pre-biotic organic matter from comets and asteroids. Nature 342: 255–257.

Anders, E. and N. Grevesse. 1989. Abundance of the elements: Meteoritic and solar. Geochimica et Cosmochimica Acta 53: 197–214.

Anderson, D.W. and E. A. Paul. 1984. Organo-mineral complexes and their study by radiocarbon dating. Soil Science Society of America Journal 48: 298–301.

Anderson, I.C. and J.S. Levine. 1987. Simultaneous field measurements of biogenic emissions of nitric oxide and nitrous oxide. Journal of Geophysical Research 92: 965–976.

Anderson, I.C., J.S. Levine, M.A. Poth and P.J. Riggan. 1988. Enhanced biogenic emissions of nitric oxide and nitrous oxide following surface biomass burning. Journal of Geophysical Research 93: 3893–3898.

Anderson, J.M., P. Ineson and S.A. Huish. 1983. Nitrogen and cation mobilization by soil fauna feeding on leaf litter and soil organic matter from deciduous woodlands. Soil Biology and Biochemistry 15: 463–467.

Anderson, J.P.E. and K.H. Domsch. 1980. Quantities of plant nutrients in the microbial biomass of selected soils. Soil Science 130: 211–216.

Andreae, M.O. 1985. The emission of sulfur to the remote atmosphere: Background paper. pp. 5–25. *In* J.N. Galloway, R.J. Charlson, M.O. Andreae, and H. Rodhe (eds.), The Biogeochemical Cycling of Sulfur and Nitrogen in the Remote Atmosphere. Reidel, Dordrecht, The Netherlands.

Andreae, M.O. 1986. The oceans as a source of biogenic gases. Oceanus 29(4): 27–35

Andreae, M.O. and H. Raemdonck. 1983. Dimethyl sulfide in the surface ocean and the marine atmosphere: A global view. Science 221: 744–747.

Andreae, M.O. and W.R. Barnard. 1984. The marine chemistry of dimethylsulfide. Marine Chemistry 14: 267–279.

Andreae, M.O., R.J. Charlson, F. Bruynseels, H. Storms, R. Van Grieken and W. Maenhaut. 1986. Internal mixture of sea salt, silicates, and excess sulfate in marine aerosols. Science 232: 1620–1623.

Andrews, M. 1986. The partitioning of nitrate assimilation between root and shoot of higher plants. Plant, Cell and Environment 9: 511–519.

Antibus, R.K., J.G. Croxdale, O.K. Miller and A.E. Linkins. 1981. Ecotomycorrhizal fungi of *Salix rotundifolia*. III. Resynthesized mycorrhizal complexes and their surface phosphatase activities. Canadian Journal of Botany 59: 2458–2465.

Antisari, L.V. and P. Sequi. 1988. Comparison of total nitrogen by four procedures and sequential determination of exchangeable ammonium, organic nitrogen, and fixed ammonium in soil. Soil Science Society of America Journal 52: 1020–1023.

Antweiler, R.C. and J.I. Drever. 1983. The weathering of a late Tertiary volcanic ash: Importance of organic solutes. Geochimica et Cosmochimca Acta 47: 623–629.

April, R., R. Newton and L.T. Coles. 1986. Chemical weathering in two Adirondack watersheds: Past and present-day rates. Geological Society of America Bulletin 97: 1232–1238.

April, R. and D. Keller. 1990. Mineralogy of the rhizosphere in forest soils of the eastern United States: Mineralogic studies of the rhizosphere. Biogeochemistry 9: 1–18.

Araujo-lima, C.A.R.M., B.R. Forsberg, R. Victoria and L. Martinelli. 1986. Energy sources for detritivorous fishes in the Amazon. Science 234: 1256–1258.

Arkley, R.J. 1963. Calculation of carbonate and water movement in soil from climatic data. Soil Science 92: 239–248.

Arkley, R.J. 1967. Climates of some great soil groups of the western United States. Soil Science 103: 389–400.

Armentano, T.V. and E.S. Menges. 1986. Patterns of change in the carbon balance of organic soils of the temperate zone. Journal of Ecology 74: 755–774.

Armentano, T.V. and C.W. Ralston. 1980. The role of temperate zone forests in the global carbon cycle. Canadian Journal of Forest Research 10: 53–60.

Art, H.W., F.H. Bormann, G.K. Voigt and G.M. Woodwell. 1974. Barrier island forest ecosystem: Role of meteorologic inputs. Science 184: 60–62.

Arther, M.A., W.E. Dean and L.M. Pratt. 1988. Geochemical and climatic effects of increased marine organic carbon burial at the Cenomanian/Turonian boundary. Nature 335: 714–717.

Aselmann, T. and P.J. Crutzen. 1989. Global distribution of natural freshwater wetlands and rice paddies, their net primary productivity, seasonality and possible methane emissions. Journal of Atmospheric Chemistry 8: 307–358.

Atjay, G.L., P. Ketner and P. Duvigneaud. 1979. Terrestrial primary production and phytomass. pp. 129–181. *In* B. Bolin, E.T. Degens, S. Kempe and P. Ketner (eds.), The Global Carbon Cycle. Wiley, New York.

Aumann, G.D. 1965. Microtine abundance and soil sodium levels. Journal of Mammalogy 46: 594–604.

Avnimelech, Y. and J.R. McHenry. 1984. Enrichment of transported sediments with organic carbon, nutrients, and clay. Soil Science Society of America Journal 48: 259–266.

Awramik, S.M., J.W. Schopf and M.R. Walter. 1983. Filamentous fossil bacteria from the Archean of western Australia. Precambrian Research 20: 357–374.

Azevedo, J. and D.L. Morgan. 1974. Fog precipitation in coastal California forests. Ecology 55: 1135–1141.

Bacastow, R. B. 1976. Modulation of atmospheric carbon dioxide by the southern oscillation. Nature 261: 116–118.

Bacastow, R and A. Bjorkstrom. 1981. Comparisons of ocean models for the carbon cycle. pp. 29–79. *In* B. Bolin (ed.), Carbon Cycle Modelling. Wiley, New York.

Bacastow, R.B., C.D. Keeling, and T.P. Whorf. 1985. Seasonal amplitude increase in atmospheric CO_2 concentration at Mauna Loa, Hawaii, 1959–1982. Journal of Geophysical Research 90: 10529–10540.

Bach, L.B., P.J. Wierenga and T.J. Ward. 1986. Estimation of the Philip infiltration parameters from rainfall simulation data. Soil Science Society of America Journal 50: 1319–1323.

Bache, B.W. 1984. The role of calcium in buffering soils. Plant, Cell and Environment 7: 391–395.

Baethgen, W.E. and M.M. Alley. 1987. Nonexchangeable ammonium nitrogen contribution to plant available nitrogen. Soil Science Society of America Journal 51: 110–115.

Baker, L.A., C.D. Pollman and J.M. Eilers. 1988. Alkalinity regulation of softwater Florida lakes. Water Resources Research 24: 1069–1082.

Baker, P.A. and M. Kastner. 1981. Constraints on the formation of sedimentary dolomite. Science 213: 214–216.

Baker, V.R. 1977. Stream-channel response to floods, with examples from central Texas. Geological Society of America Bulletin 88: 1057–1071.

Baker, W.E. 1973. Role of humic acids from Tasmanian podzolic soils in mineral degradation and metal mobilization. Geochimica et Cosmochimica Acta 37: 269–281.

Baker-Blocker, A., T.M. Donahue and K.H. Mancy. 1977. Methane flux from wetland areas. Tellus 29: 245–250.

Banin, A. and J. Navrot. 1975. Origin of Life: Clues from relations between chemical compositions of living organisms and natural environments. Science 189: 550–551.

Barber, S.A. 1962. A diffusion and mass-flow concept of soil nutrient availability. Soil Science 93: 39–49.

Barnola, J.M., D. Raynaud, Y.S. Korotkevich and C. Lorius. 1987. Vostok ice core provides 160,000-year record of atmospheric CO_2. Nature 329: 408–414.

Barrett, J.W., P.M. Solomon, R.L. de Zafra, M. Jaramillo, L. Emmons and A. Parrish. 1988. Formation of the Antarctic ozone hole by the ClO dimer mechanism. Nature 336: 455–458.

Barrie, L.A. and J.M. Hales. 1984. The spatial distributions of precipitation acidity and major ion wet deposition in North America during 1980. Tellus 36B: 333–355.

Barsdate, R.J. and V. Alexander. 1975. The nitrogen balance of Arctic tundra: Pathways, rates, and environmental implications. Journal of Environmental Quality 4: 111–117.

Bartel-Ortiz, L.M. and M.B. David. 1988. Sulfur constituents and transformations in upland and floodplain forest soils. Canadian Journal of Forest Research 18: 1106–1112.

Bartlett, K.B., D.S. Bartlett, R.C. Harriss and D.I. Sebacher. 1987. Methane emissions along a salt marsh salinity gradient. Biogeochemistry 4: 183–202.

Bartlett, K.B., P.M. Crill, D.I. Sebacher, R.C. Harriss, J.O. Wilson and J.M. Melack. 1988. Methane flux from the central Amazonian floodplain. Journal of Geophysical Research 93: 1571–1582.

Bartlett, R.J. 1986. Soil redox behavior. pp. 179–207. *In* D.L. Sparks (ed.), Soil Physical Chemistry. CRC Press, Boca Raton, Florida.

Bass Becking, L.G.M., I.R. Kaplan and D. Moore. 1960. Limits of the natural environment in terms of pH and oxidationreduction potentials. Journal of Geology 68: 243–284.

Bates, T.S., R.J. Charlson and R.H. Gammon. 1987. Evidence for the climatic role of marine biogenic sulphur. Nature 329: 319–321.

Bazely, D.R. and R.L. Jefferies. 1989. Lesser snow geese and the nitrogen economy of a grazed salt marsh. Journal of Ecology 77: 24–34.

Beadle, N.C.W. 1966. Soil phosphate and its role in molding segments of the Australian flora and vegetation, with special reference to xeromorphy and sclerophylly. Ecology 47: 992–1007.

Beck, K.C., J.H. Reuter and E.M. Perdue. 1974. Organic and inorganic geochemistry of some coastal plain rivers of the southeastern United States. Geochimica et Cosmochimica Acta 38: 341–364.

Beilke, S. and D. Lamb. 1974. On the absorption of SO_2 in ocean water. Tellus 26: 268–271.

Bell, R.A., P.V. Athey and M.R. Sommerfeld. 1986. Cryptoendolithic algal communities of the Colorado Plateau. Journal of Phycology 22: 429–435.

Bell, R.T., G.M. Ahlgren and I. Ahlgren. 1983. Estimating bacterioplankton production by measuring [^{3}H]thymidine incorporation in a eutrophic Swedish lake. Applied and Environmental Microbiology 45: 1709–1721.

Bender, M.L., G.D. Klinkhammer and D.W. Spencer. 1977. Manganese in seawater and the marine manganese balance. Deep Sea Research 24: 799–812.

Bender, M., R. Jahnke, R. Weiss, W. Martin, D.T. Heggie, J. Orchardo and T. Sowers. 1989. Organic carbon oxidation and benthic nitrogen and silica dynamics in San Clemente Basin, a continental borderland site. Geochimica et Cosmochimica Acta 53: 685–697.

Benke, A.C. and J.L. Meyer. 1988. Structure and function of a blackwater river in the southeastern U.S.A.. Verhandelingen Internationalen Vereins Limnologie 23: 1209–1218.

Benner, R., M.A. Moran and R.E. Hodson. 1986. Biogeochemical cycling of lignocellulosic carbon in marine and freshwater ecosystems: Relative contributions of procaryotes and eucaryotes. Limnology and Oceanography 31: 89–100.

Bennett, P.C., M.E. Melcer, D.I. Siegel and J. P. Hassett. 1988. The dissolution of quartz in dilute aqueous solutions of organic acids at 25°C. Geochimica et Cosmochimica Acta 52: 1521–1530.

Berg, B. 1988. Dynamics of nitrogen (^{15}N) in decomposing Scots pine (*Pinus sylvestris*) needle litter. Long-term decomposition in a Scots pine forest VI. Canadian Journal of Botany 66: 1539–1546.

Berg, P., L. Klemedtsson and T. Rosswall. 1982. Inhibitory effect of low partial pressures of acetylene on nitrification. Soil Biology and Biochemistry 14: 301–303.

Berkner, L.V. and L.C. Marshall. 1965. On the origin and rise of oxygen concentration in the Earth's atmosphere. Journal of the Atmospheric Sciences 22: 225–261.

Berliner, R., B. Jacoby and E. Zamski. 1986. Absence of *Cistus incanus* from basaltic soils in Israel: Effect of mycorrhizae. Ecology 67: 1283–1288.

Berner, R.A. 1971. Worldwide sulfur pollution of rivers. Journal of Geophysical Research 76: 6597–6600.

Berner, R.A. 1982. Burial of organic carbon and pyrite sulfur in the modern ocean: its geochemical and environmental significance. American Journal of Science 282: 451–473.

Berner, R.A. 1984. Sedimentary pyrite formation: An update. Geochimica et Cosmochimica Acta 48: 605–615.

Berner, R.A. and S. Honjo. 1981. Pelagic sedimentation of aragonite: Its geochemical significance. Science 211: 940–942.

Berner, R.A. and R. Raiswell. 1983. Burial of organic carbon and pyrite sulfur in sediments over Phanerozoic time: A new theory. Geochimica et Cosmochimica Acta 47: 855–862.

Berner, E.K. and R.A. Berner. 1987. The Global Water Cycle. Prentice Hall, Englewood Cliffs, New Jersey.

Berner, R.A. and G.P. Landis. 1988. Gas bubbles in fossil amber as possible indicators of the major gas composition of ancient air. Science 239: 1406–1409.

Berner, R.A. and A.C. Lasaga. 1989. Modeling the geochemical carbon cycle. Scientific American 260(3): 74–81.

Berner, R.A., A. C. Lasaga and R.M. Garrels. 1983. The carbonate-silicate geochemical cycle and its effect on atmospheric carbon dioxide over the past 100 million years. American Journal of Science 283: 641–683.

Bernier, B. and M. Brazeau. 1988a. Foliar nutrient status in relation to sugar maple dieback and decline in the Quebec Appalachians. Canadian Journal of Forest Research 18: 754–761.

Bernier, B. and M. Brazeau. 1988b. Magnesium deficiency symptoms associated with sugar maple dieback in a Lower Laurentians site in southeastern Quebec. Canadian Journal of Forest Research 18: 1265–1269.

Bernstein, R.E., P.R. Betzer, R.A. Feeley, R.H. Byrne, M.F. Lamb and A. F. Michaels. 1987. Acantharian fluxes and strontium to chlorinity ratios in the North Pacific Ocean. Science 237: 1490–1494.

Berresheim, H. and W. Jaeschke. 1983. The contribution of volcanoes to the global atmospheric sulfur budget. Journal of Geophysical Research 88: 3732–3740.

Bertine, K. K. and E.D. Goldberg. 1971. Fossil fuel combustion and the major sedimentary cycle. Science 173: 233–235.

Betlach, M.R. 1982. Evolution of bacterial denitrification and denitrifier diversity. Antonie van Leeuwenhoek Journal of Microbiology 48: 585–607.

Betzer, P.R., R.H. Byrne, J.G. Acker, C.S. Lewis, R.R. Jolley and R.A. Feely. 1984. The oceanic carbonate system: A reassessment of biogenic controls. Science 226: 1074–1077.

Beven, K. and P. Germann. 1982. Macropores and water flow in soils. Water Resources Research 18: 1311–1325.

Billen, G. 1975. Nitrification in the Scheldt estuary (Belgium and the Netherlands). Estuarine and Coastal Marine Science 3: 79–89.

Billings, W.D. 1950. Vegetation and plant growth as affected by chemically altered rocks in the western Great Basin. Ecology 31: 62–74.

Billings, W.D. 1987. Carbon balance of Alaskan tundra and taiga ecosystems: past, present and future. Quaternary Science Reviews 6: 165–177.

Billings, W.D., J.O. Luken, D.A. Mortensen and K.M. Peterson. 1982. Arctic tundra: A source or sink for atmospheric carbon dioxide in a changing environment? Oecologia 53: 7–11.

Billings, W.D., K.M. Peterson, J.O. Luken and D.A. Mortensen. 1984. Interaction of increasing atmospheric carbon dioxide and soil nitrogen on the carbon balance of tundra microcosms. Oecologia 65: 26–29.

Bilzi, A.F. and E.J. Ciolkosz. 1977. Time as a factor in the genesis of four soils developed in recent alluvium in Pennsylvania. Soil Science Society of America Journal 41: 122–127.

Binkley, D. 1986. Forest Nutrition Management. Wiley, New York.

Binkley, D. and D. Richter. 1987. Nutrient cycles and H^+ budgets of forest ecosystems. Advances in Ecological Research 16: 1–51.

Binkley, D. and S.C. Hart. 1989. The components of nitrogen availability assessments in forest soils. Advances in Soil Science 10: 57–112.

Birk, E.M. and P.M. Vitousek. 1986. Nitrogen availability and nitrogen use efficiency in loblolly pine stands. Ecology 67: 69–79.

Birkeland, P.W. 1978. Soil development as an indication of relative age of Quaternary deposits, Baffin Island, N.W.T., Canada. Arctic and Alpine Research 10: 733–747.

Birkeland, P.W. 1984. Soils and Geomorphology. Oxford University Press, Oxford.

Bishop, J.K.B. 1988. The barite-opal-organic carbon association in oceanic particulate matter. Nature 332: 341–343.

Blaise, T. and J. Garbaye. 1983. Effets de la fertilisation minérale sur les ectomycorhizes d'une hêtraie. Acta Oecologia Plantarum 4: 165–169.

Blake, D.R. and F.S. Rowland. 1988. Continuing worldwide increase in tropospheric methane, 1978–1987. Science 239: 1129–1131.

Blatt, H. and R.L. Jones, 1975. Proportions of exposed igneous, metamorphic, and sedimentary rocks. Geological Society of America Bulletin 86: 1085–1088.

Bloesch, J., P. Stadelmann and H. Bührer. 1977. Primary production, mineralization, and sedimentation in the euphotic zone of two Swiss lakes. Limnology and Oceanography 22: 511–526.

Bockheim, J.G. 1979. Properties and relative age of soils of southwestern Cumberland Peninsula, Baffin Island, N.W.T., Canada. Arctic and Alpine Research 11: 289–306.

Bockheim, J.G. 1980. Solution and use of chronofunctions in studying soil development. Geoderma 24: 71–85.

Boerner, R.E.J. 1984. Foliar nutrient dynamics and nutrient use efficiency of four deciduous tree species in relation to site fertility. Journal of Applied Ecology 21: 1029–1040.

Bohn, H.L., B.L. McNeal and G.A. O'Connor. 1985. Soil Chemistry. 2nd ed. Wiley, New York.

Bolan, N.S., A.D. Robson, N.J. Barrow, and L.A.G. Aylmore. 1984. Specific activity of phosphorus in mycorrhizal and non-mycorrhizal plants in relation to the availability of phosphorus to plants. Soil Biology and Biochemistry 16: 299–304.

Bolin, B., B.R. Döös, J. Jäger and R.A. Warrick (eds.). 1986. The Greenhouse Effect, Climatic Change, and Ecosystems. Wiley, New York.

Bolle, H.-J., W. Seiler and B. Bolin. 1986. Other greenhouse gases and aerosols. pp. 157–203. In B. Bolin, B.R. Doos, J. Jager, and R.A. Warrick (eds.), The Greenhouse Effect, Climatic Change, and Ecosystems. Wiley, Chichester.

Bondietti, E.A., C.F. Baes and S.B. McLaughlin. 1989. Radial trends in cation ratios in tree rings as indicators of the impact of atmospheric deposition on forests. Canadian Journal of Forest Research 19: 586–594.

Bonin, P., M. Gilewicz and J.C. Bertrand. 1989. Effects of oxygen on each step of denitrification on *Pseudomonas nautica*. Canadian Journal of Microbiology 35: 1061–1064.

Boring, L.R., C.D. Monk and W.T. Swank. 1981. Early regeneration of a clear-cut southern Appalachian forest. Ecology 62: 1244–1253.

Boring, L.R., W.T. Swank, J.B. Waide and G.S. Henderson. 1988. Sources, fates, and impacts of nitrogen inputs to terrestrial ecosystems: Review and synthesis. Biogeochemistry 6: 119–159.

Bormann, F.H. and G.E. Likens. 1979. Pattern and Process in a Forested Ecosystem. Springer-Verlag, New York.

Bormann, F.H., G.E. Likens, T.G. Siccama, R.S. Pierce and J.S. Eaton. 1974. The export of nutrients and recovery of stable conditions following deforestation at Hubbard Brook. Ecological Monographs 44: 255–277.

Bormann, B.T. and J.C. Gordon. 1984. Stand density effects in young red alder plantations: Productivity, photosynthate partitioning, and nitrogen fixation. Ecology 65: 394–402.

Born, M., H. Dörr and I. Levin. 1990. Methane consumption in aerated soils of the temperate zone. Tellus 42B: 2–8.

Boström, B., M. Jansson and C. Forsberg. 1982. Phosphorus release from lake sediments. Archiv fur Hydrobiologie Beiheft Ergebnisse der Limnologie 18: 5–59.

Bostrom, B., J.M. Andersen, S. Fleischer and M. Jansson. 1988. Exchange of phosphorus across the sediment-water interface. Hydrobiologia 170: 229–244.

Botkin, D.B. and C. R. Malone. 1968. Efficiency of net primary production based on light intercepted during the growing season. Ecology 49: 438–444.

Botkin, D.B., P.A. Jordan, A.S. Dominski, H.S. Lowendorf and G.E. Hutchinson. 1973. Sodium dynamics in a northern ecosystem. Proceedings of the National Academy of Sciences, U.S.A. 70: 2745–2748.

Botkin, D.B. and L.G. Simpson. 1990. Biomass of North American boreal forest: A step toward accurate global measures. Biogeochemistry 9: 161–174.

Boudreau, B.P. and J.T. Westrich. 1984. The dependence of bacterial sulfate reduction on sulfate concentration in marine sediments. Geochimica et Cosmochimica Acta 48: 2503–2516.

Bowden, R.D., P.A. Steudler, J.M. Melillo, and J.D. Aber. 1990. Annual nitrous oxide fluxes from temperate forest soils in the northeastern United States. Journal of Geophysical Research 95: 13997–14006.

Bowden, W.B. 1986. Gaseous nitrogen emissions from undisturbed terrestrial ecosystems: An assessment of their impacts on local and global nitrogen budgets. Biogeochemistry 2: 249–279.

Bowden, W.B. 1987. The biogeochemistry of nitrogen in freshwater wetlands. Biogeochemistry 4: 313–348.

Bowden, W.B. and F.H. Bormann. 1986. Transport and loss of nitrous oxide in soil water after forest clear-cutting. Science 233: 867–869.

Bowen, G.D. and S.E. Smith. 1981. The effects of mycorrhizas on nitrogen uptake by plants. pp. 237–247. *In* F.E. Clark and T. Rosswall (eds.), Terrestrial Nitrogen Cycles. Swedish Natural Science Research Council, Stockholm.

Bowen, H.J.M. 1966. Trace Elements in Biochemistry. Academic Press, New York.

Bowman. K.P. 1988. Global trends in total ozone. Science 239: 48–50.

Box, E. 1978. Geographical dimensions of terrestrial net and gross primary productivity. Radiation and Environmental Biophysics 15: 305–322.

Box, E.O. 1988. Estimating the seasonal carbon source-sink geography of a natural, steady-state terrestrial biosphere. Journal of Applied Meteorology 27:1109–1124.

Box, E.O., B.N. Holben and V. Kalb. 1989. Accuracy of the AVHRR vegetation index as a predictor of biomass, primary productivity and net CO_2 flux. Vegetatio 80: 71–89.

Boyle, E.A., R. Collier, A.T. Dengler, J.M. Edmond, A.C. Ng, and R.F. Stallard. 1974. On the chemical mass-balance in estuaries. Geochimica et Cosmochimica Acta 38: 1719–1728.

Boyle, E.A., J.M. Edmond and E.R. Sholkovitz. 1977. The mechanism of iron removal in estuaries. Geochimica et Cosmochimica Acta 41: 1313–1324.

Boyle, E.A., F. Sclater and J.D. Edmond. 1976. On the marine geochemistry of cadmium. Nature 263: 42–44.

Boyle, J.R. and G.K. Voigt. 1973. Biological weathering of silicate minerals. Implications for tree nutrition and soil genesis. Plant and Soil 38: 191–201.

Bradley, R.S., H.F. Diaz, J.K. Eischeid, P.D. Jones, P.M. Kelly and C.M. Goodess. 1987. Precipitation fluctuations over northern hemisphere land areas since the mid-l9th century. Science 237: 171–175.

Brand, U. 1989. Biogeochemistry of Late Paleozoic North American brachiopods and secular variation of seawater composition. Biogeochemistry 7: 159–193.

Brasseur, G. and M.H. Hitchman. 1988. Stratospheric response to trace gas perturbations: Changes in ozone and temperature distributions. Science 240: 634–637.

Bravard, S. and D. Righi. 1989. Geochemical differences in an Oxisol-Spodosol toposequence of Amazonia, Brazil. Geoderma 44: 29–42.

Bray, J.R. and E. Gorham. 1964. Litter production in forests of the world. Advances in Ecological Research 2: 101–157.

Bremner, J.M. and A.M. Blackmer. 1978. Nitrous oxide: Emission from soils during nitrification of fertilizer nitrogen. Science 199: 295–296.

Brewer, P.G., C. Goyet and D. Dyrssen. 1989. Carbon dioxide transport by ocean currents at $25°$ N latitude in the Atlantic Ocean. Science 246: 477–479.

Bricker, O.P. 1982. Redox potential: Its measurement and importance in water systems. pp. 55–83. In R.A. Minear and L.H. Keith (eds.), Water Analysis, Vol. 1. Academic Press, New York.

Brimblecombe, P., C. Hammer, H. Rodhe, A. Ryaboshapko and C. F. Boutron. 1989. Human influence on the sulphur cycle. pp. 77–121. In P. Brimblecombe and A.Y. Lein (eds.), Evolution of the Global Biogeochemical Sulphur Cycle. Wiley, New York.

Brinkmann, W.L.F. and U. De Santos. 1974. The emission of biogenic hydrogen sulfide from Amazonian floodplain lakes. Tellus 26: 261–267.

Brinson, M.M., A.E. Lugo and S. Brown. 1981. Primary productivity, decomposition and consumer activity in freshwater wetlands. Annual Review of Ecology and Systematics 12: 123–161.

Brock, T.D. 1985. Life at high temperatures. Science 230: 132–138.

Broda, E. 1975. The history of inorganic nitrogen in the biosphere. Journal of Molecular Evolution 7:87–100.

Broecker, W.S. 1973. Factors controlling CO_2 content in the oceans and atmosphere. pp. 32–50. In G.M. Woodwell and E.V. Pecan (eds.), Carbon and the Biosphere. CONF 720510, National Technical Information Service, Washington, D.C.

Broecker, W.S. 1974. Chemical Oceanography. Harcourt Brace Jovanovich, New York.

Broecker, W.S., T. Takahashi, H.J. Simpson and T.H. Peng. 1979. Fate of fossil fuel carbon dioxide and the global carbon budget. Science 206: 409–418.

Broecker, W.S. 1982. Ocean chemistry during glacial time. Geochimica et Cosmochimica Acta 46: 1689–1705.

Broecker, W.S. and T.-H. Peng. 1987. The oceanic salt pump: Does it contribute to the glacial-interglacial difference in atmospheric CO_2 content? Global Biogeochemical Cycles 1: 251–259.

Brook, G.A., M.E. Folkoff and E.O. Box. 1983. A world model of soil carbon dioxide. Earth Surface Processes and Landforms 8: 79–88.

Brookes, P.C., D.S. Powlson and D.S. Jenkinson. 1982. Measurement of microbial biomass phosphorus in soil. Soil Biology and Biochemistry 14: 319–329.

Brookes, P.C., D.S. Powlson and D.S. Jenkinson. 1984. Phosphorus in the soil microbial biomass. Soil Biology and Biochemistry 16: 169–175.

Brookes, P.C., A. Landman, G. Pruden and D.S. Jenkinson. 1985. Chloroform fumigation and the release of soil nitrogen: A rapid direct extraction method to measure microbial biomass nitrogen in soil. Soil Biology and Biochemistry 17: 837–842.

Brown, K.A. 1985. Sulphur distribution and metabolism in waterlogged peat. Soil Biology and Biochemistry 17: 39–45.

Brown, K.A. 1986. Formation of organic sulphur in anaerobic peat. Soil Biology and Biochemistry 18: 131–140.

Brown, K.A. and J.F. MacQueen. 1985. Sulphate uptake from surface water by peat. Soil Biology and Biochemistry 17: 411–420.

Brown, S. 1981. A comparison of the structure, primary productivity, and transpiration of cypress ecosystems in Florida. Ecological Monographs 51: 403–427.

Brown, S. and A.E. Lugo. 1982. The storage and production of organic matter in tropical forests and their role in the global carbon cycle. BioTropica 14: 161–187.

Brown, S. and A.E. Lugo. 1984. Biomass of tropical forests: A new estimate based on forest volumes. Science 223: 1290–1293.

Brown, S., A.E. Lugo and J. Chapman. 1986. Biomass of tropical tree plantations and its implications for the global carbon budget. Canadian Journal of Forest Research 16: 390–394.

Brugam, R.B. 1978. Human disturbance and the historical development of Linsley pond. Ecology 59: 19–36.

Brunskill, G.J. 1969. Fayetteville Green Lake, New York. II. Precipitation and sedimentation of calcite in a meromictic lake with laminated sediments. Limnology and Oceanography 14: 830–847.

Brylinsky, M. and K.H. Mann. 1973. An analysis of factors governing productivity in lakes and reservoirs. Limnology and Oceanography 18: 1–14.

Bundy, L.G. and J.M. Bremner. 1973. Inhibition of nitrification in soils. Soil Science Society of America Proceedings 37: 396–398.

Burbidge, E.M., G.R. Burbidge, W. A. Fowler and F. Hoyle. 1957. Synthesis of the elements in stars. Reviews of Modern Physics 29: 547–650.

Burford, J.R. and J.M. Bremner. 1975. Relationships between the denitrification capacities of soils and total, water-soluble and readily decomposable soil organic matter. Soil Biology and Biochemistry 7: 389–394.

Burges, A. and D.P. Drover. 1953. The rate of podzol development in sands of the Woy Woy District, N.S.W. Australian Journal of Botany 1: 83–94.

Buringh, P. 1984. Organic carbon in soils of the world. pp. 91–109. In G.M. Woodwell (ed.), The Role of Terrestrial Vegetation in the Global Carbon Cycle. Wiley, New York.

Burke, I.C. 1989. Control of nitrogen mineralization in a sagebrush steppe landscape. Ecology 70: 1115–1126.

Burke, I.C., W.A. Reiners and D.S. Schimel. 1989. Organic matter turnover in a sagebrush steppe landscape. Biogeochemistry 7: 11–31.

Burnett, W.C., M.J. Beers and K.K. Roe. 1982. Growth rates of phosphate nodules from the continental margin off Peru. Science 215: 1616–1618.

Burns, R.C. and R.W.F. Hardy. 1975. Nitrogen Fixation in Bacteria and Higher Plants. Springer-Verlag, New York.

Burns, R.G. 1982. Enzyme activity in soil: Location and a possible role in microbial ecology. Soil Biology and Biochemistry 14: 423–427.

Burns, S.J. and P.A. Baker. 1987. A geochemical study of dolomite in the Monterey Formation, California. Journal of Sedimentary Petrology 57: 128–139.

Butler, J.H., J.W. Elkins, T.M. Thompson and K.B. Egan. 1989. Tropospheric and dissolved N_2O of the west Pacific and east Indian Oceans during the El Niño southern oscillation event of 1987. Journal of Geophysical Research 94: 14865–14877.

Buyanovsky, G.A. and G.H. Wagner. 1983. Annual cycles of carbon dioxide level in soil air. Soil Science Society of America Journal 47: 1139–1145.

Cai, D.-L., F.C. Tan and J.M. Edmond. 1988. Sources and transport of particulate organic carbon in the Amazon River and estuary. Estuarine, Coastal and Shelf Science 26: 1–14.

Cairns-Smith, A.G. 1985. The first organisms. Scientific American 252(6) 90–100.

Cameron, E.M. 1982. Sulphate and sulphate reduction in early Precambrian oceans. Nature 296: 145–148.

Campbell, C.A., E.A. Paul, D.A. Rennie and K.J. McCallum. 1967. Factors affecting the accuracy of the carbon-dating method in soil humus studies. Soil Science 104:81–85.

Canfield, D.E. 1989. Sulfate reduction and oxic respiration in marine sediments: Implications for organic carbon preservation in euxinic environments. Deep Sea Research 36: 121–138.

Canfield, D.E., W.J. Green, T.J. Gardner and T. Ferdelman. 1984. Elemental residence times in Acton Lake, Ohio. Arch. Hydrobiol. 100: 501–509.

Canfield, D.E., E. Philips and C.M. Duarte. 1989. Factors influencing the abundance of blue-green algae in Florida lakes. Canadian Journal of Fisheries and Aquatic Science 46: 1232–1237.

Capone, D.G. and E.J. Carpenter. 1982. Nitrogen fixation in the marine environment. Science 217: 1140–1142.

Caraco, N.F., J.J. Cole and G.E. Likens. 1989. Evidence for sulphate-controlled phosphorus release from sediments of aquatic systems. Nature 341: 316–318.

Caraco, N.F., J.J. Cole and G.E. Likens. 1990. A comparison of phosphorus immobilization in sediments of freshwater and coastal marine systems. Biogeochemistry 9: 277–290.

Carlisle, A., A.H.F. Brown and E.J. White. 1966. The organic matter and nutrient elements in the precipitation beneath a sessile oak (*Quercus petraea*) canopy. Journal of Ecology 54: 87–98.

Carlson, P.R. and J. Forrest. 1982. Uptake of dissolved sulfide by *Spartina alterniflora:* Evidence from natural sulfur isotope abundance ratios. Science 216: 633–635.

Carlyle, J.C. and D.C. Malcolm. 1986. Larch litter and nitrogen availability in mixed larch-spruce stands. I. Nutrient withdrawal, redistribution, and leaching loss from larch foliage at senescence. Canadian Journal of Forest Research 16: 321–326.

Carpenter, S.R., J.G. Kitchell, J.R. Hodgson, P.A. Cochran, J.J. Elser, M.M. Elser, D.M. Lodge, D. Kretchmer, X. He, and C.N. von Ende. 1987. Regulation of lake primary productivity by food web structure. Ecology 68: 1863–1876.

Carr, M.H. 1987. Water on Mars. Nature 326: 30–35.

Carroll, M.A., L.E. Heidt, R.J. Cicerone and R.G. Prinn. 1986. OCS, H_2S and CS_2 fluxes from a salt water marsh. Journal of Atmospheric Chemistry 4: 375–395.

Cassagrande, D.J., K. Siefert, C. Berschinski and N. Sutton. 1977. Sulfur in peat-forming systems of the Okefenokee Swamp and Florida Everglades: Origins of sulfur in coal. Geochimica et Cosmochimica Acta 41: 161–167.

Castelle, A.J. and J.N. Galloway. 1990. Carbon dioxide dynamics in acid forest soils in Shenandoah National Park, Virginia. Soil Science Society of America Journal 54: 252–257.

Castro, M.S. and F.E. Dierberg. 1987. Biogenic hydrogen sulfide emissions from selected Florida wetlands. Water, Air, and Soil Pollution 33: 1–13.

Cavanaugh, C.M., S.L. Gardiner, M. L. Jones, H.W. Jannasch and J.B. Waterbury. 1981. Prokaryotic cells in the hydrothermal vent tube worm *Riftia pachyptila* Jones: Possible chemoautotrophic symbionts. Science 213: 340–342.

Cejudo, F.J., A. de La Torre and A. Paneque. 1984. Short-term ammonium inhibition of nitrogen fixation in *Azotobacter*. Biochemical and Biophysical Research Communications 123: 431–437.

Cess, R.D., G.L. Potter, J.P. Blanchet, G.J. Boer, S.J. Ghan, J.T. Kiehl, H. Le Treut, Z.-X. Li, X.-Z. Liang, J.F.B. Mitchell, J.-J. Morcrette, D.A. Randall, M.R. Riches, E. Roeckner, U. Schlese, A. Slingo, K.E. Taylor, W.M. Washington, R.T. Wetherald and I. Yagai. 1989. Interpretation of cloud-climate feedback as produced by 14 atmospheric general circulation models. Science 245: 513–516.

Chambers, L.A. and P.A. Trudinger. 1979. Microbiological fractionation of stable sulfur isotopes: A review and critique. Geomicrobiology Journal 1: 249–293.

Chameides, W.L., R.W. Lindsay, J. Richardson, and C.S. Kiang. 1988. The role of biogenic hydrocarbons in urban photochemical smog: Atlanta as a case study. Science 241: 1473–1475.

Chandler, R.F. 1942. The time required for podzol profile formation as evidenced by the Mendenhall glacial deposits near Juneau, Alaska. Soil Science Society of America Proceedings 7: 454–459.

Chang, S., D. DesMarais, R. Mack, S.L. Miller and G.E. Strathearn. 1983. Prebiotic organic syntheses and the origin of life. pp. 53–92. *In* J.W. Schopf (ed.), Earth's Earliest Biosphere. Princeton University Press, Princeton, New Jersey.

Chang, W.Y.B. and R.A. Moll. 1980. Prediction of hypolimnetic oxygen deficits: Problems of interpretation. Science 209: 721–722.

Chapin, F.S. 1980. The mineral nutrition of wild plants. Annual Review of Ecology and Systematics 11: 233–260.

Chapin, F.S. 1974. Morphological and physiological mechanisms of temperature compensation in phosphate absorption along a latitudinal gradient. Ecology 55: 1180–1198

Chapin, F.S., R.J. Barsdate and D. Barèl. 1978. Phosphorus cycling in Alaskan coastal tundra: A hypothesis for the regulation of nutrient cycling. Oikos 31: 189–199

Chapin, F.S. and R. A. Kedrowski. 1983. Seasonal changes in nitrogen and phosphorus fractions and autumn retranslocation in evergreen and deciduous taiga trees. Ecology 64: 376–391.

Chapin, F.S. and W.C. Oechel. 1983. Photosynthesis, respiration, and phosphate absorption by *Carex aquatilis* ecotypes along latitudinal and local environmental gradients. Ecology 64: 743–751.

Chapin, F.S., P.M. Vitousek and K. Van Cleve. 1986a. The nature of nutrient limitation in plant communities. American Naturalist 127: 48–58.

Chapin, F.S., G.R. Shaver and R.A. Kedrowski. 1986b. Environmental controls over carbon, nitrogen and phosphorus fractions in *Eriophorum vaginatum* in Alaskan tussock tundra. Journal of Ecology 74: 167–195.

Chapin, F.S. 1988. Ecological aspects of plant mineral nutrition. Advances in Mineral Nutrition 3: 161–191.

Chapman, D.J. and J.W. Schopf. 1983. Biological and biochemical effects of the development of an aerobic environment. pp. 302–320. *In* J.W. Schopf (ed.), Earth's Earliest Biosphere. Princeton University Press, Princeton, New Jersey.

Chapman, S.B., J. Hibble and C. R. Rafarel. 1975. Net aerial production by *Calluna vulgaris* on lowland heath in Britain. Journal of Ecology 63: 233–258.

Chappellaz, J., J.M. Barnola, D. Raynaud, Y.S. Korotkevich and C. Lorius. 1990. Ice-core record of atmospheric methane over the past 160,000 years. Nature 345: 127–131.

Charley, J.L. and N.E. West. 1977. Micro-patterns of nitrogen mineralization activity in soils of some shrub-dominated semi-desert ecosystems of Utah. Soil Biology and Biochemistry 9: 357–365.

Charlou, J.L., L. Dmitriev, H. Bougault and H.D. Needham. 1988. Hydrothermal CH_4 between 12°N and 15°N over the Mid-Atlantic ridge. Deep Sea Research 35: 121–131.

Charlson, R.J., J.E. Lovelock, M.O. Andreae and S.G. Warren. 1987. Oceanic phytoplankton, atmospheric sulphur, cloud albedo and climate. Nature 326: 655–661.

Chase, E.M. and F.L. Sayles. 1980. Phosphorus in suspended sediments of the Amazon river. Estuarine and Coastal Marine Science 11: 383–391.

Cherry, R.D., J.J.W. Higgo and S.W. Fowler. 1978. Zooplankton fecal pellets and element residence times in the ocean. Nature 274:246–248.

Chesworth, W. and F. Macias-Vasquez. 1985. Pe, pH, and podzolization. American Journal of Science 285: 128–146.

Chesworth, W., J. Dejou and P. Larroque. 1981. The weathering of basalt and relative mobilities of the major elements at Belbex, France. Geochimica et Cosmochimica Acta 45: 1235–1243.

Chevalier, R.A. and C. L. Sarazin. 1987. Hot gas in the Universe. American Scientist 75: 609–618.

Cho, B.C. and F. Azam. 1988. Major role of bacteria in biogeochemical fluxes in the ocean's interior. Nature 332:441–443.

Cho, C.M. 1982. Oxygen consumption and denitrification kinetics in soil. Soil Science Society of America Journal 46: 756–762.

Christensen, J.P., J.W. Murray, A.H. Devol and L. A. Codispoti. 1987. Denitrification in continental shelf sediments has major impact on the oceanic nitrogen budget. Global Biogeochemical Cycles 1: 97–116.

Christensen, N.L. 1973. Fire and the nitrogen cycle in California chaparral. Science 181: 66–68.

Christensen, N.L. 1977. Fire and soil-plant nutrient relations in a pine-wiregrass savanna on the coastal plain of North Carolina. Oecologia 31: 27–44.

Christensen, N.L. and T. MacAller. 1985. Soil mineral nitrogen transformations during succession in the Piedmont of North Carolina. Soil Biology and Biochemistry 17: 675–681.

Chyba, C.F. 1987. The cometary contribution to the oceans of primitive Earth. Nature 330: 632–635.

Chyba, C.F. 1990. Impact delivery and erosion of planetary oceans in the early inner solar system. Nature 343: 129–133.

Cicerone, R.J. 1987. Changes in stratospheric ozone. Science 237: 35–42.

Cicerone, R.J. and R.S. Oremland. 1988. Biogeochemical aspects of atmospheric methane. Global Biogeochemical Cycles 2: 299–327.

Clarkson, D.T. and J.B. Hanson. 1980. The mineral nutrition of higher plants. Annual Review of Plant Physiology 31: 239–298.

Clayton, J.L. 1976. Nutrient gains to adjacent ecosystems during a forest fire: An evaluation. Forest Science 22: 162–166.

Clymo, R.S. 1984. The limits to peat bog growth. Philosophical Transactions of the Royal Society of London 303B: 605–654.

Codispoti, L.A. and J.P. Christensen. 1985. Nitrification, denitrification and nitrous oxide cycling in the eastern tropical South Pacific Ocean. Marine Chemistry 16: 277–300.

Codispoti, L.A., G. E. Friederich, T.T. Packard, H.E. Glover, P.J. Kelley, R.W. Spinrad, R.T. Barber, J.W. Elkins, B.B. Ward, F. Lipschultz and N. Lostaunau. 1986. High nitrite levels off northern Peru: A signal of instability in the marine denitrification rate. Science 233: 1200–1202.

Cogbill, C.V. and G.E. Likens. 1974. Acid precipitation in the northeastern United States. Water Resources Research 10: 1133–1137.

Cohen, A.D. 1974. Petrography and paleoecology of Holocene peats from the Okefenokee swamp-marsh complex of Georgia. Journal of Sedimentary Petrology 44: 716–726.

Cohen, Y. and L.I. Gordon. 1979. Nitrous oxide production in the ocean. Journal of Geophysical Research 84: 347–353.

COHMAP. 1988. Climatic changes of the last 18,000 years: Observations and model simulations. Science 241: 1043–1052.

Cole, C.V. and S.R. Olsen. 1959. Phosphorus solubility in calcareous soils. I. Dicalcium phosphate activities in equilibrium solutions. Soil Science Society of America Proceedings 23: 116–118.

Cole, C.V., G.S. Innis and J.W.B. Stewart. 1977. Simulation of phosphorus cycling in semiarid grasslands. Ecology 58: 3–15.

Cole, C.V., E.T. Elliott, H.W. Hunt and D.C. Coleman. 1978. Trophic interactions in soils as they affect energy and nutrient dyanmics. V. Phosphorus transformations. Microbial Ecology 4: 381–387.

Cole, D.W. and M. Rapp. 1981. Element cycling in forest ecosystems. pp. 341–409. In D.E.

Reichle (ed.), Dynamic Properties of Forest Ecosystems. Cambridge University Press, London

Cole, J.J. and S.G. Fisher. 1979. Nutrient budgets of a temporary pond ecosystem. Hydrobiologia 63: 213–222.

Cole, J.J., S. Honjo and J. Erez. 1987. Benthic decomposition of organic matter at a deep-water site in the Panama basin. Nature 327: 703–704.

Cole, J.J., S. Findlay and M.L. Pace. 1988. Bacterial production in fresh and saltwater ecosystems: a cross-system overview. Marine Ecology Progress Series 43: 1–10.

Coleman, J.M. and E.S. Deevey. 1987. Lacustrine sediment/groundwater nutrient dynamics. Biogeochemistry 4: 3–14.

Coley, P.D., J.P. Bryant and F. S. Chapin. 1985. Resource availability and plant antiherbivore defense. Science 230: 895–899.

Conley, D.J., M.A. Quigley and C.L. Schelske. 1988. Silica and phosphorus flux from sediments: Importance of internal recycling in Lake Michigan. Canadian Journal of Fisheries and Aquatic Science 45: 1030–1035.

Conrad, R., W. Seiler and G. Bunse. 1983. Factors influencing the loss of fertilizer nitrogen into the atmosphere as N_2O. Journal of Geophysical Research 88: 6709–6718.

Conrad, R. and W. Seiler. 1988. Methane and hydrogen in seawater (Atlantic Ocean). Deep Sea Research 35: 1903–1917.

Conway, T.J., P. Tans, L.S. Waterman, K.W. Thoning, K.A. Masarie and R.H. Gammon. 1988. Atmospheric carbon dioxide measurements in the remote global troposphere, 1981–1984. Tellus 40B: 81–115.

Cook, E.A., L.R. Iverson and R.L. Graham. 1989. Estimating forest productivity with thematic mapper and biogeographical data. Remote Sensing of Environment 28: 131–141.

Cooke, R.U. and A. Warren. 1973. Geomorphology in Deserts. University of California Press, Berkeley.

Copin-Montegut, C. and G. Copin-Montegut. 1983. Stoichiometry of carbon, nitrogen, and phosphorus in marine particulate matter. Deep Sea Research 30: 31–46.

Corliss, B.H., D.G. Martinson and T. Keffer. 1986. Late Quaternary deep-ocean circulation. Geological Society of America Bulletin 97: 1106–1121.

Corliss, J.B., J. Dymond, L.I. Gordon, R.P. Von Herzen, R.D. Ballard, K. Green, D. Williams, A. Bainbridge, K. Crane and T.H. van Andel. 1979. Submarine thermal springs on the Galápagos Rift. Science 203: 1073–1083.

Cornett, R.J. and F.H. Rigler. 1979. Hypolimnetic oxygen deficits: Their prediction and interpretation. Science 205: 580–581.

Cornett, R.J. and F.H. Rigler. 1980. The areal hypolimnetic oxygen deficit: An empirical test of the model. Limnology and Oceanography 25: 672–679.

Correll, D.L. 1981. Nutrient mass balances for the watershed, headwaters intertidal zone, and basin of the Rhode River estuary. Limnology and Oceanography 26: 1142–1149.

Correll, D.L. and D. Ford. 1982. Comparison of precipitation and land runoff as sources of estuarine nitrogen. Estuarine, Coastal and Shelf Science 15: 45–56.

Cosby, B.J., G.M. Hornberger, J.N. Galloway and R.F. Wright. 1985. Modeling the effects of acid deposition: Assessment of a lumped parameter model of soil water and streamwater chemistry. Water Resources Research 21: 51–63.

Cosby, B.J., G. M. Hornberger, R.F. Wright and J.N. Galloway. 1986. Modeling the effects of acid deposition: Control of long-term sulfate dynamics by soil sulfate adsorption. Water Resources Research 22: 1283–1291.

Courchesene, F. and W.H. Hendershot. 1989. Sulfate retention in some podzolic soils of the southern Laurentians, Quebec. Canadian Journal of Soil Science 69: 337–350.

Cox, T.L., W.F. Harris, B.S. Ausmus and N.T. Edwards. 1978. The role of roots in biogeochemical cycles in an eastern deciduous forest. Pedobiologia 18: 264–271.

Craig, H., C.C. Chou, J.A. Welhan, C.M. Stevens and A. Engelkemeir. 1988. The isotopic composition of methane in polar ice cores. Science 242: 1535–1539.

Craig, H. and T. Hayward. 1987. Oxygen supersaturation in the ocean: Biological versus physical contributions. Science 235: 199–202.

Craig, P.J. 1980. Metal cycles and biological methylation. pp. 169–227. In O. Hutzinger (ed.), The Handbook of Environmental Chemistry, Vol. 1, Part A, The Natural Environment and the Biogeochemical Cycles. Springer Verlag, New York.

Crill, P.M. and C.S. Martens. 1986. Methane production from bicarbonate and acetate in an anoxic marine sediment. Geochimica et Cosmochimica Acta 50: 2089–2097.

Crill, P.M., K.B. Bartlett, R.C. Harriss, E. Gorham, E.S. Verry, D.I. Sebacher, L. Madzar and W. Sanner. 1988. Methane flux from Minnesota peatlands. Global Biogeochemical Cycles 2: 371–384.

Crisp, D.T. 1966. Input and output of minerals for an area of Pennine moorland: The importance of precipitation, drainage, peat erosion and animals. Journal of Applied Ecology 3: 327–348.

Crocker, R.L. and B.A. Dickson. 1957. Soil development on the recessional moraines of the Herbert and Mendenhall Glaciers, south-eastern Alaska. Journal of Ecology 45: 169–185.

Crocker, R.L. and J. Major. 1955. Soil development in relation to vegetation and surface age at Glacier Bay, Alaska. Journal of Ecology 43: 427–448.

Cromack, K., P. Sollins, W. C. Graustein, K. Speidel, A.W. Todd, G. Spycher, C.Y. Li and R.L. Todd. 1979. Calcium oxalate accumulation and soil weathering in mats of the hypogeous fungus Hysterangium crassum. Soil Biology and Biochemistry 11: 463–468.

Cronan, C.S. 1980. Solution chemistry of a New Hampshire subalpine ecosystem: A biogeochemical analysis. Oikos 34: 272–281.

Cronan, C.S. and G.R. Aiken. 1985. Chemistry and transport of soluble humic substances in forested watersheds of the Adirondack Park, New York. Geochimica et Cosmochimica Acta 49: 1697–1705.

Cronan, C.S., C.T. Driscoll, R.M. Newton, J.M. Kelly, C.L. Schofield, R.J. Bartlett and R. April. 1990. A comparative analysis of aluminum biogeochemistry in a northeastern and a southeastern forested watershed. Water Resources Research 26: 1413–1430.

Cross, F.A., L.H. Hardy, N.Y. Jones and R.T. Barber. 1973. Relation between total body weight and concentrations of manganese, iron, copper, zinc, and mercury in white muscle of bluefish (Pomatomus saltatrix) and a bathyl-demersal fish Antimora rostrata. Journal of the Fisheries Research Board of Canada 30: 1287–1291.

Cross, P.M. and F.H. Rigler. 1983. Phosphorus and iron retention in sediments measured by mass budget calculations and directly. Canadian Journal of Fisheries and Aquatic Science 40: 1589–1597.

Crutzen, P.J. 1983. Atmospheric interactions-Homogeneous gas reactions of C, N, and S containing compounds. pp. 67–114. In B. Bolin and R.B. Cook (eds.), The Major Biogeochemical Cycles and Their Interactions, Wiley, New York.

Crutzen, P.J. 1988. Variability in atmospheric-chemical systems. pp. 81–108. In T. Rosswall, R.G. Woodmansee and P.G. Risser (eds.), Scales and Global Change. Wiley, New York.

Crutzen, P.J., I. Aselmann and W. Seiler. 1986. Methane production by domestic animals, wild ruminants, other herbivorous fauna, and humans. Tellus 38B: 271–284.

Crutzen, P.J., A.C. Delany, J. Greenberg, P. Haagenson, L. Heidt, R. Lueb, W. Pollock, W. Seiler, A. Wartburg and P. Zimmerman. 1985. Tropospheric chemical composition measurements in Brazil during the dry season. Journal of Atmospheric Research 2: 233–256.

Cuevas, E. and E. Medina. 1986. Nutrient dynamics within Amazonian forest ecosystems. I. Nutrient flux in fine litter fall and efficiency of nutrient utilization. Oecologia 68: 466–472.

Cuevas, E. and E. Medina. 1988. Nutrient dynamics within Amazonian forests. II. Fine root growth, nutrient availability and leaf litter decomposition. Oecologia 76: 222–235.

Cuffney, T.F. 1988. Input, movement and exchange of organic matter within a subtropical coastal blackwater river–floodplain system. Freshwater Biology 19: 305–320.

Cullis, C.F. and M.M. Hirschler. 1980. Atmospheric sulphur: Natural and man-made sources. Atmospheric Environment 14: 1263–1278.

Cushon, G.H. and M.C. Feller. 1989. Asymbiotic nitrogen fixation and denitrification in a mature forest in coastal British Columbia. Canadian Journal of Forest Research 19: 1194–1200.

Dacey, J.W.H. 1981. Pressurized ventilation in the yellow waterlily. Ecology 62: 1137–1147.

Dacey, J.W.H. and B.L. Howes. 1984. Water uptake by roots controls water table movement and sediment oxidation in short *Spartina* marsh. Science 224: 487–489.

Dalal, R.C. and R.J. Mayer. 1986a. Long-term trends in fertility of soils under continuous cultivation and cereal cropping in southern Queensland. III. Distribution and kinetics of soil organic carbon in particle-size fractions. Australian Journal of Soil Research 24: 293–300.

Dalal, R.C. and R.J. Mayer. 1986b. Long-term trends in fertility of soils under continuous cultivation and cereal cropping in southern Queensland. IV. Loss of organic carbon from different density fractions. Australian Journal of Soil Research 24: 301–309.

Damman, A.W.H. 1978. Distribution and movement of elements in ombrotrophic peat bogs. Oikos 30: 480–495.

Damman, A.W.H. 1988. Regulation of nitrogen removal and retention in sphagnum bogs and other peatlands. Oikos 51: 291–305.

Daniels, L., N. Belay, B.S. Rajagopal and P.J. Weimer. 1987. Bacterial methanogenesis and growth from CO_2 with elemental iron as the sole source of electrons. Science 237: 509–511.

Dankers, N., M. Binsbergen, K. Zegers, R. Laane and M.R. van der Loeff. 1984. Transportation of water, particulate and dissolved organic and inorganic matter between a salt marsh and the Ems-Dollard estuary, The Netherlands. Estuarine, Coastal and Shelf Science 19: 143–166.

Darling, M.S. 1976. Interpretation of global differences in plant calorific values: The significance of desert and arid woodland vegetation. Oecologia 23: 127–139.

D'Arrigo, R., G.C. Jacoby and I.Y. Fung. 1987. Boreal forests and atmosphere-biosphere exchange of carbon dioxide. Nature 329: 321–323.

David, M.B., M.J. Mitchell and J.P. Nakas. 1982. Organic and inorganic sulfur constituents of a forest soil and their relationship to microbial activity. Soil Science Society of America Journal 46: 847–857.

David, M.B., J.O. Reuss and P.M. Walthall. 1988. Use of a chemical equilibrium model to understand soil chemical processes that influence soil solution and surface water alkalinity. Water, Air, and Soil Pollution 38: 71–83.

Davidson, E.A. and W.T. Swank. 1987. Factors limiting denitrification in soils from mature and disturbed southeastern hardwood forests. Forest Science 33: 135–144.

Davidson, E.A., W.T. Swank and T.O. Perry. 1986. Distinguishing between nitrification and denitrification as sources of gaseous nitrogen production in soil. Applied and Environmental Microbiology 52: 1280–1286.

Davidson, E.A. and W.T. Swank. 1986. Environmental parameters regulating gaseous nitrogen losses from two forested ecosystems via nitrification and denitrification. Applied and Environmental Microbiology 52: 1287–1292.

Davies, W.G. 1972. Introduction to Chemical Thermodynamics. W.B. Saunders, Philadelphia.

Davison, W., C. Woof and E. Rigg. 1982. The dynamics of iron and manganese in a seasonally anoxic lake: Direct measurement of fluxes using sediment traps. Limnology and Oceanography 27: 987–1003.

Dawson, G.A. 1977. Atmospheric ammonia from undisturbed land. Journal of Geophysical Research 82: 3125–3133.

De Angelis, M., N.I. Barkov and N.V. Petrov. 1987. Aerosol concentrations over the last climatic cycle (160 kyr) from an Antarctic ice core. Nature 325: 318–321.

DeBano, L.F. and C.E. Conrad. 1978. The effects of fire on nutrients in a chaparral ecosystem. Ecology 59: 489–497.

DeBano, L.F. and J.M. Klopatek. 1988. Phosphorus dynamics of pinyon-juniper soils following simulated burning. Soil Science Society of America Journal 52: 271–277.

DeBell, D.S. and C.W. Ralston. 1970. Release of nitrogen by burning light forest fuels. Soil Science Society of America Proceedings 34: 936–938.

Deevey, E.S. 1970a. Mineral cycles. Scientific American 223(3): 148–158.

Deevey, E.S. 1970b. In defense of mud. Bulletin of the Ecological Society of America 51(l): 5–8.

Deevey, E.S. 1988. Estimation of downward leakage from Florida lakes. Limnology and Oceanography 33: 1308–1320.

Degens, E.T. and K. Mopper. 1976. Factors controlling the distribution and early diagenesis of organic material in marine sediments. pp. 60–113. In J.P. Riley and R. Chester (eds.), Chemical Oceanography, 2nd ed., Vol. 6. Academic Press, New York.

Degens, E.T., H.K. Wong and S. Kempe. 1981. Factors controlling global climate of the past and the future. pp. 3–24. In G.E. Likens (ed.), Some Perspectives of the Major Biogeochemical Cycles. Wiley, New York.

De Jonge, V.N. and L.A. Villerius. 1989. Possible role of carbonate dissolution in estuarine phosphate dynamics. Limnology and Oceanography 34: 332–340.

De Kimpe, C.R. and Y. A. Martel. 1976. Effects of vegetation on the distribution of carbon, iron, and aluminum in the B horizons of Northern Appalachian Spodosols. Soil Science Society of America Journal 40: 77–80.

DeLaune, R.D., R.H. Baumann and J.G. Gosselink. 1983. Relationships among vertical accretion, coastal submergence, and erosion in a Louisiana Gulf Coast marsh. Journal of Sedimentary Petrology 53: 147–157.

Delcourt, H.R. and W.F. Harris. 1980. Carbon budget of the southeastern U.S. biota: Analysis of historical change in trend from source to sink. Science 210: 321–323.

Delmas, R. and J. Servant. 1983. Atmospheric balance of sulphur above an equatorial forest. Tellus 35B: 110–120.

DeLucia, E.H., W.H. Schlesinger and W.D. Billings. 1988. Water relations and the maintenance of Sierran conifers on hydrothermally altered rock. Ecology 69: 303–311.

DeLucia, E.H. and W.H. Schlesinger. 1991. Resource-use efficiency and drought tolerance in adjacent Great Basin and Sierran plants. Ecology 72: 51–48.

Delwiche, C.C. 1970. The nitrogen cycle. Scientific American 223(3): 136–146.

DeMaster, D.J. 1981. The supply and accumulation of silica in the marine environment. Geochimica et Cosmochimica Acta 45: 1715–1732.

Denmead, O.T., J.R. Simpson and J.R. Freney. 1974. Ammonia flux into the atmosphere from a grazed pasture. Science 185: 609–610.

Denmead, O.T., J.R. Freney and J.R. Simpson. 1976. A closed ammonia cycle within a plant canopy. Soil Biology and Biochemistry 8: 161–164.

Dethier, D.P., S.B. Jones, T. P. Feist and J.E. Ricker. 1988. Relations among sulfate, aluminum, iron, dissolved organic carbon, and pH in upland forest soils of northwestern Massachusetts. Soil Science Society of America Journal 52: 506–512.

Detwiler, R.P. 1986. Land use change and the global carbon cycle: The role of tropical soils. Biogeochemistry 2: 67–93.

Detwiler, R.P. and C.A.S. Hall. 1988. Tropical forests and the global carbon cycle. Science 239: 42–47.

Deuser, W.G., E.H. Ross and R.F. Anderson. 1981. Seasonality in the supply of sediment to the deep Sargasso Sea and implications for the rapid transfer of matter to the deep ocean. Deep Sea Research 28: 495–505.

Deuser, W.G., P.G. Brewer, T.D. Jickells and R.F. Commeau. 1983. Biological control of the removal of abiogenic particles from the surface ocean. Science 219: 388–391.

Dévai, I., L. Felföldy, T. Wittner and S. Plósz. 1988. Detection of phosphine: New aspects of the phosphorus cycle in the hydrosphere. Nature 333: 343–345.

Devol, A.H., J.E. Richey, W.A. Clark, S.L. King and L.A. Martinelli. 1988. Methane emissions to the troposphere from the Amazon floodplain. Journal of Geophysical Research 93: 1583–1592.

De Vooys, C.G.N. 1979. Primary production in aquatic environments. pp. 259–292. *In* B. Bolin, E.T. Degens, S. Kempe and P. Ketner (eds.), The Global Carbon Cycle. Wiley, New York.

Dhamala, B.R., M.J. Mitchell, and A.C. Stam. 1990. Sulfur dynamics in mineral horizons of two northern hardwood soils: A column study with ^{35}S. Biogeochemistry 10: 143–160.

Dianov-Klokov, V.I., L.N. Yurganov, E.I. Grechko and A.V. Dzhola. 1989. Spectroscopic measurements of atmospheric carbon monoxide and methane: 1. Latitudinal distribution. Journal of Atmospheric Chemistry 8: 139–151.

Dickerson R.E. 1978. Chemical evolution and the origin of life. Scientific American 239(3): 70–86.

Dickerson, R.R., G.J. Huffman, W.T. Luke, L.J. Nunnermacker, K.E. Pickering, A.C.D. Leslie, C.G. Lindsey, W.G.N. Slinn, T.J. Kelly, P.H. Daum, A.C. Delany, J.P. Greenberg, P.R. Zimmerman, J.F. Boatman, J.D. Ray and D.H. Stedman. 1987. Thunderstorms: An important mechanism in the transport of air pollutants. Science 235: 460–465.

Dickinson, R.E. and R.J. Cicerone. 1986. Future global warming from atmospheric trace gases. Nature 319: 109–115.

Dickson, B.A. and R.L. Crocker. 1953. A chronosequence of soils and vegetation near Mt. Shasta, California. II. The development of the forest floors and the carbon and nitrogen profiles of the soils. Journal of Soil Science 4: 142–154.

Dickson, R.R., E.M. Gmitrowicz and A.J. Watson. 1990. Deep-water renewal in the northern North Atlantic. Nature 344: 848–850.

Dillon, P.J. and F.H. Rigler. 1974. The phosphorus-chlorophyll relationship in lakes. Limnology and Oceanography 19: 767–773.

Dingman, S.L. and A.H. Johnson. 1971. Pollution potential of some New Hampshire lakes. Water Resources Research 7: 1208–1215.

Dixon, K.W., J.S. Pate and W.J. Bailey. 1980. Nitrogen nutrition of the tuberous sundew *Drosera erythrorhiza* Lindl with special reference to catch of arthropod fauna by its glandular leaves. Australian Journal of Botany 28: 283–297.

Dodd, J.C., C.C. Burton, R.G. Burns and P. Jeffries. 1987. Phosphatase activity associated with the roots and the rhizosphere of plants infected with vesicular—arbuscular mycorrhizal fungi. New Phytologist 107: 163–172.

Donahue, T.M., J.H. Hoffman, R.R. Hodges, and A.J. Watson. 1982. Venus was wet: A measurement of the ratio of deuterium to hydrogen. Science 216: 630–633.

Dormaar, J.F. 1979. Organic matter characteristics of undisturbed and cultivated chernozemic and solonetzic A horizons. Canadian Journal of Soil Science 59: 349–356.

Dorn, R.I. and T.M. Oberlander. 1981. Microbial origin of desert varnish. Science 213: 1245–1247.

Dregne, H.E. 1976. Soils of Arid Regions. Elsevier Scientific, Amsterdam.

Drever, J.I. 1988. The Geochemistry of Natural Waters, 2nd ed. Prentice-Hall, Englewood Cliffs, New Jersey.

Drever, J.I. and C.L. Smith. 1978. Cyclic wetting and drying of the soil zone as an influence on the chemistry of ground water in arid terrains. American Journal of Science 278: 1448–1454.

Driscoll, C.T., H. Van Breemen, and J. Mulder. 1985. Aluminum chemistry in a forested Spodosol. Soil Science Society of America Journal 49: 437–444.

Duce, R.A. 1983. Biogeochemical cycles and the air-sea exchange of aerosols. pp. 427–456. *In* B. Bolin and R.B. Cook (eds.), The Major Biogeochemical Cycles and Their Interactions. Wiley, New York.

Duce, R.A., C.K. Unni, B.J. Ray, J.M. Prospero and J.T. Merrill. 1980. Long-range atmospheric transport of soil dust from Asia to the tropical North Pacific: Temporal variability. Science 209: 1522–1524.

Ducklow, H.W., D.A. Purdie, P.J. LeB. Williams and J.M. Davies. 1986. Bacterioplankton: A sink for carbon in a coastal marine plankton community. Science 232: 865–867.

Dunn, P.H., L.F. DeBano and G.E. Eberlein. 1979. Effect of burning on chaparral soils: II. Soil microbes and nitrogen mineralization. Soil Science Society of America Journal 43: 509–514.

Du Rietz, G.E. 1949. Huvudenheter och Huvudgränser i Svensk Myrvegetation. Svensk Botanisk Tidskrift 43: 274–309.

Eamus, D. and P.G. Jarvis. 1989. The direct effects of increase in the global atmospheric CO_2 concentration on natural and commercial temperate trees and forests. Advances in Ecological Research 19: 1–55.

Edmond, J.M., C. Measures, R.E. McDuff, L.H. Chan, R. Collier, B. Grant, L. I. Gordon and J. B. Corliss. 1979. Ridge crest hydrothermal activity and the balances of the major and minor elements in the ocean: The Galapagos data. Earth and Planetary Science Letters 46: 1–18.

Edmond, J.M., E.A. Boyle, B. Grant and R.F. Stallard. 1981. The chemical mass balance in the Amazon plume. I: The nutrients. Deep Sea Research 28: 1339–1374.

Edmondson, W.T. and J.T. Lehman. 1981. The effect of changes in the nutrient income on the condition of Lake Washington. Limnology and Oceanography 26: 1–29.

Edwards, N.T. 1975. Effects of temperature and moisture on carbon dioxide evolution in a mixed deciduous forest floor. Soil Science Society of America Proceedings 39: 361–365.

Edwards, N.T. and P. Sollins. 1973. Continuous measurement of carbon dioxide evolution from partitioned forest floor components. Ecology 54: 406–412.

Edwards, N.T. and W.F. Harris. 1977. Carbon cycling in a mixed deciduous forest floor. Ecology 58: 431–437.

Edwards, P.J. 1977. Studies of mineral cycling in a montane rain forest in New Guinea. II. The production and disappearance of litter. Journal of Ecology 65: 971–992.

Edwards, P.J. 1982. Studies of mineral cycling in a montane rain forest in New Guinea. V. Rates of cycling in throughfall and litter fall. Journal of Ecology 70: 807–827.

Edwards, P.J. and P.J. Grubb. 1982. Studies of mineral cycling in a montane rain forest in New Guinea. IV. Soil characteristics and the division of mineral elements between the vegetation and soil. Journal of Ecology 70: 649–666.

Edwards, R.T. 1987. Sestonic bacteria as a food source for filtering invertebrates in two southeastern blackwater rivers. Limnology and Oceanography 32: 221–234.

Edwards, R.T. and J.L. Meyer. 1987. Metabolism of a sub-tropical low gradient blackwater river. Freshwater Biology 17: 251–263.

Edwards, R.T. and J.L. Meyer. 1986. Production and turnover of planktonic bacteria in two southeastern blackwater rivers. Applied and Environmental Microbiology 52: 1317–1323.

Eghbal, M.K., R.J. Southard and L.D. Whittig. 1989. Dynamics of evaporite distribution in soils on a fan-playa transect in the Carrizo Plain, California. Soil Science Society of America Journal 53: 898–903.

Ehhalt, D.H. 1974. The atmospheric cycle of methane. Tellus 26: 58–70.

Ehhalt, D.H. 1981. Chemical coupling of the nitrogen, sulphur, and carbon cycles in the atmosphere. pp. 81–91. *In* G.E. Likens (ed.), Some Perspectives of the Major Biogeochemical Cycles. Wiley, New York.

Ehrlich, H.L. 1975. The formation of ores in the sedimentary environment of the deep sea with microbial participation: The case for ferromanganese concretions. Soil Science 119: 36–41.

Ehrlich, H.L. 1982. Enhanced removal of Mn^{2+} from seawater by marine sediments and clay minerals in the presence of bacteria. Canadian Journal of Microbiology 28: 1389–1395.

Eichner, M.J. 1990. Nitrous oxide emissions from fertilized soils: Summary of available data. Journal of Environmental Quality 19: 272–280.

Eisele, K.A., D.S. Schimel, L.A. Kapustka and W.J. Parton. 1989. Effects of available P and N:P ratios on non-symbiotic dinitrogen fixation in tallgrass prairie soils. Oecologia 79: 471–474.

Elliott, E.T. 1986. Aggregate structure and carbon, nitrogen, and phosphorus in native and cultivated soils. Soil Science Society of America Journal 50: 627–633.

Elwood, J.W., J.D. Newbold, A.F. Trimble and R.W. Stark. 1981. The limiting role of phosphorus in a woodland stream ecosystem: Effects of P enrichment on leaf decomposition and primary producers. Ecology 62: 146–158.

Emanuel, W.R., H.H. Shugart and M.P. Stevenson. 1985a. Climatic change and the broad-scale distribution of terrestrial ecosystem complexes. Climatic Change 7: 29–43.

Emanuel, W.R., I.Y.-S. Fung, G.G. Killough, B. Moore and T.-H. Peng. 1985b. Modeling the global carbon cycle and changes in the atmospheric carbon dioxide levels. pp. 141–173. In J.R. Trabalka (ed.), Atmospheric Carbon Dioxide and the Global Carbon Cycle. U.S. Department of Energy, Washington, D.C.

Emerson, S., K. Fischer, C. Reimers and D. Heggie. 1985. Organic carbon dynamics and preservation in deep-sea sediments. Deep Sea Research 32: 1–21.

Emsley, J. 1980. The phosphorus cycle. pp. 147–167. In O. Hutzinger (ed.), The Handbook of Environmental Chemistry, Vol. I, Part A, The Natural Environment and the Biogeochemical Cycles. Springer-Verlag, New York.

Environmental Protection Agency. 1986. Effects of ozone on natural ecosystems and their components. pp. 7–1 to 7–66. Air Quality Criteria for Ozone and Other Photochemical Oxidants. EPA-600–8–84–020cF, Research Triangle Park, N.C.

Eppley, R.W. and B.J. Peterson. 1979. Particulate organic matter flux and planktonic new production in the deep ocean. Nature 282: 677–680.

Epstein, C.B. and M. Oppenheimer. 1986. Empirical relation between sulphur dioxide emissions and acid deposition derived from monthly data. Nature 323: 245–247.

Erickson, D.J., J.S. Ghan and J.E. Penner. 1990. Global ocean-to-atmosphere dimethyl sulfide flux. Journal of Geophysical Research 95: 7543–7552.

Eriksson, E. 1960. The yearly circulation of chloride and sulfur in nature; meteorological, geochemical and pedological implications. Part II. Tellus 12: 63–109.

Ertel, J.R., J.I. Hedges, A.H. Devol, J.E. Richey and M.G. Ribeiro. 1986. Dissolved humic substances of the Amazon river system. Limnology and Oceanography 31: 739–754.

Esser, G., I. Aselmann, and H. Lieth. 1982. Modelling the carbon reservoir in the system compartment 'litter,' pp. 39–58. In Mitteilungen aus dem Geologisch-Palantologischen Institut der Universitat Hamburg, Volume 52. University of Hamburg, Germany.

Esser, G. 1987. Sensitivity of global carbon pools and fluxes to human and potential climatic impacts. Tellus 39B: 245–260.

Evans, J.R. 1989. Photosynthesis and nitrogen relationships in leaves of C_3 plants. Oecologia 78: 9–19.

Evans, L.J. 1980. Podzol development north of Lake Huron in relation to geology and vegetation. Canadian Journal of Soil Science 60: 527–539.

Evans, L.J. and B.H. Cameron. 1979. A chronosequence of soils developed from granitic morainal material, Baffin Island, N.W.T. Canadian Journal of Soil Science 59: 203–210.

Fahey, T.J. 1983. Nutrient dynamics of aboveground detritus in lodgepole pine (*Pinus contorta* ssp. *latifolia*) ecosystems, southeastern Wyoming. Ecological Monographs 53: 51–72.

Fahey, T.J. and J.B. Yavitt. 1988. Soil solution chemistry in lodgepole (*Pinus contorta* spp. *latifolia*) ecosystems, southeastern Wyoming, USA. Biogeochemistry 6: 91–118.

Fairbanks, R.G. 1989. A 17,000-year glacio-eustatic sea level record: Influence of glacial melting rates on the Younger Dryas event and deep-ocean circulation. Nature 342: 637–642.

Fanale, F.P. 1971. A case for catastrophic early degassing of the Earth. Chemical Geology 8: 79–105.

Fanning, K.A. 1989. Influence of atmospheric pollution on nutrient limitation in the ocean. Nature 339: 460–463.

Farquhar, G.D., R. Wetselaar and P.M. Firth. 1979. Ammonia volatilization from senescing leaves of maize. Science 203: 1257–1258.

Farquhar, G.D., K.T. Hubick, A.G. Condon and R.A. Richards. 1989. Carbon isotope fractionation and plant water-use efficiency. pp. 21–40. *In* P.W. Rundel, J.R. Ehleringer, and K.A. Nagy (eds.), Stable Isotopes in Ecological Research. Springer-Verlag, New York.

Fassbinder, J.W.E., H. Stanjek and H. Vali. 1990. Occurrence of magnetic bacteria in soil. Nature 343: 161–163.

Faulkner, S.P., W.H. Patrick and R.P. Gambrell. 1989. Field techniques for measuring wetland soil parameters. Soil Science Society of America Journal 53: 883–890.

Federer, C.A. 1983. Nitrogen mineralization and nitrification: Depth variation in four New England forest soils. Soil Science Society of America Journal 47: 1008–1014.

Federer, C.A. and J.W. Hornbeck. 1985. The buffer capacity of forest soils in New England. Water, Air, and Soil Pollution 26: 163–173.

Feller, M.C. and J.P. Kimmins. 1979. Chemical characteristics of small streams near Haney in southwestern British Columbia. Water Resources Research 15: 247–258.

Ferek, R.J. and M.O. Andreae. 1984. Photochemical production of carbonyl sulphide in marine surface waters. Nature 307: 148–150.

Ferek, R.J., R.B. Chatfield, and M.O. Andreae. 1986. Vertical distribution of dimethylsulphide in the marine atmosphere. Nature 320: 514–516.

Field, C. and H.A. Mooney. 1986. The photosynthesis-nitrogen relationship in wild plants. pp. 25–55. *In* T.J. Givnish (ed.), On the Economy of Plant Form and Function. Cambridge University Press, Cambridge.

Field, C., J. Merino and H.A. Mooney. 1983. Compromises between water-use efficiency and nitrogen-use efficiency in five species of California evergreens. Oecologia 60: 384–389.

Fife, D.N and E.K.S. Nambiar. 1984. Movement of nutrients in *Radiata* pine needles in relation to the growth of shoots. Annals of Botany 54: 303–314.

Firestone, M.K. 1982. Biological denitrification. pp. 289–326. *In* F.J. Stevenson (ed.), Nitrogen in Agricultural Soils. American Society of Agronomy, Madison, Wisconsin.

Firestone, M.K., R.B. Firestone and J.M. Tiedje. 1980. Nitrous oxide from soil denitrification: Factors controlling its biological production. Science 208: 749–751.

Fisher, R.F. 1972. Spodosol development and nutrient distribution under *Hydnaceae* fungal mats. Soil Science Society of America Proceedings 36: 492–495.

Fisher, R.F. 1977. Nitrogen and phosphorus mobilization by the fairy ring fungus, *Marasmius oreades* (Bolt.) Fr. Soil Biology and Biochemistry 9: 239–241.

Fisher, S.G. 1977. Organic matter processing by a stream-segment ecosystem: Fort River, Massachusetts, U.S.A. Internationale Revue der Gesamten Hydrobiologie 62: 701–727.

Fisher, S.G. and G.E. Likens. 1973. Energy flow in Bear Brook, New Hampshire: An integrative approach to stream ecosystem metabolism. Ecological Monographs 43: 421–439.

Fisher, S.G. and W.L. Minckley. 1978. Chemical characteristics of a desert stream in flash flood. Journal of Arid Environments 1: 25–33.

Fisher, S.G. and N.B. Grimm. 1985. Hydrologic and material budgets for a small Sonoran desert watershed during three consecutive cloudburst floods. Journal of Arid Environments 9: 105–118.

Fisher, T.R., L.W. Harding, D.W. Stanley and L.G. Ward. 1988. Phytoplankton, nutrients, and turbidity in the Chesapeake, Delaware, and Hudson estuaries. Estuarine, Coastal and Shelf Science 27: 61–93.

Fishman, J. and E.V. Browell. 1988. Comparison of satellite total ozone measurements with the distribution of tropospheric ozone obtained by an airborne UV-DIAL system over the Amazon Basin. Tellus 40B: 393–407.

Fitzgerald, J.W., T.L. Andrew and W.T. Swank. 1984. Availability of carbon-bonded sulfur for mineralization in forest soils. Canadian Journal of Forest Research 14: 839–843.

Fitzgerald, J.W., T.C. Strickland and J.T. Ash. 1985. Isolation and partial characterization of forest floor and soil organic sulfur. Biogeochemistry 1: 155–167.

Flaig, W., H. Beutelspacher and E. Rietz. 1975. Chemical composition and physical properties of humic substances. pp. 1–211. In J.E. Gieseking (ed.), Soil Components, Vol. 1, Organic Components. Springer-Verlag, New York.

Fleisher, Z., A. Kenig, I. Ravina and J. Hagin. 1987. Model of ammonia volatilization from calcareous soils. Plant and Soil 103: 205–212.

Flint, R.F. 1971. Glacial and Quaternary Geology. Wiley, New York.

Floret, C., R. Pontanier and S. Rambal. 1982. Measurement and modelling of primary production and water use in a south Tunisian steppe. Journal of Arid Environments 5: 77–90.

Force, A., D.K. Killinger, W.E. DeFeo and N. Menyuk. 1985. Laser remote sening of atmospheric ammonia using a CO_2 lidar system. Applied Optics 24: 2837–2841.

Foster, R.C. 1981. Polysaccharides in soil fabrics. Science 214: 665–667.

Fowler, W.A. 1984. The quest for the origin of the elements. Science 226: 922–935.

Fox, G.E., E. Stackebrandt, R.B. Hespell, J. Gibson, J. Maniloff, T.A. Dyer, R.S. Wolfe, W.E. Balch, R.S. Tanner, L.J. Magrum, L.B. Zablen, R. Blakemore, R. Gupta, L. Bonen, B.J. Lewis, D.A. Stahl, K.R. Luehrsen, K.N. Chen and C.R. Woese. 1980. The phylogeny of prokaryotes. Science 209: 457–463.

Fox, L.E. 1983. The removal of dissolved humic acid during estuarine mixing. Estuarine, Coastal and Shelf Science 16: 431–440.

Fox, R.F. 1988. Energy and the Evolution of Life. W.H. Freeman Company, New York.

Francis, A.J., J.M. Slater and C.J. Dodge. 1989. Denitrification in deep subsurface sediments. Geomicrobiology Journal 7: 103–116.

Frangi, J.L. and A.E. Lugo. 1985. Ecosystem dynamics of a subtropical floodplain forest. Ecological Monographs 55: 351–369.

Franzblau, E. and C.J. Popp. 1989. Nitrogen oxides produced from lightning. Journal of Geophysical Research 94: 11089–11104.

Franzmeier, D.P., E.P. Whiteside and M.M. Mortland. 1963. A chronosequence of podzols in northern Michigan, III. Mineralogy, micromorphology, and net changes during soil formation. Michigan Agriculture Experiment Station Quarterly Bulletin 46: 37–57.

Fraser, P.J., R.A. Rasmussen, J.W. Creffield, J.R. French and M.A.K. Khalil. 1986. Termites and global methane—another assessment. Journal of Atmospheric Chemistry 4: 295–310.

Freney, J.R., J.R. Simpson and O.T. Denmead. 1983. Volatilization of ammonia. pp. 1–32. In J.R. Freney and J.R. Simpson (eds.), Gaseous Loss of Nitrogen from Plant-Soil Systems. Martinus Nijhoff, The Hague.

Frey, R.W. and P.B. Basan. 1985. Coastal salt marshes. pp. 225–301. In R.A. Davis (ed.), Coastal Sedimentary Environments, 2nd ed. Springer-Verlag, New York.

Fridovich, I. 1975. Superoxide dismutases. Annual Review of Biochemistry 44: 147–159.

Friedli, H., H. Lotscher, H. Oeschger, U. Siegenthaler and B. Stauffer. 1986. Ice core record of the $^{13}C/^{12}C$ ratio of atmospheric CO_2 in the past two centuries. Nature 324: 237–238.

Friedmann, E.I. 1982. Endolithic microorganisms in the Antarctic cold desert. Science 215: 1045–1053.

Froelich, P.N. 1984. Interactions of the marine phosphorus and carbon cycles. pp. 141–176. In B. Moore and M.N. Dastoor (eds.), The Interaction of Global Biochemical Cycles. Jet Propulsion Laboratory Publication 84–21, Pasadena, California.

Froelich, P.N. 1988. Kinetic control of dissolved phosphate in natural rivers and estuaries: A primer on the phosphate buffer mechanism. Limnology and Oceanography 33: 649–668.

Froelich, P.N., M.L. Bender, N.A. Luedtke, G.R. Heath and T. DeVries. 1982. The marine phosphorus cycle. American Journal of Science 282: 474–511.

Fruchter, J.S., D.E. Robertson, J.C. Evans, K.B. Olsen, E.A. Lepel, J.C. Laul, K.H. Abel, R.W. Sanders, P.O. Jackson, N.S. Wogman, R.W. Perkins, H.H. Van Tuyl, R.H. Beauchamp, J.W. Shade, J.L. Daniel, R.L. Erikson, G.A. Sehmel, R.N. Lee, A.V. Robinson, O.R. Moss, J.K. Briant and W.C. Cannon. 1980. Mount St. Helens ash from the 18 May 1980 eruption: Chemical, physical, mineralogical, and biological properties. Science 209: 1116–1125.

Fuhrman, J.A. and F. Azam. 1982. Thymidine incorporation as a measure of heterotrophic bacterioplankton production in marine surface waters: Evaluation and field results. Marine Biology 66: 109–120.

Fuhrman, J.A., H.W. Ducklow, D.L. Kirchman, J. Hudak, G.B. McManus and J. Kramer. 1986. Does adenine incorporation into nucleic acids measure total microbial production? Limnology and Oceanography 31: 627–636.

Fuller, R.D., M.J. Mitchell, H.R. Krouse, B.J. Wyskowski and C.T. Driscoll. 1986. Stable sulfur isotope ratios as a tool for interpreting ecosystem sulfur dynamics. Water, Air, and Soil Pollution 28: 163–171.

Fung, I.Y., C.J. Tucker and K.C. Prentice. 1987. Application of advanced very high resolution radiometer vegetation index to study atmosphere-biosphere exchange of CO_2. Journal of Geophysical Research 92: 2999–3015.

Gächter, R., J.S. Meyer and A. Mares. 1988. Contribution of bacteria to release and fixation of phosphorus in lake sediments. Limnology and Oceanography 33: 1542–1558.

Gallagher, J.L. and F.G. Plumley. 1979. Underground biomass profiles and productivity in Atlantic coastal marshes. American Journal of Botany 66: 156–161.

Gallardo, A. and W.H. Schlesinger. 1990. Estimating microbial biomass nitrogen using the fumigation-incubation and fumigation-extraction methods in a warm-temperate forest soil. Soil Biology and Biochemistry 22: 927–932.

Galloway, J.N., G. E. Likens, and E.S. Edgerton. 1976. Acid precipitation in the northeastern United States: pH and acidity. Science 194: 722–724.

Galloway, J.N. and G.E. Likens. 1979. Atmospheric enhancement of metal deposition in Adirondack lake sediments. Limnology and Oceanography 24: 427–433.

Galloway, J.N., G.E. Likens and M.E. Hawley. 1984. Acid precipitation: Natural versus anthropogenic components. Science 226: 829–831.

Galloway, J.N. and D.M. Whelpdale. 1980. An atmospheric sulfur budget for eastern North America. Atmospheric Environment 14: 409–417.

Galloway, J.N. and D.M. Whelpdale. 1987. WATOX-86 overview and western North Atlantic ocean S and N atmospheric budgets. Global Biogeochemical Cycles 1: 261–281.

Gammon, R.H., E.T. Sundquist and P.J. Fraser. 1985. History of carbon dioxide in the atmosphere. pp. 25–62. In J.R. Trabalka (ed.), Atmospheric Carbon Dioxide and the Global Carbon Cycle. U.S. Department of Energy, Er-0239, Washington, D.C.

Ganor, E. and Y. Mamane. 1982. Transport of Saharan dust across the eastern Mediterranean. Atmospheric Environment 16: 581–587.

Gardner, W.S., T.F. Nalepa and J.M. Malczyk. 1987. Nitrogen mineralization and denitrification in Lake Michigan sediments. Limnology and Oceanography 32: 1226–1238.

Gardner, W.S. and D.W. Menzel. 1974. Phenolic aldehydes as indicators of terrestrially derived organic matter in the sea. Geochimica et Cosmochimica Acta 38: 813–822.

Garrels, R.M., F.T. MacKenzie and C. Hunt. 1975. Chemical Cycles and the Global Environment. W. Kaufmann, Los Altos, California.

Garrels, R.M. and A. Lerman. 1981. Phanerozoic cycles of sedimentary carbon and sulfur. Proceedings of the National Academy of Sciences U.S.A. 78: 4652–4656.

Garrels, R.M. and F.T. MacKenzie. 1967. Origin of the chemical compositions of some springs and lakes. pp. 222–242. In W. Stumm (ed.), Equilibrium Concepts in Natural Water Systems. American Chemical Society, Washington, D.C.

Garrels, R.M. and F.T. MacKenzie. 1971. Evolution of Sedimentary Rocks. W.W. Norton, New York.

Gates, D.M. 1985. Global biospheric response to increasing atmospheric carbon dioxide concentration. pp. 171–184. In B.R. Strain and J.D. Cure (eds.), Direct Effects of Increasing Carbon Dioxide on Vegetation. DOE/ER-0238. U.S. Department of Energy, Washington, D.C.

Gates, W.L. 1976. Modeling the ice-age climate. Science 191: 1138–1144.

Gatz, D.F. and A.N. Dingle. 1971. Trace substances in rain water: Concentration variations during convective rains, and their interpretation. Tellus 23: 14–27.

Gause, G.E. 1934. The Struggle for Existence. Hafner, New York.

Gensel, P.G. and H.N. Andrews. 1987. The evolution of early land plants. American Scientist 75: 478–489.

George, G.N., R.C. Prince and S.P. Cramer. 1989. The manganese site of the photosynthetic water-splitting enzyme. Science 243: 789–791.

Georgii, H.-W. and D. Wötzel. 1970. On the relation between drop size and concentration of trace elements in rainwater. Journal of Geophysical Research 75: 1727–1731.

Gersper, P.L. and N. Holowaychuk. 1971. Some effects of stem flow from forest canopy trees on chemical properties of soils. Ecology 52: 691–702.

Gholz, H.L. 1982. Environmental limits on aboveground net primary production, leaf area, and biomass in vegetation zones of the Pacific Northwest. Ecology 63: 469–481.

Gholz, H.L., R.F. Fisher and W.L. Pritchett. 1985. Nutrient dynamics in slash pine plantation ecosystems. Ecology 66: 647–659.

Gibbs, R.J. 1970. Mechanisms controlling world water chemistry. Science 170: 1088–1090.

Giblin, A.E. 1988. Pyrite formation in marshes during early diagenesis. Geomicrobiology Journal 6: 77–97.

Gifford, G.F. and F.E. Busby 1973. Loss of particulate organic materials from semiarid watersheds as a result of extreme hydrologic events. Water Resources Research 9: 1443–1449.

Gijsman, A.J. 1990. Nitrogen nutrition of Douglas fir (*Pseudotsuda menziesii*) on strongly acid sandy soil. I. Growth, nutrient uptake, and ionic balance. Plant and Soil 126: 53–61.

Gile, L.H., F.F. Peterson and R.B. Grossman. 1966. Morphological and genetic sequences of carbonate accumulation in desert soils. Soil Science 101: 347–360.

Gilmore, A.R., G.Z. Gertner and G.L. Rolfe. 1984. Soil chemical changes associated with roosting birds. Soil Science 138: 158–163.

Glass, S.J. and M.J. Matteson. 1973. Ion enrichment in aerosols dispersed from bursting bubbles in aqueous salt solutions. Tellus 25: 272–280.

Glover, H.E., B.B. Prézelin, L. Campbell, M. Wyman and C. Garside. 1988. A nitrate-dependent Synechococcus bloom in surface Sargasso sea water. Nature 331: 161–163.

Glynn, P.W. 1988. El Niño-Southern Oscillation 1982–1983: Nearshore population, community, and ecosystem responses. Annual Review of Ecology and Systematics 19: 309–345.

Godbold, D.L., E. Fritz and A. Hüttermann. 1988. Aluminum toxicity and forest decline. Proceedings of the National Academy of Sciences, U.S.A. 85: 3888–3892.

Goldan, P.D., W.C. Kuster, D.L. Albritton and F.C. Fehsenfeld. 1987. The measurement of natural sulfur emissions from soils and vegetation: Three sites in the eastern United States revisited. Journal of Atmospheric Chemistry 5: 439–467.

Goldan, P.D., R. Fall, W. C. Kuster and F.C. Fehsenfeld. 1988. Uptake of COS by growing vegetation: A major tropospheric sink. Journal of Geophysical Research 93: 14186–14192.

Goldberg, A.B., P.J. Maroulis, L.A. Wilner and A.R. Bandy. 1981. Study of H₂S emissions from a salt water marsh. Atmospheric Environment 15: 11–18.

Goldberg, D.E. 1982. The distribution of evergreen and deciduous trees relative to soil type: An example from the Sierra Madre, Mexico, and a general model. Ecology 63: 942–951.

Goldberg, D.E. 1985. Effects of soil pH, competition, and seed predation on the distributions of two tree species. Ecology 66: 503–511.

Goldich, S.S. 1938. A study in rock-weathering. Journal of Geology 46: 17–58.

Goldman, C.R. 1988. Primary productivity, nutrients, and transparency during the early onset of eutrophication in ultra-oligotrophic Lake Tahoe, California-Nevada. Limnology and Oceanography 33:1321–1333.

Goldman, J.C. and P.M. Gilbert. 1982. Comparative rapid ammonium uptake by four species of marine phytoplankton. Limnology and Oceanography 27: 814–827.

Goldstein, R.M., T.P. Barnett and H.A. Zebker. 1989. Remote sensing of ocean currents. Science 246: 1282–1285.

Golley, F.B. 1972. Energy flux in ecosystems. pp. 69–90. In J.A. Wiens (ed.), Ecosystem Structure and Function. Oregon State University Press, Corvallis, Oregon.

Goodroad, L.L. and D.R. Keeney. 1984. Nitrous oxide emission from forest, marsh, and prairie ecosystems. Journal of Environmental Quality 13: 448–452.

Gorham, E. 1957. The development of peat lands. Quarterly Review of Biology 32:145–166.

Gorham, E. 1961. Factors influencing supply of major ions to inland waters, with special reference to the atmosphere. Geological Society of America Bulletin 72: 795–840.

Gorham, E., P.M. Vitousek and W.A. Reiners. 1979. The regulation of chemical budgets over the course of terrestrial ecosystem succession. Annual Review of Ecology and Systematics 10: 53–84.

Gorham, E., F.B. Martion and J.T. Litzau. 1984. Acid rain: Ionic correlations in the eastern United States, 1980–1981. Science 225: 407–409.

Gornitz, V., S. Lebedeff and J. Hansen. 1982. Global sea level trend in the past century. Science 215: 1611–1614.

Gosz, J.R., R.T. Holmes, G.E. Likens and F.H. Bormann. 1978. The flow of energy in a forest ecosystem. Scientific American 238(3): 92–102.

Gosz, J.R. 1981. Nitrogen cycling in coniferous ecosystems. pp. 405–426. In F.E. Clark and T. Rosswall (eds.), Terrestrial Nitrogen Cycles. Swedish Natural Science Research Council, Stockholm.

Goudie, A.S. 1978. Dust storms and their geomorphological implications. Journal of Arid Environments 1: 291–310.

Goward, S.N., C.J. Tucker and D.G. Dye. 1985. North American vegetation patterns observed with the NOAA-7 advanced very high resolution radiometer. Vegetatio 64: 3–14.

Graedel, T.E., G.W. Kammlott and J.P. Franey. 1981. Carbonyl sulfide: Potential agent of atmospheric sulfur corrosion. Science 212: 663–665.

Graham, W.F. and R.A. Duce. 1979. Atmospheric pathways of the phosphorus cycle. Geochimica et Cosmochimica Acta 43: 1195–1208.

Grandstaff, D.E. 1986. The dissolution rate of forsteritic olivine from Hawaiian beach sand. pp. 41–59. In S.M. Colman and D.P. Dethier (eds.), Rates of Chemical Weathering of Rocks and Minerals. Academic Press, Orlando, Florida.

Granhall, U. 1981. Biological nitrogen fixation in relation to environmental factors and functioning of natural ecosystems. pp. 131–144. *In* F.E. Clark and T. Rosswall (eds.), Terrestrial Nitrogen Cycles. Swedish Natural Science Research Council, Stockholm.

Grassle, J.F. 1985. Hydrothermal vent animals: Distribution and biology. Science 229: 713–717.

Graustein, W.C., K. Cromack and P. Sollins. 1977. Calcium oxalate: Occurrence in soils and effect on nutrient and geochemical cycles. Science 198: 1252–1254.

Gray, J.T. and W.H. Schlesinger. 1981. Nutrient cycling in Mediterranean type ecosystems. pp. 259–285. In P.C. Miller (ed.), Resource Use by Chaparral and Matorral. Springer-Verlag, New York.

Gray, J.T. 1982. Community structure and productivity in *Ceanothus* chaparral and coastal sage scrub of southern California. Ecological Monographs 52: 415–435.

Gray, J.T. 1983. Nutrient use by evergreen and deciduous shrubs in southern California. I. Community nutrient cycling and nutrient-use efficiency. Journal of Ecology 71:21–41.

Greenberg, J.P., P.R. Zimmerman, L. Heidt and W. Pollock. 1984. Hydrocarbon and carbon monoxide emissions from biomass burning in Brazil. Journal of Geophysical Research 89: 1350–1354.

Greenberg, J.P. and P.R. Zimmerman. 1984. Nonmethane hydrocarbons in remote tropical, continental, and marine atmospheres. Journal of Geophysical Research 89: 4767–4778.

Greenland, D.J. 1971. Interactions between humic and fulvic acids and clays. Soil Science 111: 34–41.

Gregor, B. 1970. Denudation of the continents. Nature 228: 273–275.

Gregory, G.L., E.V. Browell and L.S. Warren. 1988. Boundary layer ozone: An airborne survey above the Amazon Basin. Journal of Geophysical Research 93: 1452–1468.

Grier, C.C. 1975. Wildfire effects on nutrient distribution and leaching in a coniferous forest ecosystem. Canadian Journal of Forest Research 5: 599–607.

Grier, C.C. and S.W. Running. 1977. Leaf area of mature northwestern coniferous forests: Relation to site water balance. Ecology 58: 893–899.

Griffin, T.M., M.C. Rabenhorst, and D.S. Fanning. 1989. Iron and trace metals in some tidal marsh soils of the Chesapeake Bay. Soil Science Society of America Journal 53: 1010–1019.

Griffith, E.J., C. Ponnamperuma and N.W. Gabel. 1977. Phosphorus, A key to life on the primitive Earth. Origins of Life 8: 71–85.

Grinspoon, D.H. 1987. Was Venus wet? Deuterium reconsidered. Science 238: 1702–1704.

Groffman, P.M. and J.M. Tiedje. 1989. Denitrification in north temperate forest soils: Spatial and temporal patterns at the landscape and seasonal scales. Soil Biology and Biochemistry 21: 613–620.

Gross, M.G. 1982. Oceanography, 2nd ed. Prentice-Hall, Englewood Cliffs, New Jersey.

Grusbaugh, J.W. and R. V. Anderson. 1989. Upper Mississippi River: Seasonal and flood-plain forest influences on organic matter transport. Hydrobiologia 174: 235–244.

Gutschick. V.P. 1981. Evolved strategies in nitrogen acquisition by plants. American Naturalist 118: 607–637.

Hackley, K.C. and T.F. Anderson. 1986. Sulfur isotopic variations in low-sulfur coals from the Rocky Mountain region. Geochimica et Cosmochimica Acta 50: 1703–1713.

Haering, K.C., M.C. Rabenhorst and D.S. Fanning. 1989. Sulfur speciation in some Chesapeake Bay tidal marsh soils. Soil Science Society of America Journal 53: 500–505.

Hahn, J. 1974. The North Atlantic Ocean as a source of atmospheric N_2O. Tellus 26: 160–168.

Hahn, J. 1980. Organic constituents of natural aerosols. Annals of the New York Academy of Sciences 338:359–376.

Hahn, J. 1981. Nitrous oxide in the oceans. pp. 191–277. *In* C.C. Delwiche (ed.), Denitrification, Nitrification, and Atmospheric Nitrous Oxide. Wiley, New York.

Haines, E.B. 1977. The origins of detritus in Georgia salt marsh estuaries. Oikos 29: 254–260.

Hanks, T.C. and D.L. Anderson. 1969. The early thermal history of the Earth. Physics of the Earth and Planetary Interiors 2: 19–29.

Hansen, J., D. Johnson, A. Lacis, S. Lebedeff, P. Lee, D. Rind and G. Russell. 1981. Climatic impact of increasing atmospheric carbon dioxide. Science 213: 957–966.

Harden, J.W. 1988. Genetic interpretations of elemental and chemical differences in a soil chronosequence, California. Geoderma 43: 179–193.

Harley, J.L. and S.E. Smith. 1983. Mycorrhizal Symbiosis. Academic Press, New York.

Harmon, M.E., J.F. Franklin, F.J. Swanson, P. Sollins, S.V. Gregory, J.D. Lattin, N.H. Anderson, S.P. Cline, N.G.Aumen, J.R. Sedell, G.W. Lienkaemper, K. Cromack and K.W. Cummins. 1986. Ecology of coarse woody debris in temperate ecosystems. Advances in Ecological Research 15: 133–302.

Harmon, M.E., W.K. Ferrell and J.F. Franklin. 1990. Effects on carbon storage of conversion of old-growth forests to young forests. Science 247: 699–702.

Harrington, J.B. 1987. Climatic change: A review of causes. Canadian Journal of Forest Research 11: 1313–1339.

Harris, S.A. 1971. Podsol development on volcanic ash deposits in the Talamanea Range, Costa Rica. pp. 191–209. In D.H. Yaalon (ed.), Paleopedology: Origin, Nature and Dating of Paleosols. International Society of Soil Science, Jerusalem.

Harrison, A.F. 1982. ^{32}P method to compare rates of mineralization of labile organic phosphorus in woodland soils. Soil Biology and Biochemistry 14: 337–341.

Harrison, R.B., D.W. Johnson and D.E. Todd. 1989. Sulfate adsorption and desorption reversibility in a variety of forest soils. Journal of Environmental Quality 18: 419–426.

Harriss, R.C., D.I. Sebacher and F.P. Day. 1982. Methane flux in the Great Dismal Swamp. Nature 297: 673–674.

Harriss, R.C., D.I. Sebacher, K.B. Bartlett, D.S. Bartlett and P.M. Crill. 1988. Sources of atmospheric methane in the South Florida environment. Global Biogeochemical Cycles 2: 231–243.

Harvey, L.D.D. 1988. Climatic impact of ice-age aerosols. Nature 334: 333–335.

Hatcher, B.G. and K.H. Mann. 1975. Above-ground production of marsh cordgrass (*Spartina alterniflora*) near the northern end of its range. Journal of the Fisheries Research Board of Canada 32: 83–87.

Haynes, R.J. and K.M. Goh. 1978. Ammonium and nitrate nutrition of plants. Biological Reviews 53: 465–510.

Heath, D.F. 1988. Non-seasonal changes in total column ozone from satellite observations, 1970–86. Nature 332: 219–227.

Hedges, J.I. and P.L. Parker. 1976. Land-derived organic matter in surface sediments from the Gulf of Mexico. Geochimica et Cosmochimica Acta 40: 1019–1029.

Hedges, J.I., W.A. Clark and G.L. Cowie. 1988. Organic matter sources to the water column and surficial sediments of a marine bay. Limnology and Oceanography 33: 1116–1136.

Hedley, M.J., P.H. Nye and R.E. White. 1982a. Plant-induced changes in the rhizosphere of rape (*Brassica napus* var. emerald) seedlings. II. Origin of the pH change. New Phytologist 91: 31–44.

Hedley, M.J., J.W.B. Stewart and B.S. Chauhan. 1982b. Changes in inorganic and organic soil phosphorus fractions induced by cultivation practices and by laboratory incubations. Soil Science Society of America Journal 46: 970–976.

Hedlin, L.O., G.E. Likens and F.H. Bormann. 1987. Decrease in precipitation acidity resulting from decreased SO_4^{2-} concentration. Nature 325: 244–246.

Hegg, D.A., L.F. Radke, P.V. Hobbs, R.A. Rasmussen and P.J. Riggan. 1990. Emissions of some trace gases from biomass fires. Journal of Geophysical Research 95: 5669–5675.

Hemond, H.F. 1980. Biogeochemistry of Thoreau's Bog, Concord, Massachusetts. Ecological Monographs 50: 507–526.

Hemond, H.F. 1983. The nitrogen budget of Thoreau's Bog. Ecology 64: 99–109.

Henderson, G.S., W.T.Swank, J.B. Waide and C.C. Grier. 1978. Nutrient budgets of Appalachian and Cascade Region watersheds: A comparison. Forest Science 24: 385–397.

Henrichs, S.M. and W.S. Reeburgh. 1987. Anaerobic mineralization of marine sediment organic matter: Rates and the role of anaerobic processes in the oceanic carbon economy. Geomicrobiology Journal 5: 191–237.

Herron, M.M., C.C. Langway, H.W. Weiss and J.H. Cragin. 1977. Atmospheric trace metals and sulfate in the Greenland ice sheet. Geochimica et Cosmochimica Acta 41: 915–920.

Hess, S.L., R.M. Henry, C.B. Leovy, J.A. Ryan, J.E. Tillman, T.E. Chamberlain, H.L. Cole, R.G. Dutton, G.C. Greene, W.E. Simon and J.L. Mitchell. 1976. Preliminary meteorological results on Mars from the Viking I Lander. Science 193: 788–791.

Hester, K. and E. Boyle. 1982. Water chemistry control of cadmium content in recent benthic foraminifera. Nature 298: 260–262.

Hewitt, C.M. and R.M. Harrison. 1985. Tropospheric concentrations of the hydroxyl radical—A review. Atmospheric Environment 19: 545–554.

Hidy, G.M. 1970. Theory of diffusive and impactive scavenging. pp. 355–371. In R.J. Englemann and W.G.N.Slinn (eds.), Precipitation Scavenging (1970). U.S. Atomic Energy Commission, Division of Technical Information, Oak Ridge, Tennessee.

Hidy, G.M. and J.R. Brock. 1971. An assessment of the global sources of tropospheric aerosols. pp. 1088–1097. In H. M. Englund and W.T. Berry (eds.), Proceedings of the Second International Clean Air Congress. Academic Press, New York.

Hingston, F.J., R.J. Atkinson, A. M. Posner and J.P. Quirk. 1967. Specific adsorption of anions. Nature 215: 1459–1461.

Hoffman, D.J. 1990. Increase in the stratospheric background sulfuric acid aerosol mass in the past 10 years. Science 248: 996–1000.

Hoffman, D.J. and J.M. Rosen. 1980. Stratospheric sulfuric acid layer: Evidence for an anthropogenic component. Science 208: 1368–1370.

Hoffman, D.J. and J.M. Rosen. 1983. Sulfuric acid droplet formation and growth in the stratosphere after the 1982 eruption of El Chichón. Science 222: 325–327.

Hole, F.D. 1981. Effects of animals on soil. Geoderma 25:75–112.

Holland, H.D. 1965. The history of ocean water and its effect on the chemistry of the atmosphere. Proceedings of the National Academy of Sciences U.S.A. 53: 1173–1183.

Holland, H.D. 1978. The Chemistry of the Atmosphere and Oceans. Wiley, New York.

Holland, H.D. 1984. The Chemical Evolution of the Atmosphere and Oceans. Princeton University Press, Princeton.

Holland, H.D., B. Lazar and M. McCaffrey. 1986. Evolution of the atmosphere and oceans. Nature 320: 27–33.

Hollinger, D.Y. 1986. Herbivory and the cycling of nitrogen and phosphorus in isolated California oak trees. Oecologia 70: 291–297.

Holmes, R.T., T.W. Sherry and F.W. Sturges. 1986. Bird community dynamics in a temperate deciduous forest: Long-term trends at Hubbard Brook. Ecological Monographs 56: 201–220.

Holser, W.T. and I.R. Kaplan. 1966. Isotope geochemistry of sedimentary sulfates. Chemical Geology 1: 93–135.

Holser, W.T., J.B. Maynard and K.M. Cruikshank. 1989. Modelling the natural cycle of sulphur through Phanerozoic time. pp. 21–56. In P. Brimblecombe and A.Y. Lein (eds.), Evolution of the Global Biogeochemical Sulphur Cycle. Wiley, New York.

Honeycutt, C.W., R.D. Heil and C.V. Cole. 1990. Climatic and topographic relations of three Great Plains soils. I. Soil morphology. Soil Science Society of America Journal 54: 469–475.

Honjo, S., S.J. Manganini and J.J. Cole. 1982. Sedimentation of biogenic matter in the deep ocean. Deep Sea Research 29: 609–625.

Hooper, F.F. and L.S. Morris. 1982. Mat-water phosphorus exchange in an acid bog lake. Ecology 63: 1411–1421.

Horne, A.J. and C.R. Goldman. 1974. Suppression of nitrogen fixation by blue-green algae in a eutrophic lake with trace additions of copper. Science 183: 409–411.

Horne, A.J. and D.L. Galat. 1985. Nitrogen fixation in an oligotrophic, saline desert lake: Pyramid Lake, Nevada. Limnology and Oceanography 30: 1229–1239.

Horowitz, N.H. 1977. The search for life on Mars. Scientific American 237(5): 52–61.

Horrigan, S.G., J.P. Montoya, J.L. Nevins, and J.J. McCarthy. 1990. Natural isotopic composition of dissolved inorganic nitrogen in the Chesapeake Bay. Estuarine, Coastal and Shelf Science 30: 393–410.

Hosker, R.P. and S.E. Lindberg. 1982. Review: Atmospheric deposition and plant assimilation of gases and particles. Atmospheric Environment 16: 889–910.

Houghton, R.A., J.E. Hobbie, J.M. Melillo, B. Moore, B.J. Peterson, G.R. Shaver and G.M. Woodwell. 1983. Changes in the carbon content of terrestrial biota and soils between 1860 and 1980: A net release of CO_2 to the atmosphere. Ecological Monographs 53: 235–262.

Houghton, R.A., R.D. Boone, J.R. Fruci, J.E. Hobbie, J.M. Melillo, C.A.Palm, B.J. Peterson, G.R. Shaver, G.M. Woodwell, B. Moore, D.L. Skole and N. Myers. 1987. The flux of carbon from terrestrial ecosystems to the atmosphere in 1980 due to changes in land use: geographic distribution of the global flux. Tellus 39B: 122–139.

Houghton, R.A. 1987. Biotic changes consistent with the increased seasonal amplitude of atmospheric CO_2 concentrations. Journal of Geophysical Research 92: 4223–4230.

Hovis, W.A., D.K. Clark, F. Anderson, R.W. Austin, W.H. Wilson, E.T. Baker, D. Ball, H.R. Gordon, J.L. Mueller, S.Z. El-Sayed, B. Sturm, R.C. Wrigley and C.S. Yentsch. 1980. Nimbus-7 Coastal Zone Color Scanner: system description and initial imagery. Science 210: 60–63.

Howard, J.A. and C. W. Mitchell. 1985. Phytogeomorphology. Wiley, New York.

Howarth, R.W. 1979. Pyrite: Its rapid formation in a salt marsh and its importance in ecosystem metabolism. Science 203: 49–51.

Howarth, R.W. 1984. The ecological significance of sulfur in the energy dynamics of salt marsh and coastal marine sediments. Biogeochemistry 1: 5–27.

Howarth, R.W. 1988. Nutrient limitation of net primary production in marine ecosystems. Annual Review of Ecology and Systematics 19: 89–110.

Howarth, R.W. and J.J. Cole. 1985. Molybdenum availability, nitrogen limitation, and phytoplankton growth in natural waters. Science 229: 653–655.

Howarth, R.W., R. Marino, J. Lane and J.J. Cole. 1988a. Nitrogen fixation in freshwater, estuarine, and marine ecosystems. I. Rates and importance. Limnology and Oceanography 33: 669–687.

Howarth, R.W., R. Marino and J.J. Cole. 1988b. Nitrogen fixation in freshwater, estuarine, and marine ecosystems. 2. Biogeochemical controls. Limnology and Oceanography 33: 688–701.

Howeler, R.H. and D.R. Bouldin. 1971. The diffusion and consumption of oxygen in submerged soils. Soil Science Society of America Proceedings 35: 202–208.

Howell, D.G. and R. W. Murray. 1986. A budget for continental growth and denudation. Science 233: 446–449.

Howes, B.L., J.W.H. Dacey and G.M. King. 1984. Carbon flow through oxygen and sulfate reduction pathways in salt marsh sediments. Limnology and Oceanography 29: 1037–1051.

Howes, B.L., J.W.H. Dacey and J.M. Teal. 1985. Annual carbon mineralization and belowground production of *Spartina alterniflora* in a New England salt marsh. Ecology 66: 595–605.

Huang, P.M. 1988. Ionic factors affecting aluminium transformations and the impact on soil and environment sciences. Advances in Soil Science 8: 1–78.

Huber, R., M. Kurr, H.W. Jannasch and K.O. Stetter. 1989. A novel group of abyssal methanogenic archaebacteria (*Methanopyrus*) growing at 110°C. Nature 342: 833–834.

Hurley, J.P., D.E. Armstrong, G.J. Kenoyer and C.J. Bowser. 1985. Ground water as a silica source for diatom production in a precipitation-dominated lake. Science 227: 1576–1578.

Husar, R.B. and J.D. Husar. 1985. Regional sulfur runoff. Journal of Geophysical Research 90: 1115–1125.

Huss-Danell, K. 1986. Nitrogen in shoot litter, root litter, and root exudates from nitrogen-fixing *Alnus incana*. Plant and Soil 91: 43–49.

Hutchinson, G.E. 1938. On the relation between oxygen deficit and the productivity and typology of lakes. Internationale Revue der Gesamten Hydrobiologie Hydrographie 36: 336–355.

Hutchinson, G.L., R. J. Millington and D. B. Peters. 1972. Atmospheric ammonia: Absorption by plant leaves. Science 175: 771–772.

Hutchinson, J.N. 1980. The record of peat wastage in the East Anglian Fenlands at Holme Post, 1848–1978 A.D. Journal of Ecology 68: 229–249.

Hydes, D.J. 1979. Aluminum in seawater: Control by inorganic processes. Science 205: 1260–1262.

Hyman, M. R. and P.M. Wood. 1983. Methane oxidation by *Nitrosomonas europaea*. Biochemical Journal 212: 31–37.

Idso, S.B. and A.J. Brazel. 1984. Rising atmospheric carbon dioxide concentrations may increase streamflow. Nature 312: 51–53.

Ingestad, T. 1979a. Mineral nutrient requirements of *Pinus silvestris* and *Picea abies* seedlings. Physiologica Plantarum 45: 373–380.

Ingestad, T. 1979b. Nitrogen stress in birch seedlings. II. N, K, P, Ca, and Mg nutrition. Physiologica Plantarum 45: 149–157.

Ingestad, T. 1982. Relative addition rate and external concentration: Driving variables used in plant nutrition research. Plant, Cell and Environment 5: 443–453.

Inn, E.C.Y., J.F. Vedder, E.P. Condon and D. O'Hara. 1981. Gaseous constituents in the plume from eruptions of Mount St. Helens. Science 211: 821–823.

Inskeep, W.P. and P.R. Bloom. 1986. Kinetics of calcite precipitation in the presence of water-soluble organic ligands. Soil Science Society of America Journal 50: 1167–1172.

Irwin, J.G. and M.L.Williams. 1988. Acid rain: Chemistry and transport. Environmental Pollution 50: 29–59.

Isaksen, I.S.A. and O. Hov. 1987. Calculation of trends in the tropospheric concentration of O_3, OH, CO, CH_4 and NO_x. Tellus 39B: 271–285.

Ittekkot, V. and R. Arain. 1986. Nature of particulate organic matter in the river Indus, Pakistan. Geochimica et Cosmochimica Acta 50: 1643–1653.

Ittekkot, V. and S. Zhang. 1989. Pattern of particulate nitrogen transport in world rivers. Global Biogeochemical Cycles 3: 383–391.

Ivanov, M.V. 1983. Major fluxes of the global biogeochemical cycle of sulphur. pp. 449–463. *In* M.V. Ivanov and J.R. Freney (eds.), The Global Biogeochemical Sulphur Cycle. Wiley, New York.

Ivanov, M.V., Y.A. Grinenko and A.P. Rabinovich. 1983. Sulphur flux from continents to oceans. pp. 331–356. *In* M.V. Ivanov and J.R. Freney (eds.), The Global Biogeochemical Sulphur Cycle. Wiley, New York.

Iverson, R.L., F.L Nearhoof and M.O. Andreae. 1989. Production of dimethylsulfonium propionate and dimethylsulfide by phytoplankton in estuarine and coastal waters. Limnology and Oceanography 34: 53–67.

Jackson, L.E., J.P. Schimel and M.K. Firestone. 1989. Short-term partitioning of ammonium and nitrate between plants and microbes in an annual grassland. Soil Biology and Biochemistry 21: 409–415.

Jackson, R.B., J.H. Manwaring and M.M. Caldwell. 1990. Rapid physiological adjustment of roots to localized soil enrichment. Nature 344: 58–60.

Jacobs, S.S. 1986. The polar ice sheets: A wild card in the deck? Oceanus 29(4): 50–54.

Jacobs, T.C. and J.W. Gilliam. 1985. Riparian losses of nitrate from agricultural drainage waters. Journal of Environmental Quality 14: 472–478.

Jacoby, G.C. 1986. Long-term temperature trends and a positive departure from the climate-growth response since the 1950s in high elevation lodgepole pine from California. pp. 81–83. *In* C. Rosenzweig and R. Dickinson (eds.), Climate-Vegetation Interactions. University Corporation for Atmospheric Research, Boulder, Colorado.

James, B.R. and S.J. Riha. 1986. pH buffering in forest soil organic horizons: Relevance to acid precipitation. Journal of Environmental Quality 15: 229–234.

Jannasch, H.W. 1989. Sulphur emission and transformations at deep sea hydrothermal vents. pp. 181–190. *In* P. Brimblecombe and A.Y. Lein (eds.), Evolution of the Global Biogeochemical Sulphur Cycle. Wiley, New York.

Jannasch, H.W. and C.O. Wirsen. 1979. Chemosynthetic primary production at East Pacific sea floor spreading centers. BioScience 29: 592–598.

Jannasch, H.W. and M.J. Mottl. 1985. Geomicrobiology of deep-sea hydrothermal vents. Science 229: 717–725.

Janos, D.P. 1980. Vesicular-arbuscular mycorrhizae affect lowland tropical rain forest plant growth. Ecology 61: 151–162.

Jaramillo, V.J. and J.K. Detling. 1988. Grazing history, defoliation, and competition: Effects on shortgrass production and nitrogen accumulation. Ecology 69: 1599–1608.

Jeffries, D.S., R.G. Semkin, R. Neureuther and M. Seymour. 1988. Ion mass budgets for lakes in the Turkey Lakes watershed, June 1981-May 1983. Canadian Journal of Fisheries and Aquatic Science 45(Suppl.): 47–58.

Jenkinson, D.S. and D.S. Powlson. 1976a. The effects of biocidal treatments on metabolism in soil. I. Fumigation with chloroform. Soil Biology and Biochemistry 8: 167–177.

Jenkinson, D.S. and D.S. Powlson. 1976b. The effects of biocidal treatments on metabolism in soil. V. A method for measuring soil biomass. Soil Biology and Biochemistry 8: 209–213.

Jenkinson, D.S. and J.H. Rayner. 1977. The turnover of soil organic matter in some of the Rothamsted classical experiments. Soil Science 123: 298–305.

Jennings, J.N. 1983. Karst landforms. American Scientist 71: 578–586.

Jenny, H. 1980. The Soil Resource. Springer-Verlag, New York.

Jensen, E.S., B.T. Christensen and L.H. Sorensen. 1989. Mineral-fixed ammonium in clay- and silt-size fractions of soils incubated with 15N-ammonium sulphate for five years. Biology and Fertility of Soils 8: 298–302.

Johansson, C., H. Rodhe and E. Sanhueza. 1988. Emission of NO in a tropical savanna and a cloud forest during the dry season. Journal of Geophysical Research 93: 7180–7192.

Johnson, D.W., D.W. Cole, S.P. Gessel, M.J. Singer and R.V. Minden. 1977. Carbonic acid leaching in a tropical, temperate, subalpine, and northern forest soil. Arctic and Alpine Research 9: 329–343.

Johnson, D.W. 1984. Sulfur cycling in forests. Biogeochemistry 1: 29–43.

Johnson, D.W., G.S. Henderson, D.D. Huff, S.E. Lindberg, D.D. Richter, D.S. Shriner, D.E. Todd and J. Turner. 1982. Cycling of organic and inorganic sulphur in a chestnut oak forest. Oecologia 54: 141–148.

Johnson, D.W., G.S. Henderson and D.E. Todd. 1988. Changes in nutrient distribution in forests and soils of Walker Branch watershed, Tennessee, over an eleven-year period. Biogeochemistry 5: 275–293.

Johnson, D.W. and D.W. Cole. 1980. Anion mobility in soils: Relevance to nutrient transport from forest ecosystems. Environment International 3: 79–90.

Johnson, D.W., G.S. Henderson and D.E. Todd. 1981. Evidence of modern accumulations of adsorbed sulfate in an east Tennessee forested Ultisol. Soil Science 132: 422–426.

Johnson, D.W. and D.E. Todd. 1983. Relationships among iron, aluminum, carbon, and sulfate in a variety of forest soils. Soil Science Society of America Journal 47: 792–800.

Johnson, D.W., D.W. Cole, H. Van Miegroet and F.W. Horng. 1986. Factors affecting anion movement and retention in four forest soils. Soil Science Society of America Journal 50: 776–783.

Johnson, K.S., C.L. Beehler, C. M. Sakamoto-Arnold and J.J. Childress. 1986. In situ measurements of chemical distributions in a deep-sea hydrothermal vent field. Science 231: 1139–1141.

Johnson, N.M., G.E. Likens, F.H. Bormann and R.S. Pierce. 1968. Rate of chemical weathering of silicate minerals in New Hampshire. Geochimica et Cosmochimica Acta 32: 531–545.

Johnson, N.M., G.E. Likens, F.H. Bormann, D.W. Fisher and R.S. Pierce. 1969. A working model for the variation in stream water chemistry at the Hubbard Brook Experimental Forest, New Hampshire. Water Resources Research 5: 1353–1363.

Johnson, N.M. 1971. Mineral equilibria in ecosystem geochemistry. Ecology 52: 529–531.

Johnson, N.M., R.C. Reynolds and G.E. Likens. 1972. Atmospheric sulfur: Its effect on the chemical weathering of New England. Science 177: 514–516.

Johnson, N.M., C.T. Driscoll, J.S. Eaton, G.E. Likens and W.H. McDowell. 1981. 'Acid rain,' dissolved aluminum and chemical weathering at the Hubbard Brook Experimental Forest, New Hampshire. Geochimica et Cosmochimica Acta 45: 1421–1437.

Johnson, W.C. and D.M. Sharpe. 1983. The ratio of total to merchantable forest biomass and its application to the global carbon budget. Canadian Journal of Forest Research 13: 372–383.

Jonasson, S., J.P. Bryant, F.S. Chapin and M. Andersson. 1986. Plant phenolics and nutrients in relation to variations in climate and rodent grazing. American Naturalist 128: 394–408.

Jones, J.M. and B.N. Richards. 1977. Effect of reforestation on turnover of 15N-labelled nitrate and ammonium in relation to changes in soil microflora. Soil Biology and Biochemistry 9: 383–392.

Jones, M.J. 1973. The organic matter content of the savanna soils of west Africa. Journal of Soil Science 24: 42–53.

Jones, P.D., T.M.L. Wigley, and P.B. Wright. 1986. Global temperature variations between 1861 and 1984. Nature 322: 430–434.

Jones, R.L. and H.C. Hanson. 1985. Mineral Licks, Geophagy, and Biogeochemistry of North American Ungulates. Iowa State University Press, Ames.

Jones, R.D. and R.Y. Morita. 1983. Methane oxidation by *Nitrosococcus oceanus* and *Nitrosomonas europaea*. Applied and Environmental Microbiology 45: 401–410.

Jordan, C.F. 1971. A world pattern in plant energetics. American Scientist 59: 425–433.

Jordan, M. and G.E. Likens. 1975. An organic carbon budget for an oligotrophic lake in New Hampshire, U.S.A. Verhandlungen der Internationalen Vereins Limnologie 19: 994–1003.

Jørgensen, B.B. 1977. The sulfur cycle of a coastal marine sediment (Limfjorden, Denmark). Limnology and Oceanography 22: 814–832.

Jorgensen, J.R., C.G. Wells and L. J. Metz. 1980. Nutrient changes in decomposing loblolly pine forest floor. Soil Science Society of America Journal 44: 1307–1314.

Jørgensen, K.S. 1989. Annual pattern of denitrification and nitrate ammonification in estuarine sediment. Applied and Environmental Microbiology 55: 1841–1847.

Jørgensen, K.S., H.B. Jensen and J. Sørensen. 1984. Nitrous oxide production from nitrification and denitrification in marine sediment at low oxygen concentrations. Canadian Journal of Microbiology 30: 1073–1078.

Juang, F.H.T. and N.M. Johnson. 1967. Cycling of chlorine through a forested watershed in New England. Journal of Geophysical Research 72: 5641–5647.

Junge, C.E. and R.T. Werby. 1958. The concentration of chloride, sodium, potassium, calcium, and sulfate in rain water over the United States. Journal of Meteorology 15: 417–425.

Junge, C.E. 1960. Sulfur in the atmosphere. Journal of Geophysical Research 65: 227–237.

Junge, C.E. 1974. Residence time and variability of tropospheric trace gases. Tellus 26: 477–488.

Junge, C.E. and R.T. Werby. 1958. The concentration of chloride, sodium, potassium, calcium, and sulfate in rain water over the United States. Journal of Meteorology 25: 417–425.

Jurinak, J.J., L. M. Dudley, M.F. Allen, and W.G. Knight. 1986. The role of calcium oxalate in the availability of phosphorus in soils of semiarid regions: A thermodynamic study. Soil Science 142: 255–261.

Kadeba, O. 1978. Organic matter status of some savanna soils of northern Nigeria. Soil Science 125: 122-127.

Kaplan, I.R. 1975. Stable isotopes as a guide to biogeochemical processes. Proceedings of the Royal Society of London 189B: 183–211.

Kaplan, W., I. Valiela and J.M. Teal. 1979. Denitrification in a salt marsh ecosystem. Limnology and Oceanography 24: 726–734.

Kaplan, W.A., S.C. Wofsy, M. Keller and J.M. da Costa. 1988. Emission of NO and deposition of O3 in a tropical forest system. Journal of Geophysical Research 93: 1389–1395.

Kasting, J.F. and S.M. Richardson. 1985. Seafloor hydrothermal activity and spreading rates: The Eocene carbon dioxide greenhouse revisited. Geochimica et Cosmochimica Acta 49: 2541–2544.

Kasting, J.F., O. B. Toon and J.B. Pollack. 1988. How climate evolved on the terrestrial planets. Scientific American 258(2): 90–97.

Kasting, J.F. and J.C.G. Walker. 1981. Limits on oxygen concentration in the prebiological atmosphere and the rate of abiotic fixation of nitrogen. Journal of Geophysical Research 86: 1147–1158.

Kaufman, Y.J., C.J. Tucker, and I. Fung. 1990. Remote sensing of biomass burning in the tropics. Journal of Geophysical Research 95: 9927–9939.

Keeling, C.D. 1983. The global carbon cycle: What we know and could know from atmospheric, biospheric, and oceanic observations. pp. II.3–62. Proceedings: Carbon Dioxide Research Conference: Carbon Dioxide, Science and Consensus. CONF 820970. U.S. Department of Energy, Washington, D.C.

Keeling, C.D. 1986. Atmospheric CO_2 concentrations—Mauna Loa Observatory, Hawaii 1958–1986. NDP-001/Rl Carbon Dioxide Information Center, Oak Ridge National Laboratory, Oak Ridge Tennessee.

Keeling, C.D., S.C. Piper and M. Heimann. 1989. A three-dimensional model of atmospheric CO_2 transport based on observed winds: 4. Mean annual gradients and interannual variations. pp. 305–363. In D.H. Peterson, (ed.), Aspects of Climate Variability in the Pacific and Western Americas. American Geophysical Union, Washington, D.C.

Keeling, C.D., A.F. Carter and W.G. Mook. 1984. Seasonal, latitudinal, and secular variations in the abundance and isotopic ratios of atmospheric CO_2. 2. Results from oceanographic cruises in the tropical Pacific Ocean. Journal of Geophysical Research 89: 4615–4628.

Keeney, D.R. 1980. Prediction of soil nitrogen availability in forest ecosystems: A literature reveiw. Forest Science 26: 159–171.

Keffer, T., D.G. Martinson and B.H. Corliss. 1988. The position of the Gulf Stream during Quaternary glaciations. Science 241: 440–442.

Keller, M., T.J. Goreau, S.C. Wofsy, W.A. Kaplan and M.B. McElroy. 1983. Production of nitrous oxide and consumption of methane by forest soils. Geophysical Research Letters 10: 1156–1159.

Keller, M., W.A. Kaplan and S.C. Wofsy. 1986. Emissions of N_2O, CH4, and CO_2 from tropical forest soils. Journal of Geophysical Research 91: 11791–11802.

Keller, M., W.A. Kaplan, S.C. Wofsy, and J.M. DaCosta. 1988. Emission of N_2O from tropical forest soils: Response to fertilization with $NH4^+$, $NO3^-$ and $PO4^{3-}$. Journal of Geophysical Research 93:1600–1604.

Keller, M., M.E. Mitre and R.F. Stallard. 1990. Consumption of atmospheric methane in soils of central Panama: Effects of agricultural development. Global Biogeochemical Cycles 4: 21–27.

Kelly, C.A. and D.P. Chynoweth. 1981. The contribution of temperature and of the input of organic matter in controlling rates of sediment methanogenesis. Limnology and Oceanography 26: 891–897.

Kelly, D.P. and N.A. Smith. 1990. Organic sulfur compounds in the environment. Advances in Microbial Ecology 11: 345–385.

Kempe, S. 1988. Estuaries—Their natural and anthropogenic changes. pp. 251–285. *In* T. Rosswall, R.G. Woodmansee and R.G. Risser (eds.), Scales and Global Change. Wiley, London.

Khalil, M.A.K. and R.A. Rasmussen. 1983. Increase and seasonal cycles of nitrous oxide in the earth's atmosphere. Tellus 35B: 161–169.

Khalil, M.A.K. and R.A. Rasmussen. 1984. Global sources, lifetimes and mass balances of carbonyl sulfide (OCS) and carbon disulfide (CS_2) in the Earth's atmosphere. Atmospheric Environment 18: 1805–1813.

Khalil, M.A.K. and R. A. Rasmussen. 1985. Causes of increasing atmospheric methane: Depletion of hydroxyl radicals and the rise of emissions. Atmospheric Environment 19: 397–407.

Khalil, M.A.K. and R.A. Rasmussen. 1988. Carbon monoxide in the Earth's atmosphere: Indications of a global increase. Nature 332: 242–244.

Khalil, M.A.K. and R.A. Rasmussen. 1989. Climate-induced feedbacks for the global cycles of methane and nitrous oxide. Tellus 41B: 554–559.

Khalil, M.A.K. and R.A. Rasmussen. 1990. Constraints on the global sources of methane and an analysis of recent budgets. Tellus 42B: 229–236.

Khalil, M.A.K., R.A. Rasmussen, J.R.J. French and J.A. Holt. 1990. The influence of termites on atmospheric trace gases: CH_4, CO_2, $CHCl_3$, N_2O, CO, H_2, and light hydrocarbons. Journal of Geophysical Research 95: 3619–3634.

Kieffer, H.H. 1976. Soil and surface temperatures at the Viking landing sites. Science 194: 1344–1346.

Kiene, R.P. and T.S. Bates. 1990. Biological removal of dimethyl sulphide from sea water. Nature 345: 702–704.

Kilham, P. 1971. A hypothesis concerning silica and the freshwater planktonic diatoms. Limnology and Oceanography 16: 10–18.

Kilham, P. 1982. Acid precipitation: Its role in the alkalization of a lake in Michigan. Limnology and Oceanography 27: 856–867.

Killingbeck, K.T. 1985. Autumnal resorption and accretion of trace metals in gallery forest trees. Ecology 66: 283–286.

Killough, G.G. and W.R. Emanuel. 1981. A comparison of several models of carbon turnover in the ocean with respect to their distributions of transit time and age, and responses to atmospheric CO_2 and ^{14}C. Tellus 33: 274–290.

Kim, K-R. and H. Craig. 1990. Two-isotope characterization of N_2O in the Pacific ocean and constraints on its origin in deep water. Nature 347: 58–61.

King, G.M. 1988. Patterns of sulfate reduction and the sulfur cycle in a South Carolina salt marsh. Limnology and Oceanography 33: 376–390.

King, G.M. 1990. Regulation by light of methane emissions from a wetland. Nature 345: 513–515.

King, G.M. and W.J. Wiebe. 1978. Methane release from soils of a Georgia salt marsh. Geochimica et Cosmochimica Acta 42: 343–348.

Kira, T. and T. Shidei. 1967. Primary production and turnover of organic matter in different forest ecosystems of the western Pacific. Japanese Journal of Ecology 17: 70–87.

Kirchhoff, V.W.J.H. 1988. Surface ozone measurements in Amazonia. Journal of Geophysical Research 93: 1469–1476.

Klappa, C.F. 1980. Rhizoliths in terrestrial carbonates: Classification, recognition, genesis and significance. Sedimentology 27: 613–629.

Kling, G.W. 1988. Comparative transparency, depth of mixing, and stability of stratification in lakes of Cameroon, West Africa. Limnology and Oceanography 33: 27–40.

Klinkhammer, G.P. 1980. Early diagenesis in sediments from the eastern equatorial Pacific. II. Pore water metal results. Earth and Planetary Science Letters 49: 81–101.

Klopatek, J.M. 1987. Nitrogen mineralization and nitrification in mineral soils of pinyon-juniper ecosystems. Soil Science Society of America Journal 51: 453–457.

Knauss, J.A. 1978. Introduction to Physical Oceanography. Prentice Hall, Englewood Cliffs, New Jersey

Knight, D.H., T.J. Fahey and S.W. Running. 1985. Water and nutrient outflow from contrasting lodgepole pine forests in Wyoming. Ecological Monographs 55: 29–48.

Knittle, E. and R. Jeanloz. 1987. Synthesis and equation of state of $(Mg,Fe)SiO_3$ perovskite to over 100 gigapascals. Science 235: 668–670.

Knoll, A.H. and E.S. Barghoorn. 1975. Precambrian eukaryotic organisms: A reassessment of the evidence. Science 190: 52–54.

Knowles, R. 1982. Denitrification. Microbiological Reviews 46: 43–70.

Kodama, H. and M. Schnitzer. 1977. Effect of fulvic acid on the crystallization of Fe(III) oxides. Geoderma 19: 279–291.

Kodama, H. and M. Schnitzer. 1980. Effect of fulvic acid on the crystallization of aluminum hydroxides. Geoderma 24: 195–205.

Koerselman, W., H. De Caluwe and W.H. Kieskamp. 1989. Denitrification and dinitrogen fixation in two quaking fens in the Vechtplassen area, The Netherlands. Biogeochemistry 8: 153–165.

Kohlmaier, G.H., E. Siré, A. Janecek, C.D. Keeling, S.C. Piper and R. Revelle. 1989. Modelling the seasonal contribution of a CO_2 fertilization effect of the terrestrial vegetation to the amplitude increase in atmospheric CO_2 at Mauna Loa Observatory. Tellus 41B: 487–510.

Kramer, P.J. 1981. Carbon dioxide concentration, photosynthesis, and dry matter production. BioScience 31: 29–33.

Kramer, P.J. 1982. Water and plant productivity of yield. pp. 41–47. In M. Rechcigl (ed.), Handbook of Agricultural Productivity. CRC Press, Boca Raton, Florida.

Kratz, T.K. and C.B. DeWitt. 1986. Internal factors controlling peatland-lake ecosystem development. Ecology 67: 100–107.

Kristjansson, J.K. and P. Schönheit. 1983. Why do sulfate-reducing bacteria outcompete methanogenic bacteria for substrates? Oecologia 60: 264–266.

Kroehler, C.J. and A.E. Linkins. 1988. The root surface phosphatases of Eriophorum vaginatum: Effects of temperature, pH, substrate concentration and inorganic phosphorus. Plant and Soil 105: 3–10.

Krouse, H.R. and R.G.L. McCready. 1979. Reductive reactions in the sulfur cycle. pp. 315–368. In P.A. Trudinger and D.J. Swaine (eds.), Biogeochemical Cycling of Mineral-Forming Elements. Elsevier Scientific, Amsterdam.

Krumbein, W.E. 1971. Manganese-oxidizing fungi and bacteria in recent shelf sediments of the Bay of Biscay and the North Sea. Naturwissenschaften 58: 56–57.

Krumbein, W.E. 1979. Calcification by bacteria and algae. pp. 47–68. In P.A. Trudinger and D.J. Swaine (eds.), Biogeochemical Cycling of Mineral-Forming Elements. Elsevier Scientific, Amsterdam.

Kuivila, K.M., J.W. Murray, A.H. Devol, M.E. Lidstrom and C.E. Reimers. 1988. Methane cycling in the sediments of Lake Washington. Limnology and Oceanography 33: 571–581.

Kuivila, K.M., J.W. Murray, A.H. Devol, and P.C. Novelli. 1989. Methane production, sulfate reduction and competition for substrates in the sediments of Lake Washington. Geochimica et Cosmochimica Acta 53: 409–416.

Kump, L.R. and R.M. Garrels. 1986. Modeling atmospheric O_2 in the global sedimentary redox cycle. American Journal of Science 286: 337–360.

Kunishi, H.M. 1988. Sources of nitrogen and phosphorus in an estuary of the Chesapeake Bay. Journal of Environmental Quality 17: 185–188.

Lacis, A., J. Hansen, P. Lee, T. Mitchell and S. Lebedeff. 1981. Greenhouse effect of trace gases, 1970–1980. Geophysical Research Letters 8: 1035–1038.

Ladd, J.N., J.M. Oades and M. Amato. 1981. Microbial biomass formed from ^{14}C, ^{15}N-labelled plant material decomposing in soils in the field. Soil Biology and Biochemistry 13: 119–126.

Lajtha, K. 1987. Nutrient reabsorption efficiency and the response to phosphorus fertilization in the desert shrub *Larrea tridentata*. (DC.) Cov. Biogeochemistry 4: 265–276.

Lajtha, K. and S.H. Bloomer. 1988. Factors affecting phosphate sorption and phosphate retention in a desert ecosystem. Soil Science 146: 160–167.

Lajtha, K. and W.H. Schlesinger. 1986. Plant response to variations in nitrogen availability in a desert shrubland ecosystem. Biogeochemistry 2: 29–37.

Lajtha, K. and W.H. Schlesinger. 1988. The biogeochemistry of phosphorus cycling and phosphorus availability along a desert soil chronosequence. Ecology 69: 24–39.

Lajtha, K. and W.G. Whitford. 1989. The effect of water and nitrogen amendments on photosynthesis, leaf demography, and resource-use efficiency in *Larrea tridentata*, a desert evergreen shrub. Oecologia 80: 341–348.

Lal, D. 1977. The oceanic microcosm of particles. Science 198: 997–1009.

Lalisse-Grundmann, G., B. Brunel and A. Chalamet. 1988. Denitrification in a cultivated soil: Optimal glucose and nitrate concentrations. Soil Biology and Biochemistry 20: 839–844.

LaMarche, V.C., D.A. Graybill, H.C. Fritts and M.R. Rose. 1984. Increasing atmospheric carbon dioxide: Tree ring evidence for growth enhancement in natural vegetation. Science 225: 1019–1021.

Lamb, B., A. Guenther, D. Gay and H. Westberg. 1987. A national inventory of biogenic hydrocarbon emissions. Atmospheric Environment 21: 1695–1705.

Lamb, D. 1985. The influence of insects on nutrient cycling in eucalypt forests: A beneficial role? Australian Journal of Ecology 10: 1–5.

Lambert, R.L., G.E. Lang and W.A. Reiners. 1980. Loss of mass and chemical change in decaying boles of a subalpine balsam fir forest. Ecology 61: 1460–1473.

Lang, G.E. and R.T.T. Forman. 1978. Detrital dynamics in a mature oak forest: Hutcheson Memorial Forest, New Jersey. Ecology 59: 580–595.

Lang, G.E., W.A. Reiners and R.K. Heier. 1976. Potential alteration of precipitation chemistry by epiphytic lichens. Oecologia 25: 229–241.

Lantzy, R.J. and F. T. MacKenzie. 1979. Atmospheric trace metals: Global cycles and assessment of man's impact. Geochimica et Cosmochimica Acta 43: 511–525.

Lapeyrie, F., G.A. Chilvers and C.A. Bhem. 1987. Oxalic acid synthesis by the mycorrhizal fungus *Paxillus involutus* (Battsch. ex. Fr.). Fr. New Phytologist 106: 139–146.

Lasenby, D.C. 1975. Development of oxygen deficits in 14 southern Ontario lakes. Limnology and Oceanography 20: 993–999.

Lashof, D.A. and D.R. Ahuja. 1990. Relative contributions of greenhouse gas emissions to global warming. Nature 344: 529–531.

Lauenroth, W.K. and W.C. Whitman. 1977. Dynamics of dry matter production in a mixed-grass prairie in western North Dakota. Oecologia 27: 339–351.

Laurmann, J.A. 1979. Market penetration characteristics for energy production and atmospheric carbon dioxide growth. Science 205: 896–898.

Lawless, J.G. and N. Levi. 1979. The role of metal ions in chemical evolution: Polymerization of alanine and glycine in a cation-exchanged clay environment. Journal of Molecular Evolution 13: 281–286.

Lawson, D.R. and J.W. Winchester. 1979. A standard crustal aerosol as a reference for elemental enrichment factors. Atmospheric Environment 13: 925–930.

LaZerte, B.D. 1983. Stable carbon isotope ratios: Implications for the source of sediment carbon and for phytoplankton carbon assimilation in Lake Memphremagog, Quebec. Canadian Journal of Fisheries and Aquatic Science 40: 1658–1666.

Leahey, A. 1947. Characteristics of soils adjacent to the MacKenzie River in the Northwest Territories of Canada. Soil Science Society of America Proceedings 12: 458–461.

Lean, D.R.S. 1973. Phosphorus dynamics in lake water. Science 179:678–680.

Lean, J. and D.A. Warrilow. 1989. Simulation of the regional climatic impact of Amazon deforestation. Nature 342: 411–413.

Leavitt, S.W. and A. Long. 1988. Stable carbon isotope chronologies from trees in the southwestern United States. Global Biogeochemical Cycles 2: 189–198.

Lee, J.A. and G.R. Stewart. 1978. Ecological aspects of nitrogen assimilation. Advances in Botanical Research 6: 1–43.

Legrand, M. and R.J. Delmas. 1987. A 220-year continuous record of volcanic H_2SO_4 in the Antarctic ice sheet. Nature 327: 671–676.

Lehman, J.T. 1980. Release and cycling of nutrients between planktonic algae and herbivores. Limnology and Oceanography 25: 620–632.

Lehman, J.T. 1988. Hypolimnetic metabolism in Lake Washington: Relative effects of nutrient load and food web structure on lake productivity. Limnology and Oceanography 33: 1334–1347.

Lein, A.Y. 1984. Anaerobic consumption of organic matter in modern marine sediments. Nature 312: 148–150.

Lennon, J.M., J.D. Aber and J.M. Melillo. 1985. Primary production and nitrogen allocation of field grown sugar maples in relation to nitrogen availability. Biogeochemistry 1: 135–154.

Leschine, S.B., K. Holwell and E. Canale-Parola. 1988. Nitrogen fixation by anaerobic cellulolytic bacteria. Science 242: 1157–1159.

Levin, S.A. 1989. Challenges in the development of a theory of community and ecosystem structure and function. pp. 242–255. In J. Roughgarden, R.M. May and S.A. Levin (eds.), Perspectives in Ecological Theory, Princeton University Press, Princeton, New Jersey.

Levine, J.S., T.R. Augustsson, I.C. Anderson, J. M. Hoell and D.A. Brewer. 1984. Tropospheric sources of NO_x: Lightning and biology. Atmospheric Environment 18: 1797–1804.

Levine, J.S., W.R. Cofer, D.I. Sebacher, E.L. Winstead, S. Sebacher and P.J. Boston. 1988. The effects of fire on biogenic soil emissions of nitric oxide and nitrous oxide. Global Biogeochemical Cycles 2: 445–449.

Levine, S.N., M.P. Stainton and D.W. Schindler. 1986. A radiotracer study of phosphorus cycling in a eutrophic Canadian Shield Lake, Experimental Lake 227, northwestern Ontario. Canadian Journal of Fisheries and Aquatic Science 43: 366–378.

Levine, S.N. and D.W. Schindler. 1989. Phosphorus, nitrogen, and carbon dynamics of Lake 303 during recovery from eutrophication. Canadian Journal of Fisheries and Aquatic Science 46: 2–10.

Lewis, M.R., W.G. Harrison, N.S. Oakey, D. Hebert and T. Platt. 1986. Vertical nitrate fluxes in the oligotrophic ocean. Science 234: 870–873.

Lewis, W.M. 1974. Effects of fire on nutrient movement in a South Carolina pine forest. Ecology 55: 1120–1127.

Lewis, W.M. 1981. Precipitation chemistry and nutrient loading by precipitation in a tropical watershed. Water Resources Research 17: 169–181.

Lewis, W.M. 1986. Nitrogen and phosphorus runoff losses from a nutrient-poor tropical moist forest. Ecology 67: 1275–1282.

Lewis, W.M. 1988. Primary production in the Orinoco River. Ecology 69: 679–692.

Lewis, W.M. and M.C. Grant. 1979. Relationships between stream discharge and yield of dissolved substances from a Colorado Mountain watershed. Soil Science 128: 353–363.

Lewis, W.M., S.K. Hamilton, S. L. Jones and D.D. Runnels. 1987. Major element chemistry, weathering and elements yields for the Caura River drainage, Venezuela. Biogeochemistry 4: 159–181.

Li, W.K.W., D.V. Subba Rao, W.G. Harrison, J.C. Smith, J.J. Cullen, B. Irwin and T. Platt. 1983. Autotrophic picoplankton in the tropical ocean. Science 219: 292–295.

Li, Y.-H. 1972. Geochemical mass balance among lithosphere, hydrosphere, and atmosphere. American Journal of Science 272: 119–137.

Li, Y.-H. 1981. Geochemical cycles of elements and human perturbation. Geochimica et Cosmochimica Acta 45: 2073–2084.

Lieth, H. 1975. Modeling the primary productivity of the world. pp. 237–263. In H. Lieth and R.H. Whittaker (eds.), Primary Productivity of the Biosphere. Springer-Verlag, New York.

Lightfoot, D.C. and W.G. Whitford. 1987. Variation in insect densities on desert creosotebush: Is nitrogen a factor? Ecology 68: 547–557.

Likens, G.E. 1975a. Nutrient flux and cycling in freshwater ecosystems. pp. 314–348. In F.G. Howell, J.B. Gentry and M.H. Smith (eds.), Mineral Cycling in Southeastern Ecosystems. National Technical Information Service, Springfield, Virginia.

Likens, G.E. 1975b. Primary production of inland aquatic ecosystems. pp. 185–202. In H. Lieth and R.H. Whittaker (eds.), Primary Productivity of the Biosphere. Springer-Verlag, New York.

Likens, G.E., F.H. Bormann, N.M. Johnson, D.W. Fisher and R.S. Pierce. 1970. Effects of forest cutting and herbicide treatment on nutrient budgets in the Hubbard Brook watershed-ecosystem. Ecological Monographs 40: 23–47.

Likens, G.E. and F.H. Bormann. 1974. Linkages between terrestrial and aquatic ecosystems. BioScience 24: 447–456.

Likens, G.E., F. H. Bormann, R.S. Pierce, J.S. Eaton and N.M. Johnson. 1977. Biogeochemistry of a Forested Ecosystem. Springer-Verlag, New York.

Likens, G.E., F.H. Bormann and N.M. Johnson. 1981. Interactions between major biogeochemical cycles in terrestrial ecosystems. pp. 93–112. In G.E. Likens (ed.), Some Perspectives of the Major Biogeochemical Cycles. Wiley, New York.

Likens, G.E., F.H. Bormann, R.S. Pierce, J.S. Eaton, and R.E. Munn. 1984. Long-term trends in precipitation chemistry at Hubbard Brook, New Hampshire. Atmospheric Environment 18: 2641–2647.

Lilley, M.D., M.A. de Angelis, and L.I. Gordon. 1982. CH_4, H_2, CO and N_2O in submarine hydrothermal vent waters. Nature 300: 48–50.

Linak, W.P., J.A. McSorley, R.E. Hall, J.V. Ryan, R.K. Srivastava, J.O.L. Wendt and J.B. Mereb. 1990. Nitrous oxide emissions from fossil fuel combustion. Journal of Geophysical Research 95: 7533–7541.

Lindberg, S.E. and G.M. Lovett. 1985. Field measurements of particle dry deposition rates to foliage and inert surfaces in a forest canopy. Environmental Science and Technology 19: 238–244.

Lindberg, S.E. and R.R. Turner. 1988. Factors influencing atmospheric deposition, stream export, and landscape accumulation of trace metals in forested watersheds. Water, Air, and Soil Pollution 39: 123–156.

Lindberg, S.E., G.M. Lovett, D.D. Richter and D.W. Johnson. 1986. Atmospheric deposition and canopy interactions of major ions in a forest. Science 231: 141–145.

Lindberg, S.E. and C.T. Garten. 1988. Sources of sulphur in forest canopy throughfall. Nature 336: 148–151.

Lindqvist, Q and H. Rodhe. 1985. Atmospheric mercury—A review. Tellus 37B: 136–159.

Lindsay, W.L. and E.C. Moreno. 1960. Phosphate phase equilibria in soils. Soil Science Society of America Proceedings 24: 177–182.

Lindsay, W.L. and P.L.G. Vlek. 1977. Phosphate minerals. pp. 639–672. In J.B. Dixon and S.B. Weed (eds.), Minerals in Soil Environments. Soil Science Society of America, Madison, Wisconsin.

Lindsay, W.L. 1979. Chemical Equilibria in Soils. Wiley, New York.

Liss, P.S. and P.G. Slater. 1974. Flux of gases across the air-sea interface. Nature 247: 181–184.

Liss, P.S. 1983. The exchange of biogeochemically important gases across the air-sea interface. pp. 411–426. *In* B. Bolin and R.B. Cook (eds.), The Major Biogeochemical Cycles and Their Interactions. Wiley, New York.

Littke, W.R., C.S. Bledsoe and R.L. Edmonds. 1984. Nitrogen uptake and growth *in vitro* by *Hebeloma crustuliniforme* and other Pacific Northwest mycorrhizal fungi. Canadian Journal of Botany 62: 647–652.

Liu, T.S., X.F. Gu, Z.S. An, and Y.X. Fan. 1981. The dust fall in Beijing, China on April 18, 1980. pp. 149–157. *In* T.L. Pewe (ed.), Desert Dust: Origin, Characteristics, and Effect on Man. Geological Society of America, Special Paper 186, Boulder, Colorado.

Livingston, G.P., P.M. Vitousek and P.A.Matson. 1988. Nitrous oxide flux and nitrogen transformations across a landscape gradient in Amazonia. Journal of Geophysical Research 93: 1593–1599.

Livingston, D.A. 1963. Chemical composition of rivers and lakes. U.S. Geological Survey, Professional Paper 440G, 64 pp.

Lobert, J.M., D.H. Scharffe, W.M. Hao, and P.J. Crutzen. 1990. Importance of biomass burning in the atmospheric budgets of nitrogen-containing gases. Nature 346: 552–554.

Logan, J.A. 1983. Nitrogen oxides in the troposphere: Global and regional budgets. Journal of Geophysical Research 88: 10785–10807.

Logan, J.A. 1985. Tropospheric ozone; Seasonal behavior, trends, and anthropogenic influence. Journal of Geophysical Research 90: 10463–10482.

Logan, J.A., M.J. Prather, S.C. Wofsy and M.B. McElroy. 1981. Tropospheric chemistry: A global perspective. Journal of Geophysical Research 86: 7210–7254.

Lohrmann, R. and L.E. Orgel. 1973. Prebiotic activation processes. Nature 244: 418–420.

Longmore, M.E., B.M. O'Leary, C.W. Rose and A.L Chandica. 1983. Mapping soil erosion and accumulation with the fallout isotope caesium-137. Australian Journal of Soil Research 21: 373–385.

Lonsdale, W.M. 1988. Predicting the amount of litterfall in forests of the world. Annals of Botany 61: 319–324.

Lovelock, J.E., R.J. Maggs and R.A. Rasmussen. 1972. Atmospheric dimethyl sulphide and the natural sulphur cycle. Nature 237: 452–453.

Lovelock, J.E. and M. Whitfield. 1982. Life span of the biosphere. Nature 296: 561–563.

Lovelock, J.E. 1979. Gaia: A New Look at Life on Earth. Oxford University Press, Oxford.

Lovelock, J. 1988. The Ages of Gaia. W.W. Norton, New York.

Lovett, G.M., W.A. Reiners, and R.K. Olson. 1982. Cloud droplet deposition in subalpine balsam fir forests: Hydrological and chemical inputs. Science 218: 1303–1304.

Lovett, G.M. and S.E. Lindberg. 1986. Dry deposition of nitrate to a deciduous forest. Biogeochemistry 2: 137–148.

Lovley, D.R. and E.J.P. Phillips. 1987. Competitive mechanisms for inhibition of sulfate reduction and methane production in the zone of ferric iron reduction in sediments. Applied and Environmental Microbiology 53: 2636–2641.

Lovley, D.R. and E.J.P. Phillips. 1988a. Novel mode of microbial energy metabolism: Organic carbon oxidation coupled to dissimilatory reduction of iron or manganese. Applied and Environmental Microbiology 54: 1472–1480.

Lovley, D.R. and E.J.P. Phillips. 1988b. Manganese inhibition of microbial iron reduction in anaerobic sediments. Geomicrobiology Journal 6: 145–155.

Lovley, D.R. and E.J.P. Phillips. 1989. Requirement for a microbial consortium to completely oxidize glucose in Fe(III)-reducing sediments. Applied and Environmental Microbiology 55: 3234–3236.

Lovley, D.R. and M.J. Klug. 1986. Model for the distribution of sulfate reduction and methanogenesis in freshwater sediments. Geochimica et Cosmochimica Acta 50: 11–18.

Lowrance, R., R. Todd, J. Fail, O. Hendrickson, R. Leonard and L. Asmussen. 1984. Riparian forests as nutrient filters in agricultural watersheds. Bioscience 34: 374–377.

Lowry, B., D. Lee and C. Hébant. 1980. The origin of land plants: A new look at an old problem. Taxon 29: 183–197.

Löye-Pilot, M.D., J. M. Martin and J Morelli. 1986. Influence of Saharan dust on the rain acidity and atmospheric input to the Mediterranean. Nature 321: 427–428.

Lui, K.-K. and I.R. Kaplan. 1984. Denitrification rates and availability of organic matter in marine environments. Earth and Planetary Science Letters 68: 88–100.

Lui, K.-K. and I.R. Kaplan. 1989. The eastern tropical Pacific Ocean as a source of ^{15}N-enriched nitrate in seawater off southern California. Limnology and Oceanography 34: 820–830.

Luizão, F., P. Matson, G. Livingston, R. Luizão, and P. Vitousek. 1989. Nitrous oxide flux following tropical land clearing. Global Biogeochemical Cycles 3: 281–285.

Lull, H.W. and W.E. Sopper. 1969. Hydrologic effects from urbanization of forested watersheds in the Northeast. U.S. Department of Agriculture, Northeast Forest Experiment Station Research Paper NE-146, Upper Darby, Pennsylvania.

Lupton, J.E. and H. Craig. 1981. A major helium-3 source at 15°S on the east Pacific Rise. Science 214: 13–18.

Luther, F.M. and R.D. Cess. 1985. Review of the recent carbon dioxide-climate controversy. pp. 321–335. In M.C. MacCraken and F. M. Luther (eds.), Projecting the Climatic Effects of Increasing Carbon Dioxide. U.S. Department of Energy, Er-0237, Washington, D.C.

Luther, G.W., T.M. Church, J.R. Scudlark and M. Cosman. 1986. Inorganic and organic sulfur cycling in salt-marsh pore waters. Science 232: 746–749.

Luxmoore, R.J., T. Grizzard and R.H. Strand. 1981. Nutrient translocation in the outer canopy and understory of an eastern deciduous forest. Forest Science 27: 505–518.

Lvovitch, M.I. 1973. The global water balance. EOS 54: 28–42.

MacCracken, M.C. 1985. Carbon dioxide and climate change: Background and overview. pp. 1–23. In M.C. MacCracken and F.M. Luther (eds.), Projecting the Climatic Effects of Increasing Carbon Dioxide. U.S. Department of Energy, Er-0237, Washington, D.C.

MacDonald, G.J. 1990. Role of methane clathrates in past and future climates. Climatic Change 16: 247–281.

MacDonald, N.W. and J.B. Hart. 1990. Relating sulfate adsorption to soil properties in Michigan forest soils. Soil Science Society of America Journal 54: 238–245.

Mach, D.L., A. Ramirez and H.D. Holland. 1987. Organic phosphorus and carbon in marine sediments. American Journal of Science 287: 429–441.

MacIntyre, F. 1974. The top millimeter of the ocean. Scientific American 230(5): 62–77.

MacKenzie, F.T. and R.M. Garrels. 1966. Chemical mass balance between rivers and oceans. American Journal of Science 264: 507–525.

MacKenzie, F.T., M. Stoffyn and R. Wollast. 1978. Aluminum in seawater: Control by biological activity. Science 199: 680–682.

Mackney, D. 1961. A podzol development sequence in oakwoods and heath in central England. Journal of Soil Science 12: 23–40.

Malaney, R.A. and W.A. Fowler. 1988. The transformation of matter after the big bang. American Scientist 76: 472–477.

Malcolm, R.E. 1983. Assessment of phosphatase activity in soils. Soil Biology and Biochemistry 15: 403–408.

Manabe, S. and R.T. Wetherald. 1980. On the distribution of climate change resulting from an increase in CO_2 content of the atmosphere. Journal of the Atmospheric Sciences 37: 99–118.

Manabe, S. and R.T. Wetherald. 1986. Reduction in summer soil wetness induced by an increase in atmospheric carbon dioxide. Science 232: 626–628.

Mancinelli, R.L. and C.P. McKay. 1988. The evolution of nitrogen cycling. Origins of Life and Evolution of the Biosphere 18: 311–325.

Mann, L.K. 1986. Changes in soil carbon storage after cultivation. Soil Science 142: 279–288.

Mantoura, R.F.C. and E.M.S. Woodward. 1983. Conservative behaviour of riverine dissolved organic carbon in the Severn Estuary: Chemical and geochemical implications. Geochimica et Cosmochimica Acta 47: 1293–1309.

Marion, G.M., W.H. Schlesinger and P.J. Fonteyn. 1985. CALDEP: A regional model for soil $CaCO_3$ (caliche) deposition in southwestern deserts. Soil Science 139: 468–481.

Marion, G.M. and C.H. Black. 1987. The effect of time and temperature on nitrogen mineralization in Arctic tundra soils. Soil Science Society of America Journal 51: 1501–1508.

Marion, G.M. and C.H. Black. 1988. Potentially available nitrogen and phosphorus along a chaparral fire cycle chronosequence. Soil Science Society of America Journal 52: 1155–1162.

Marks, P.L. and F.H. Bormann. 1972. Revegetation following forest cutting: Mechanisms for return ot steady-state nutrient cycling. Science 176: 914–915.

Maroulis, P.J. and A.R. Bandy. 1977. Estimate of the contribution of biologically produced dimethyl sulfide to the global sulfur cycle. Science 196: 647–648.

Marrs, R.H., J. Proctor, A. Heaney and M.D. Mountford. 1988. Changes in soil nitrogen-mineralization and nitrification along an altitudinal transect in tropical rain forest in Costa Rica. Journal of Ecology 76: 466–482.

Martens, C.S., N.E. Blair, C.D. Green and D. J. Des Marais. 1986. Seasonal variations in the stable carbon isotopic signature of biogenic methane in a coastal sediment. Science 233: 1300–1303.

Martens, C.S. and J.V. Klump. 1984. Biogeochemical cycling in an organic-rich coastal marine basin. 4. An organic carbon budget for sediments dominated by sulfate reduction and methanogenesis. Geochimica et Cosmochimica Acta 48: 1987–2004.

Martin, C.W. and R.D. Harr. 1988. Precipitation and streamwater chemistry from undisturbed watersheds in the Cascade Mountains of Oregon. Water, Air, and Soil Pollution 42: 203–219.

Martin, J.H., G.A. Knauer, D.M. Karl and W.W. Broenkow. 1987. VERTEX: Carbon cycling in the northeast Pacific. Deep Sea Research 34: 267–285.

Martin, J.H. and R.M. Gordon. 1988. Northeast Pacific iron distributions in relation to phytoplankton productivity. Deep Sea Research 35: 177–196.

Martin, J.H., R. M. Gordon, S. Fitzwater and W.W. Broenkow. 1989. VERTEX: Phytoplankton/iron studies in the Gulf of Alaska. Deep Sea Research 36: 649–680.

Martin, J.-M. and M. Meybeck. 1979. Elemental mass-balance of material carried by major world rivers. Marine Chemistry 7: 173–206.

Martinéz, L.A., M.W. Silver, J.M. King and A.L. Alldredge. 1983. Nitrogen fixation by floating diatom mats: A source of new nitrogen to oligotrophic ocean waters. Science 221: 152–154.

Marx, D.H., A.B. Hatch and J.F. Mendicino. 1977. High soil fertility decreases sucrose content and susceptibility of loblolly pine roots to ectomycorrhizal infection by *Pisolithus tinctorius*. Canadian Journal of Botany 55: 1569–1574.

Marumoto, T., J.P.E. Anderson and K.H. Domsch. 1982. Mineralization of nutrients from soil microbial biomass. Soil Biology and Biochemistry 14: 469–475.

Mason, B. 1966. Principles of Geochemistry. Wiley, New York.

Matson, P.A. and P.M. Vitousek. 1987. Cross-system comparisons of soil nitrogen transformations and nitrous oxide flux in tropical forest ecosystems. Global Biogeochemical Cycles 1: 163–170.

Matson, P.A. and P.M. Vitousek. 1990. Ecosystem approach to a global nitrous oxide budget. BioScience, 40:667–672.

Matson, P.A., P.M. Vitousek, J.J. Ewel, M.J. Mazzarino and G.P. Robertson. 1987. Nitrogen transformations following tropical forest felling and burning on a volcanic soil. Ecology 68: 491–502.

Matthews, E. and I. Fung. 1987. Methane emission from natural wetlands: Global distribution, area, and environmental characteristics of sources. Global Biogeochemical Cycles 1: 61–86.

Mattson, W.J. 1980. Herbivory in relation to plant nitrogen content. Annual Review of Ecology and Systematics 11: 119–161.

Mattson, W.J. and N.D. Addy. 1975. Phytophagous insects as regulators of forest primary production. Science 190: 515–522.

Mayewski, P.A., W.B. Lyons, M.J. Spencer, M. Twickler, W. Dansgaard, B. Koci, C.I. Davidson and R.E. Honrath. 1986. Sulfate and nitrate concentrations from a south Greenland ice core. Science 232: 975–977.

Mazumder, A. and M.D. Dickman. 1989. Factors affecting the spatial and temporal distribution of phototrophic sulfur bacteria. Archive de Hydrobiologia 116: 209–226.

McCarthy, J.J. and J.C. Goldman. 1979. Nitrogenous nutrition of marine phytoplankton in nutrient-depleted waters. Science 203: 670–672.

McColl, J.G. and D.F. Grigal. 1975. Forest fire: Effects on phosphorus movement to lakes. Science 188: 1109–1111.

McDiffett, W.F., A.W. Beidler, T.F. Dominick and K.D. McCrea. 1989. Nutrient concentration–stream discharge relationships during storm events in a first-order stream. Hydrobiologia 179: 97–102.

McDowell, W.H. and T. Wood. 1984. Podzolization: Soil processes control dissolved organic carbon concentrations in stream water. Soil Science 137: 23–32.

McDowell, W.H. and G.E. Likens. 1988. Origin, composition, and flux of dissolved organic carbon in the Hubbard Brook Valley. Ecological Monographs 58: 177–195.

McElroy, M.B. 1983. Marine biological controls on atmospheric CO_2 and climate. Nature 302: 328–329.

McElroy, M.B., Y. L. Yung and A. O. Nier. 1976. Isotopic composition of nitrogen: Implications for the past history of Mars' atmosphere. Science 194: 70–72.

McElroy, M.B., M.J. Prather and J.M. Rodriguez. 1982. Escape of hydrogen from Venus. Science 215: 1614–1615.

McElroy, M.B. and R.J. Salawitch. 1989. Changing composition of the global stratosphere. Science 243: 763–770.

McFadden, L.D. and D. M. Hendricks. 1985. Changes in the content and composition of pedogenic iron oxyhydroxides in a chronosequence of soils in southern California. Quaternary Research 23: 189–204.

McGill, W.B. and C.V. Cole. 1981. Comparative aspects of cycling of organic C, N, S and P through soil organic matter. Geoderma 26: 267–286.

McKelvey, V.E. 1980. Seabed minerals and the law of the sea. Science 209: 464–472.

McLaughlin, S.B. and G.E. Taylor. 1981. Relative humidity: Important modifier of pollutant uptake by plants. Science 211: 167–169.

McNaughton, S.J. 1988. Mineral nutrition and spatial concentrations of African ungulates. Nature 334: 343–345.

McNaughton, S.J. and F.S. Chapin. 1985. Effects of phosphorus nutrition and defoliation on C_4 graminoids from the Serengeti plains. Ecology 66: 1617–1629.

McSween, H.Y. 1989. Chondritic meteorites and the formation of planets. American Scientist 77: 146–153.

Meade, R.H., T. Dunne, J.E. Richey, U. de M. Santos and E. Salati. 1985. Storage and remobilization of suspended sediment in the Lower Amazon River of Brazil. Science 228: 488–490.

Meentemeyer, V. 1978a. Climatic regulation of decomposition rates of organic matter in terrestrial ecosystems. pp. 779–789. *In* D.C. Adriano and I.L. Brisbin (eds.), Environmental Chemistry and Cycling Processes. CONF 760429, National Technical Information Service, Springfield, Virginia.

Meentemeyer, V. 1978b. Macroclimate and lignin control of litter decomposition rates. Ecology 59: 465–472.

Meentemeyer, V., E. O. Box and R. Thompson. 1982. World patterns and amounts of terrestrial plant litter production. BioScience 32: 125–128.

Megraw, S.R. and R. Knowles. 1987. Active methanotrophs suppress nitrification in a humisol. Biology and Fertility of Soils 4: 205–212.

Meier, M.F. 1984. Contribution of small glaciers to global sea level. Science 226: 1418–1421.

Melillo, J.M., J.D. Aber and J.F. Muratore. 1982. Nitrogen and lignin control of hardwood leaf litter decomposition dynamics. Ecology 63: 621–626.

Melillo, J.M., J.D. Aber, P.A. Steudler, and J.P. Schimel. 1983. Denitrification potentials in a successional sequence of northern hardwood forest stands. pp. 217–228. *In* R. Hallberg (ed.), Environmental Biogeochemistry. Swedish Natural Science Research Council, Stockholm.

Melin, J., H. Nômmik, U. Lohm and J. Flower-Ellis. 1983. Fertilizer nitrogen budget in a Scots pine ecosystem attained by using root-isolated plots and ^{15}N tracer technique. Plant and Soil 74: 249–263.

Mendelssohn, I.A., K.L. McKee and W.H. Patrick. 1981. Oxygen deficiency in *Spartina alterniflora* roots: Metabolic adaptation to anoxia. Science 214: 439–441.

Mengel, K. and H.W. Scherer. 1981. Release of nonexchangeable (fixed) soil ammonium under field conditions during the growing season. Soil Science 131: 226–232.

Meybeck, M. 1977. Dissolved and suspended matter carried by rivers; Composition, time and space variations, and world balance. pp. 25–32. *In* H.L. Golterman (ed.), Interactions between Sediments and Freshwater. Dr. W. Junk, Publishers, The Hague.

Meybeck, M. 1979. Concentrations des eaux fluviales en elements majeurs et apports en solution aux oceans. Revue de Geologie Dynamique et de Geographe Physique 21: 215–246.

Meybeck, M. 1982. Carbon, nitrogen, and phosphorus transport by world rivers. American Journal of Science 282: 401–450.

Meyer, C. 1985. Ore metals through geologic history. Science 227: 1421–1428.

Meyer, J.L. 1979. The role of sediments and bryophytes in phosphorus dynamics in a headwater stream ecosystem. Limnology and Oceanography 24: 365–375.

Meyer, J.L. 1980. Dynamics of phosphorus and organic matter during leaf decomposition in a forest stream. Oikos 34: 44–53.

Meyer, J.L. and G.E. Likens. 1979. Transport and transformation of phosphorus in a forest stream ecosystem. Ecology 60: 1255–1269.

Meyer, J.L., G.E. Likens and J. Sloane. 1981. Phosphorus, nitrogen, and organic carbon flux in a headwater stream. Archiv fuer Hydrobiologie 91: 28–44.

Meyer, J.L., R.T. Edwards and R. Risley. 1987. Bacterial growth on dissolved organic carbon from a blackwater river. Microbial Ecology 13: 13–29.

Meyer, J.L., W.H. McDowell, T.L. Bott, J.W. Elwood, C. Ishizaki, J.M. Melack, B.L. Peckarsky, B.J. Peterson and P.A. Rublee. 1988. Elemental dynamics in streams. Journal of the North American Benthological Society 7: 410–432.

Michaelis, W., V. Ittekkot and E.T. Degens. 1986. River inputs into oceans. pp. 37–52. *In* P. Lasserre and J.-M. Martin (eds.), Biogeochemical Processes at the Land-Sea Boundary. Elsevier Science, Amsterdam.

Michel, R.L. and H.E. Suess. 1975. Bomb tritium in the Pacific Ocean. Journal of Geophysical Research 80: 4139–4152.

Middleton, K.R. and G.S. Smith. 1979. A comparison of ammoniacal and nitrate nutrition of perennial ryegrass through a thermodynamic model. Plant and Soil 53: 487–504.

Migdisov, A.A., A.B. Ronov and V.A. Grinenko. 1983. The sulphur cycle in the lithosphere.

pp. 25–127. *In* M.V. Ivanov and J.R. Freney (eds.), The Global Biogeochemical Sulphur Cycle. Wiley, New York.

Miller, H.G., J. M. Cooper and J.D. Miller. 1976. Effects of nitrogen supply on nutrients in litter fall and crown leaching in a stand of corsican pine. Journal of Applied Ecology 13: 233–248.

Miller, N.L., D.L. Sisterson and Y.P. Liaw. 1989. A normalization and comparison of theoretical, laboratory, and field estimates of global nitrogen fixation by lightning. EOS 70(43): 1005.

Miller, S.L. 1957. The formation of organic compounds on the primitive Earth. Annals of the New York Academy of Sciences 69: 260–275.

Miller, W.R. and J.I. Drever. 1977. Chemical weathering and related controls on surface water chemistry in the Absaroka Mountains, Wyoming. Geochimica et Cosmochimica Acta 41: 1693–1702.

Milliman, J.D. and R.H. Meade. 1983. World-wide delivery of river sediment to the oceans. Journal of Geology 91: 1–21.

Mispagel, M.E. 1978. The ecology and bioenergetics of the acridid grasshopper, *Bootettix punctatus*, on cresotebush, *Larrea tridentata*, in the northern Mojave desert. Ecology 59: 779–788.

Mitchell, J.F.B. 1983. The hydrological cycle as simulated by an atmospheric general circulation model. pp. 429–446. *In* A. Street-Perrott and M. Beran (eds.), Variations in the Global Water Budget. D. Reidel, Hingham, Massachusetts.

Mitchell, M.J., M.B. David, D.G. Maynard and S.A. Telang. 1986. Sulfur constituents in soils and streams of a watershed in the Rocky Mountains of Alberta. Canadian Journal of Forest Research 16: 315–320.

Mitchell, M.J., C.T. Driscoll, R.D. Fuller, M.B. David and G.E. Likens. 1989. Effect of whole-tree harvesting on the sulfur dynamics of a forest soil. Soil Science Society of America Journal 53: 933–940.

Mitsch, W.J., C.L. Dorge and J.R. Wiemhoff. 1979. Ecosystem dynamics and a phosphorus budget of an alluvial cypress swamp in southern Illinois. Ecology 60: 1116–1124.

Mitsch, W.J. and J.G. Gosselink. 1986. Wetlands. Van Nostrand Reinhold, New York.

Mizutani, H., H. Hasegawa and E. Wada. 1986. High nitrogen isotope ratio for soils of seabird rookeries. Biogeochemistry 2: 221–247.

Mizutani, H. and E. Wada. 1988. Nitrogen and carbon isotope ratios in seabird rookeries and their ecological implications. Ecology 69: 340–349.

Moeller, J.R., G.W. Minshall, K.W. Cummins, R.C. Petersen, C.E. Cushing, J.R. Sedell, R.A. Larson and R.L. Vannote. 1979. Transport of dissolved organic carbon in streams of differing physiographic characteristics. Organic Geochemistry 1: 139–150.

Mohren, G.M.J., J. Van den Berg and F.W. Burger. 1986. Phosphorus deficiency induced by nitrogen input in Douglas fir in the Netherlands. Plant and Soil 95: 191–200.

Molina, M.J., T.-L. Tso, L.T. Molina and F.C.-Y. Wang. 1987. Antarctic stratospheric chemistry of chlorine nitrate, hydrogen chloride, and ice: Release of active chlorine. Science 238: 1253–1257.

Möller, D. 1984. On the global natural sulphur emission. Atmospheric Environment 18: 29–39.

Molofsky, J., E.S. Menges, C.A.S. Hall, T.V. Armentano and K.A. Ault. 1984. The effects of land use alteration on tropical carbon exchange. pp. 181–194. *In* T.N. Veziroglu (ed.), The Biosphere: Problems and Solutions. Elsevier Science, Amsterdam.

Monk, C.D. 1966. An ecological significance of evergreenness. Ecology 47: 504–505.

Mooney, H.A., P.M. Vitousek and P.A. Matson. 1987. Exchange of materials between terrestrial ecosystems and the atmosphere. Science 238: 926–932.

Moore, T.R. and R. Knowles. 1989. The influence of water table levels on methane and carbon dioxide emissions from peatland soils. Canadian Journal of Soil Science 69: 33–38.

Moran, S.B. and R.M. Moore. 1988. Evidence from mesocosm studies for biological removal of dissolved aluminium from sea water. Nature 335: 706–708.

Mortimer, C.H. 1941. The exchange of dissolved substances between mud and water in lakes. I & II. Journal of Ecology 29: 280–329.

Mortimer, C.H. 1942. The exchange of dissolved substances between mud and water in lakes. III & IV. Journal of Ecology 30: 147–201.

Morowitz, H.J. 1968. Energy Flow in Biology: Biological Organization as a Problem in Thermal Physics. Academic Press, New York.

Morrow, P.A. and V.C. LaMarche. 1978. Tree ring evidence for chronic insect suppression of productivity in subalpine *Eucalyptus*. Science 201: 1244–1246.

Mosier, A.R., W.D. Guenzi and E.E. Schweizer. 1986. Field denitrification estimation by nitrogen-15 and acetylene inhibition techniques. Soil Science Society of America Journal 50: 831–833.

Mowbray, T. and W.H. Schlesinger. 1988. The buffer capacity of organic soils of the Bluff Mountain fen, North Carolina. Soil Science 146: 73–79.

Moyers, J.L., L.E. Ranweiler, S.B. Hopf and N.E. Korte. 1977. Evaluation of particulate trace species in southwest desert atmosphere. Environmental Science and Technology 11: 789–795.

Mroz, G.D., M.F. Jurgensen, A.E. Harvey and M.J. Larsen. 1980. Effects of fire on nitrogen in forest floor horizons. Soil Science Society of America Journal 44: 395–400.

Mulholland, P.J. 1981. Deposition of riverborne organic carbon in floodplain wetlands and deltas. pp. 142–172. *In* G.E. Likens, F.T. MacKenzie, J.E. Richey, J.R. Sedell and K.K. Turekian (eds.), Flux of Organic Carbon by Rivers to the Oceans. CONF-8009140, U.S. Department of Energy, Washington, D.C.

Mulholland, P.J. and E. J. Kuenzler. 1979. Organic carbon export from upland and forested wetland watersheds. Limnology and Oceanography 24: 960–966.

Mulholland, P.J. and J.A. Watts. 1982. Transport of organic carbon to the oceans by rivers of North America: A synthesis of existing data. Tellus 34: 176–186.

Munger, J.W. and S. J. Eisenreich. 1983. Continental-scale variations in precipitation chemistry. Environmental Science and Technology 17:32–42.

Murphy, T.P. and B.G. Brownlee. 1981. Ammonia volatilization in a hypertrophic prairie lake. Canadian Journal of Fisheries and Aquatic Science 38: 1035–1039.

Murray, J.W. and V. Grundmanis. 1980. Oxygen consumption in pelagic marine sediments. Science 209: 1527–1530.

Murray, J.W. and K.M. Kuivila. 1990. Organic matter diagenesis in the northeast Pacific: Transition from aerobic red clay to suboxic hemipelagic sediments. Deep Sea Research 37: 59–80.

Muzio, L.J. and J.C. Kramlich. 1988. An artifact in the measurement of N_2O from combustion sources. Geophysical Research Letters 15: 1369–1372.

Myers, C.R. and K.H. Nealson. 1988. Bacterial manganese reduction and growth with manganese oxide as the sole electron acceptor. Science 240: 1319–1321.

Myrold, D.D., P.A. Matson and D.L. Peterson. 1989. Relationships between soil microbial properties and aboveground stand characteristics of conifer forests in Oregon. Biogeochemistry 8: 265–281.

Nadelhoffer, K.J., J.D. Aber and J.M. Melillo. 1984. Seasonal patterns of ammonium and nitrate uptake in nine temperate forest ecosystems. Plant and Soil 80: 321–335.

Nadelhoffer, K.J. and B. Fry. 1988. Controls on natural nitrogen-15 and carbon-13 abundances in forest soil organic matter. Soil Science Society of America Journal 52: 1633–1640.

Naiman, R.J. 1982. Characteristics of sediment and organic carbon export from pristine boreal forest watersheds. Canadian Journal of Fisheries and Aquatic Science 39: 1699–1718.

Naiman, R.J. and J.R. Sedell. 1981. Stream ecosystem research in a watershed perspective.

Verhandlungen der Internationalen Vereins Theoretische Angewandte Limnologie 21: 804–811.

Neftel, A., H. Oeschger, J. Schwander, B. Stauffer and R. Zumbrunn. 1982. Ice core sample measurements give atmospheric CO_2 content during the past 40,000 yr. Nature 295: 220–223.

Nelson, D.W. 1982. Gaseous losses of nitrogen other than through denitrification. pp. 327–363. In F.J. Stevenson (ed.), Nitrogen in Agricultural Soils. American Society of Agronomy, Madison, Wisconsin.

Nettleton, W.D., J.E. Witty, R.E. Nelson and J.W. Hawley. 1975. Genesis of argillic horizons in soils of desert areas of the southwestern United States. Soil Science Society of America Proceedings 39: 919–926.

Neukum, G. 1977. Lunar cratering. Philosophical Transactions of the Royal Society of London 285A: 267–272.

Newbold, J.D., R.V. O'Neill, J.W. Elwood and W. Van Winkle. 1982. Nutrient spiralling in streams: Implications for nutrient limitation and invertebrate activity. American Naturalist 120: 628–652.

Newbold, J.D., J.W. Elwood, R.V. O'Neill and A.L. Sheldon. 1983. Phosphorus dynamics in a woodland stream ecosystem: A study of nutrient spiralling. Ecology 64: 1249–1265.

Newman, E.I. and R.E. Andrews. 1973. Uptake of phosphorus and potassium in relation to root growth and root density. Plant and Soil 38: 49–69.

Nichols, J.D. 1984. Relation of organic carbon to soil properties and climate in the southern Great Plains. Soil Science Society of America Journal 48: 1382–1384.

Nicholson, S.E. 1988. Land surface atmosphere interaction: Physical processes and surface changes and their impact. Progress in Physical Geography 12: 36–65.

Nielsen, D.R., M. T. Van Genuchten and J.W. Biggar. 1986. Water flow and solute transport processes in the unsaturated zone. Water Resources Research 22S: 89–108.

Nielsen, H. 1974. Isotopic composition of the major contributors to atmospheric sulfur. Tellus 26: 213–221.

Nixon, S.W. 1980. Between coastal marshes and coastal waters— a review of twenty years of speculation and research on the role of salt marshes in estuarine productivity and water chemistry. pp. 437–525. In P. Hamilton and K.B. MacDonald (eds.), Estuarine and Wetland Processes. Plenum Press, New York.

Nixon, S.W. 1987. Chesapeake Bay nutrient budgets—A reassessment. Biogeochemistry 4: 77–90.

Nodvin, S.C., C.T. Driscoll and G.E. Likens. 1988. Soil processes and sulfate loss at the Hubbard Brook Experimental Forest. Biogeochemistry 5: 185–199.

Nozette, S. and J.S. Lewis. 1982. Venus: Chemical weathering of igneous rocks and buffering of atmospheric composition. Science 216: 181–183.

Nriagu, J.O. and R.D. Coker. 1978. Isotopic composition of sulfur in precipitation within the Great Lakes Basin. Tellus 30: 365–373.

Nriagu, J.O. 1989. A global assessment of natural sources of atmospheric trace metals. Nature 338: 47–49.

Nriagu, J.O. and D.A. Holdway. 1989. Production and release of dimethyl sulfide from the Great Lakes. Tellus 41B: 161–169.

Nriagu, J.O., D.A. Holdway and R.D. Coker. 1987. Biogenic sulfur and the acidity of rainfall in remote areas of Canada. Science 237: 1189–1192.

Nye, P.H. 1977. The rate-limiting step in plant nutrient absorption from soil. Soil Science 123: 292–297.

Nye, P.H. 1981. Changes in pH across the rhizosphere induced by roots. Plant and Soil 61: 7–26.

Oades, J.M. 1988. The retention of organic matter in soils. Biogeochemistry 5: 35–70.

O'Brien, B.J. and J.D. Stout. 1978. Movement and turnover of soil organic matter as indicated by carbon isotope measurements. Soil Biology and Biochemistry 10: 309–317.

O'Connell, A.M. 1988. Nutrient dynamics in decomposing litter in karri (*Eucalyptus diversicolor* F. Muell.) forests of south-western Australia. Journal of Ecology 76: 1186–1203.

Odum, E.P. 1969. The strategy of ecosystem development. Science 164: 262–270.

Officer, C.B., R.B. Biggs, J.L. Talf, L.E. Cronin, M.A. Tyler and W.R. Boynton. 1984. Chesapeake Bay anoxia: Origin, development, and significance. Science 223: 22–27.

Oglesby, R.T. 1977. Phytoplankton summer standing crop and annual productivity as functions of phosphorus loading and various physical factors. Journal of the Fisheries Research Board of Canada 34: 2255–2270.

O'Leary, M.H. 1988. Carbon isotopes in photosynthesis. BioScience 38: 328–336.

Olson, J.S. 1963. Energy storage and the balance of producers and decomposers in ecological systems. Ecology 44: 322–331.

Olson, J.S., J.A. Watts and L.J. Allison. 1983. Carbon in Live Vegetation of Major World Ecosystems. DOE/NBB-0037. National Technical Information Service, Washington, D.C.

Olson, J.S., R.M. Garrels, R.A. Berner, T.V. Armentano, M.I. Dyer and D.H. Yaalon. 1985. The natural carbon cycle. pp. 175–213. *In* J.R. Trabalka (ed.), Atmospheric Carbon Dioxide and the Global Carbon Cycle. U.S. Department of Energy, Washington, D.C.

Olson, R.K., W.A. Reiners, C.S. Cronan and G.E. Lang. 1981. The chemistry and flux of throughfall and stemflow in subalpine balsam fir forests. Holarctic Ecology 4: 291–300.

Olson, R.K. and W.A. Reiners. 1983. Nitrification in subalpine balsam fir soils: Tests for inhibitory factors. Soil Biology and Biochemistry 15: 413–418.

Olson, S.R., C.V. Cole, F.S. Watanabe and L.A. Dean. 1954. Estimation of available phosphorus in soils by extraction with sodium bicarbonate. United States Department of Agriculture, Circular 939, Washington, D.C.

Olsson, M. and P.-A. Melkerud. 1989. Chemical and mineralogical changes during genesis of a podzol from till in southern Sweden. Geoderma 45: 267–287.

O'Neill, R.V. and D.L. DeAngelis. 1981 Comparative productivity and biomass relations of forest ecosystems. pp. 411–449. *In* D.E. Reichle (ed.), Dynamic Properties of Forest Ecosystems. Cambridge University Press, Cambridge.

Oort, A.H. 1970. The energy cycle of the Earth. Scientific American 223(3): 54–63.

Oppenheimer, M., C.B. Epstein and R.E. Yuhnke. 1985. Acid deposition, smelter emissions, and the linearity issue in the western United States. Science 229: 859–862.

Orians, K.J. and K.W. Bruland. 1985. Dissolved aluminium in the central North Pacific. Nature 316: 427–429.

Orians, K.J. and K.W. Bruland. 1986. The biogeochemistry of aluminum in the Pacific ocean. Earth and Planetary Science Letters 78: 397–410.

Oremland, R.S., J.T. Hollibaugh, A.S. Maest, T.S. Presser, L.G. Miller and C.W. Culbertson. 1989. Selenate reduction to elemental selenium by anaerobic bacteria in sediments and culture: Biogeochemical significance of a novel sulfate-independent respiration. Applied and Environmental Microbiology 55: 2333–2343.

Osmond, C.B., K. Winter and H. Ziegler. 1982. Functional significance of different pathways of CO2 fixation in photosynthesis. pp. 479–547. *In* A. Person and M.H. Zimmerman (eds.), Encylopedia of Plant Physiology, Vol. 12B. Springer-Verlag, New York.

Ostlund, H.G. 1983. Tritium and radiocarbon, TTO western North Atlantic Section, GEOSECS Reoccupation. Rosentiel School of Marine and Atmospheric Sciences, University of Miami, Florida.

Ostman, N.L. and G.T. Weaver. 1982. Autumnal nutrient transfers by retranslocation, leaching and litter fall in a chestnut oak forest in southern Illinois. Canadian Journal of Forest Research 12: 40–51.

Ottow, J.C.G. 1971. Iron reduction and gley formation by nitrogen-fixing *Clostridia*. Oecologia 6: 164–175.

Oudot, C., C. Andrie and Y. Montel. 1990. Nitrous oxide production in the tropical Atlantic ocean. Deep Sea Research 37: 183–202.

Owen, D.F. and R.G. Wiegert. 1976. Do consumers maximize plant fitness? Oikos 27: 488–492.

Owen, T. and K. Biemann. 1976. Composition of the atmosphere at the surface of Mars: Detection of argon-36 and preliminary analysis. Science 193: 801–803.

Owen, T., J.P. Maillard, C. de Bergh and B.L. Lutz. 1988. Deuterium on Mars: The abundance of HDO and the value of D/H. Science 240: 1767–1770.

Owen, R.M. and D.K. Rea. 1985. Sea-floor hydrothermal activity links climate to tectonics: The Eocene carbon dioxide greenhouse. Science 227: 166–169.

Oyama, V.I., G.C. Carle, F. Woeller, and J.B. Pollack. 1979. Venus lower atmospheric composition: Analysis by gas chromotography. Science 203: 802–805.

Paerl, H.W. 1985. Enhancement of marine primary production by nitrogen-enriched acid rain. Nature 315: 747–749.

Paerl, H.W., K.M. Crocker and L.E. Prufert. 1987. Limitation of N_2 fixation in coastal marine waters: Relative importance of molybdenum, iron, phosphorus, and organic matter availability. Limnology and Oceanography 32: 525–536.

Paerl, H.W. and B.M. Bebout. 1988. Direct measurement of O_2-depleted microzones in marine *Oscillatoria:* Relation to N_2 fixation. Science 241: 442–445.

Paerl, H.W. and R.G. Carlton. 1988. Control of nitrogen fixation by oxygen depletion in surface-associated microzones. Nature 332: 260–262

Palm, C.A., R. A. Houghton, J.M. Melillo, and D.L. Skole. 1986. Atmospheric carbon dioxide from deforestation in southeast Asia. Biotropica 18: 177–188.

Palmer, F.E., J.T. Staley, R.G.E. Murray, T. Counsell and J.B. Adams. 1986. Identification of manganese-oxidizing bacteria from desert varnish. Geomicrobiology Journal 4: 343–360.

Parfitt, R.L. and R. St. C. Smart. 1978. The mechanism of sulfate adsorption on iron oxides. Soil Science Society of America Journal 42: 48–50.

Parfitt, R.L. and C.W. Childs. 1988. Estimation of forms of Fe and Al: A review, and analysis of contrasting soils by dissolution and Moessbauer methods. Australian Journal of Soil Research 26: 121–144.

Parker, G. G. 1983. Throughfall and stemflow in the forest nutrient cycle. Advances in Ecological Research 13:57–133.

Parker, R.S. and B.M. Troutman. 1989. Frequency distribution for suspended sediment loads. Water Resources Research 25: 1567–1574.

Parkin, T.B., A.J. Sexstone and J.M. Tiedje. 1985. Comparison of field denitrification rates determined by acetylene-based soil core and nitrogen-15 methods. Soil Science Society of America Journal 49: 94–99.

Parkin, T.B. 1987. Soil microsites as a source of denitrification variability. Soil Science Society of America Journal 51: 1194–1199.

Parkin, T.B., J.L. Starr and J.J. Meisinger. 1987. Influence of sample size on measurement of soil denitrification. Soil Science Society of America Journal 51: 1492–1501.

Parrington, J.R., W.H. Zoller and N.K. Aras. 1983. Asian dust: Seasonal transport to the Hawaiian islands. Science 220: 195–197.

Parton, W.J., D.S. Schimel, C.V. Cole and D.S. Ojima. 1987. Analysis of factors controlling soil organic matter levels in Great Plains grasslands. Soil Science Society of America Journal 51: 1173–1179.

Parton, W.J., A.E. Mosier and D.S. Schimel. 1988a. Rates and pathways of nitrous oxide production in a shortgrass steppe. Biogeochemistry 6: 45–58.

Parton, W.J., J.W.B. Stewart and C.V. Cole. 1988b. Dynamics of C, N, P and S in grassland soils: A model. Biogeochemistry 5: 109–131.

Pastor, J., J.D. Aber, C.A. McClaugherty, and J.M. Melillo. 1984. Aboveground production and N and P cycling along a nitrogen mineralization gradient on Blackhawk Island, Wisconsin. Ecology 65: 256–268.

Pastor, J. and W.M. Post. 1986. Influence of climate, soil moisture, and succession on forest carbon and nitrogen cycles. Biogeochemistry 2: 3–27.

Pastor, J.M. and W. M. Post. 1988. Response of northern forests to CO_2-induced climate change. Nature 334: 55–58.

Patrick, W.H. and M.E. Tusneem. 1972. Nitrogen loss from flooded soil. Ecology 53: 735–737.

Patrick, W.H. and R.A. Khalid. 1974. Phosphate release and sorption by soils and sediments: Effect of aerobic and anaerobic conditions. Science 186: 53–55.

Pearson, J.A., D.H. Knight and T.J. Fahey. 1987. Biomass and nutrient accumulation during stand development in Wyoming lodgepole pine forests. Ecology 68: 1966–1973.

Pedro, G., M. Jamagne and J.C. Begon. 1978. Two routes in genesis of strongly differentiated acid soils under humid, cool-temperate conditions. Geoderma 20: 173–189.

Peirson, D.H., P.A. Cawse, L. Salmon and R.S. Cambray. 1973. Trace elements in the atmospheric environment. Nature 241:252–256.

Peltier, L. 1950. The geographic cycle in periglacial regions as it is related to climatic geomorphology. Annals of the Association of American Geographers 40: 214–236.

Peltier, W.R. and A.M. Tushingham. 1989. Global sea level rise and the greenhouse effect: Might they be connected? Science 244: 806–810.

Pennington, W. 1981. Records of a lake's life in time: The sediments. Hydrobiologia 79: 197–219.

Penzias, A.A. 1979. The origin of the elements. Science 205: 549–554.

Perdue, E.M., K.C. Beck, and J.H. Reuter. 1976. Organic complexes of iron and aluminium in natural waters. Nature 260: 418–420.

Peterjohn, W.T. and D.L. Correll. 1984. Nutrient dynamics in an agricultural watershed: Observations on the role of a riparian forest. Ecology 65: 1466–1475.

Peterson, B.J. 1980. Aquatic primary productivity and the ^{14}C-CO_2 method: A history of the productivity problem. Annual Review of Ecology and Systematics 11: 359–385.

Peterson, B.J. 1981. Perspectives on the importance of the oceanic particulate flux in the global carbon cycle. Ocean Science and Engineering 6: 71–108.

Peterson, B.J. and J.M. Melillo. 1985. The potential storage of carbon caused by eutrophication of the biosphere. Tellus 37B:117–127.

Peterson, B.J., R.W. Howarth and R.H. Garritt. 1986. Sulfur and carbon isotopes as tracers of salt-marsh organic matter flow. Ecology 67: 865–874.

Peterson, B.J. and R.W. Howarth. 1987. Sulfur, carbon, and nitrogen isotopes used to trace organic matter flow in the salt-marsh estuaries of Sapelo Island, Georgia. Limnology and Oceanography 32: 1195–1213.

Peterson, D.L., M.A. Spanner, S.W. Running and K.T. Teuber. 1987. Relationship of thematic mapper simulator data to leaf area index of temperate coniferous forests. Remote Sensing of Environment 22: 323–341.

Petit, J.R., L. Mounier, J. Jouzel, Y.S. Korotkevich, V.I. Kotlyakov and C. Lorius. 1990. Palaeoclimatological and chronological implications of the Vostok core dust record. Nature 343: 56–58.

Philander, G. 1989. El Niño and La Niña. American Scientist 77: 451–459.

Phillips, F.M., J.L. Mattick, T.A. Duval, D. Elmore and P.W. Kubik. 1988. Chlorine 36 and tritium from nuclear weapons fallout as tracers for long-term liquid and vapor movement in desert soils. Water Resources Research 24: 1877–1891.

Pierce, R.S., J.W. Hornbeck, G.E. Likens and F.H. Bormann. 1970. Effects of elimination of vegetation on stream water quantity and quality. pp. 311–328. Proceedings of the International Association of Scientific Hydrology, Wellington, New Zealand.

Pigott, E.D. and K. Taylor. 1964. The distribution of some woodland herbs in relation to the supply of nitrogen and phosphorus in the soil. Journal of Ecology 52(Suppl.): 175–185.

Pinto, J.P., G.R. Gladstone and Y.L. Yung. 1980. Photochemical production of formaldehyde in Earth's primitive atmosphere. Science 210: 183–185.

Pirozynski, K.A. and D.W. Malloch. 1975. The origin of land plants: A matter of mycotrophism. Biosystems 6: 153–164.

Platt, T. and D.V. Subba Rao. 1975. Primary production of marine microphytes. pp. 249–280. In J.P. Cooper (ed.), Photosynthesis and Productivity in Different Environments. Cambridge University Press, Cambridge.

Platt, T. and M.R. Lewis. 1987. Estimation of phytoplankton production by remote sensing. Advances in Space Research 7: 131–135.

Platt, T. and S. Sathyendranath. 1988. Organic primary production: Estimation by remote sensing at local and regional scales. Science 241: 1613–1620.

Platt, U., M. Rateike, W. Junkermann, J. Rudolph and D.H. Ehhalt. 1988. New tropospheric OH measurements. Journal of Geophysical Research 93: 5159–5166.

Pletscher, D.H., F.H. Bormann and R.S. Miller. 1989. Importance of deer compared to other vertebrates in nutrient cycling and energy flow in a northern hardwood ecosystem. American Midland Naturalist 121: 302–311.

Pomeroy, L.R., W.M. Darley, E.L. Dunn, J.L. Gallagher, E.B. Haines, and D.M. Whitney. 1981. Primary production. pp. 39–67. In L.R. Pomeroy and R.G. Wiegert (eds.), The Ecology of a Salt Marsh. Springer-Verlag, New York.

Pomeroy, L.R. and D. Deibel. 1986. Temperature regulation of bacterial activity during the spring bloom in Newfoundland coastal waters. Science 233: 359–361.

Ponnamperuma, F.N. 1972. The chemistry of submerged soils. Advances in Agronomy 24: 29–96.

Ponnamperuma, F.N., E.M. Tianco and T.A. Loy. 1967. Redox equilibria in flooded soils: 1. The iron hydroxide systems. Soil Science 103: 374–382.

Porter, K.G. 1976. Enhancement of algal growth and productivity by grazing zooplankton. Science 192: 1332–1334.

Post, W.M., W. R. Emanuel, P.J. Zinke and A.G. Stangenberger. 1982. Soil carbon pools and world life zones. Nature 298: 156–159.

Post, W.M., J. Pastor, P.J. Zinke and A.G. Stangenberger. 1985. Global patterns of soil nitrogen storage. Nature 317: 613–616.

Postgate, J.R., H.M. Kent and R.L. Robson. 1988. Nitrogen fixation by Desulfovibrio. pp. 457–471. In J.A. Cole and S.J. Ferguson (eds.), The Nitrogen and Sulphur Cycles. Cambridge University Press, Cambridge.

Potter, G.L., H.W. Ellsaesser, M.C. MacCracken and F. M. Luther. 1975. Possible climatic impact of tropical deforestation. Nature 258: 697–698.

Potts, M.J. 1978. Deposition of air-bourne salt on Pinus radiata and the underlying soil. Journal of Applied Ecology 15: 543–550.

Powell, T.M., J.E. Cloern and L.M. Huzzey. 1989. Spatial and temporal variability in South San Francisco Bay (USA). I. Horizontal distributions of salinity, suspended sediments, and phytoplankton biomass and productivity. Estuarine, Coastal and Shelf Science 28: 583–597.

Prather, M.J. 1985. Continental sources of halocarbons and nitrous oxide. Nature 317: 221–225.

Premuzic, E.T., C.M. Benkovitz, J.S. Gaffney and J.J. Walsh. 1982. The nature and distribution of organic matter in the surface sediments of the worlds oceans and seas. Organic Geochemistry 4: 63–77.

Press, F. and R. Siever. 1986. Earth, 4th ed. W. H. Freeman, New York.

Prezelin, B.B. and B.A. Boczar. 1986. Molecular bases of cell absorption and fluorescence in phytoplankton: Potential applications to studies in optical oceanography. pp. 349–464. In F.E. Round and D.J. Chapman (eds.), Progress in Phycological Research. Elsevier Science Publishers, New York.

Price, N.M. and F.M.M. Morel. 1990. Cadmium and cobalt substitution for zinc in a marine diatom. Nature 334: 658–660.

Prinn, R., D. Cunnold, R. Rasmussen, P. Simmonds, F. Alyea, A. Crawford, P. Fraser and R.

Rosen. 1987. Atmospheric trends in methylchloroform and the global average for the hydroxyl radical. Science 238: 945–950.

Probst, J.L. and Y. Tardy. 1987. Long range streamflow and world continental runoff fluctuations since the beginning of this century. Journal of Hydrology 94: 289–311.

Prospero, J.M. and D.L. Savoie. 1989. Effect of continental sources on nitrate concentrations over the Pacific Ocean. Nature 339: 687–689.

Protz, R., G.J. Ross, I.P. Martini and J. Terasmae. 1984. Rate of podzolic soil formation near Hudson Bay, Ontario. Canadian Journal of Soil Science 64: 31–49.

Protz, R., G.J. Ross, M.J. Shipitalo and J. Terasmae. 1988. Podzolic soil development in the southern James Bay lowlands, Ontario. Canadian Journal of Soil Science 68: 287–305.

Pye, K. 1987. Aeolian Dust and Dust Deposits. Academic Press, London.

Quade, J., T.E. Cerling and J.R. Bowman. 1989. Systematic variations in the carbon and oxygen isotopic composition of pedogenic carbonate along elevation transects in the southern Great Basin, United States. Geological Society of America Bulletin 101: 464–475.

Qualls, R.G. 1984. The role of leaf litter nitrogen immobilization in the nitrogen budget of a swamp stream. Journal of Environmental Quality 13: 640–644.

Quay, P.D., S.L. King, J.M. Landsdown and D.O. Wilbur. 1988. Isotopic composition of methane released from wetlands: Implications for the increase in atmospheric methane. Global Biogeochemical Cycles 2: 385–397.

Quinn, P.K., R.J. Charlson and W.H. Zoller. 1987. Ammonia, the dominant base in the remote marine troposphere: A review. Tellus 39B: 413–425.

Rabinowitz, J., J. Flores, R. Krebsbach and G. Rogers. 1969. Peptide formation in the presence of linear or cyclic polyphosphates. Nature 224: 795–796.

Rahn, K.A. and D.H. Lowenthal. 1984. Elemental tracers of distant regional pollution aerosols. Science 223: 132–139.

Raich, J.W. and K.J. Nadelhoffer. 1989. Belowground carbon allocation in forest ecosystems: Global trends. Ecology 70: 1346–1354.

Raison, R.J. 1979. Modification of the soil environment by vegetation fires, with particular reference to nitrogen transformations: A review. Plant and Soil 51: 73–108.

Raison, R.J., P.K. Khanna and P.V. Woods. 1985. Mechanisms of element transfer to the atmosphere during vegetation fires. Canadian Journal of Forest Research 15: 132–140.

Raison, R.J., M.J. Connell and P.K. Khanna. 1987. Methodology for studying fluxes of soil mineral-N in situ. Soil Biology and Biochemistry 19: 521–530.

Raiswell, R. and R.A. Berner. 1986. Pyrite and organic matter in Phanerozoic normal marine shales. Geochimica et Cosmochimica Acta 50: 1967–1976.

Ralph, B.J. 1979. Oxidative reactions in the sulfur cycle. pp. 369–400. In P.A. Trudinger and D.J. Swaine (eds.), Biogeochemical Cycling of Mineral-Forming Elements. Elsevier Scientific, Amsterdam.

Ramanathan, V. 1988. The greenhouse theory of climatic change: A test by an inadvertent global experiment. Science 240: 293–299.

Ramanathan, V., R.J. Cicerone, H.B. Singh and J.T. Kiehl. 1985. Trace gas trends and their potential role in climate change. Journal of Geophysical Research 90: 5547–5566.

Ramanathan, V., R.D. Cess, E.F. Harrison, P. Minnis, B.R. Barkstrom, E. Ahmad and D. Hartmann. 1989. Cloud-radiative forcing and climate: Results from the Earth radiation budget experiment. Science 243: 57–63.

Rank, D.M., P.A. Pinto, S.E. Woosley, J.B. Bregman, F.C. Witteborn, T.S. Axelrod and M. Cohen. 1988. Nickel, argon and cobalt in the infrared spectrum of SN1987A: The core becomes visible. Nature 331: 505–506.

Raper, C.D., D.L. Osmond, M. Wann and W.W. Weeks. 1978. Interdependence of root and shoot activities in determining nitrogen uptake rate of roots. Botanical Gazette 139: 289–294.

Rasmussen, R.A. 1974. Emission of biogenic hydrogen sulfide. Tellus 26: 254–260.

Rasmussen, R.A. and M.A.K. Khalil. 1986. Atmospheric trace gases: Trends and distributions over the last decade. Science 232: 1623–1624.

Raval, A. and V. Ramanathan. 1989. Observational determination of the greenhouse effect. Nature 342: 758–761.

Raven, J.A., A.A. Franco, E. Lino de Jesus and J. Jacob-Neto. 1990. H^+ extrusion and organic-acid synthesis in N_2-fixing symbioses involving vascular plants. New Phytologist 114: 369–389.

Raynaud, D., J. Chapellaz, J.M. Barnola, Y.S. Korotkevich and C. Lorius. 1988. Climatic and CH_4 cycle implications of glacial-interglacial CH_4 change in the Vostok ice core. Nature 333: 655–657.

Reader, R.J. and J.M. Stewart. 1972. The relationship between net primary production and accumulation for a peatland in southeastern Manitoba. Ecology 53: 1024–1037.

Reddy, K.J., W.L. Lindsay, S.M. Workman and J.I. Drever. 1990. Measurement of calcite ion activity products in soils. Soil Science Society of America Journal 54: 67–71.

Reddy, K.R. and W.H. Patrick. 1975. Effect of alternate aerobic and anaerobic conditions on redox potential, organic matter decomposition and nitrogen loss in a flooded soil. Soil Biology and Biochemistry 7: 87–94.

Reddy, K.R. and W.H. Patrick. 1976. Effect of frequent changes in aerobic and anaerobic conditions on redox potential and nitrogen loss in a flooded soil. Soil Biology and Biochemistry 8: 491–495.

Redfield, A.C. 1958. The biological control of chemical factors in the environment. American Scientist 46: 205–221.

Redfield, A.C., B.H. Ketchum and F.A. Richards. 1963. The influence of organisms on the composition of sea-water. pp. 26–77. In M.N. Hill (ed.), The Sea, Vol. 2. Wiley, New York.

Rees, C.E., W.J. Jenkins and J. Monster. 1978. The sulphur isotopic composition of ocean water sulphate. Geochimica et Cosmochimica Acta 42: 377–381.

Reich, P.B. and A.W. Schoettle. 1988. Role of phosphorus and nitrogen in photosynthetic and whole plant carbon gain and nutrient use efficiency in eastern white pine. Oecologia 77: 25–33.

Reich, P.B., M.B. Walters and T.J. Tabone. 1989. Response of Ulmus americana seedlings to varying nitrogen and water status. 2. Water and nitrogen use efficiency in photosynthesis. Tree Physiology 5: 173–184.

Reichle, D.E., B.E. Dinger, N.T. Edwards, W.F. Harris and P. Sollins. 1973a. Carbon flow and storage in a forest ecosystem. pp. 345–365. In G.M. Woodwell and E.V. Pecan (eds.), Carbon and the Biosphere. CONF 720510, National Technical Information Service, Washington, D.C.

Reichle, D.E., R.A. Goldstein, R.I. Van Hook and G.J. Dodson. 1973b. Analysis of insect consumption in a forest canopy. Ecology 54: 1076–1084.

Reiners, W.A. and R.O. Anderson. 1968. CO_2 concentrations in forests along a topographic gradient. American Midland Naturalist 80: 111–117.

Reiners, W.A. 1972. Structure and energetics of three Minnesota forests. Ecological Monographs 42: 71–94.

Reiners, W.A. 1986. Complementary models for ecosystems. American Naturalist 127: 59–73.

Reiners, W.A., L.L. Strong, P.A. Matson, I.C. Burke and D.S. Ojima. 1989. Estimating biogeochemical fluxes across sagebrush-steepe landscapes with thematic mapper imagery. Remote Sensing of Environment 28: 121–129.

Revelle, R.R. 1983. Methane hydrates in continental slope sediments and increasing atmospheric carbon dioxide. pp. 252–261. In Changing Climate. National Academy Press, Washington, D.C.

Reuss, J.O. 1980. Simulation of soil nutrient losses resulting from rainfall acidity. Ecological Modelling 11: 15–38.

Reuss, J.O. and D.W. Johnson. 1986. Acid Deposition and the Acidification of Soils and Waters. Springer-Verlag, New York.

Reuss, J.O., B.J. Cosby and R.F. Wright. 1987. Chemical processes governing soil and water acidification. Nature 329: 27–32.

Rice, E.L. and S.K. Pancholy 1972. Inhibition of nitrification by climax ecosystems. American Journal of Botany 59: 1033–1040.

Rich, P.H. and R.G. Wetzel. 1978. Detritus in the lake ecosystem. American Naturalist 112: 57–71.

Richardson, C.J. 1985. Mechanisms controlling phosphorus retention capacity in freshwater wetlands. Science 228: 1424–1427.

Richey, J.E. 1983. The phosphorus cycle. pp. 51–56. In B. Bolin and R.B. Cook (eds.), The Major Biogeochemical Cycles and Their Interactions. Wiley, New York.

Richey, J.E., A.H. Devol, S.C. Wofsy, R. Victoria and M.N.G. Riberio. 1988. Biogenic gases and the oxidation and reduction of carbon in Amazon River and floodplain waters. Limnology and Oceanography 33: 551–561.

Richey, J.E., J.I. Hedges, A.H. Devol, P.D. Quay, R. Victoria, L. Martinelli, and B.R. Forsberg. 1990. Biogeochemistry of carbon in the Amazon River. Limnology and Oceanography 35: 352–371.

Richter, D.D., C.W. Ralston and W. R. Harms. 1982. Prescribed fire: Effects on water quality and forest nutrient cycling. Science 215: 661–663.

Richter, D.D. and L.I. Babbar. 199-. Soil diversity in the tropics. Advances in Ecological Research, in press.

Ridley, W.P., L.J. Dizikes, and J.M. Wood. 1977. Biomethylation of toxic elements in the environment. Science 197: 329–332.

Riggan, P.J., R.N. Lockwood and E.N. Lopez. 1985. Deposition and processing of airborne nitrogen pollutants in Mediterranean-type ecosystems of southern California. Environmental Science and Technology 19: 781–789.

Risenhoover, K.L. and R.O. Peterson. 1986. Mineral licks as a sodium source for Isle Royale moose. Oecologia 71: 121–126.

Risley, L.S. and D.A. Crossley. 1988. Herbivore-caused greenfall in the southern Appalachians. Ecology 69: 1118–1127.

Robbins, C.S., J.R. Sauer, R.S. Greenberg and S. Droege. 1989. Population declines in North American birds that migrate to the neotropics. Proceedings of the National Academy of Sciences, USA 86: 7658–7662.

Robbins, C.W., R.J. Wagenet and J.J. Jurinak. 1980. A combined salt transport-chemical equilibrium model for calcareous and gypsiferous soils. Soil Science Society of America Journal 44: 1191–1194.

Roberts, T.L., J.W.B. Stewart and J.R. Bettany. 1985. The influence of topography on the distribution of organic and inorganic soil phosphorus across a narrow environmental gradient. Canadian Journal of Soil Science 65: 651–665.

Robertson, G.P. 1982a. Nitrification in forested ecosystems. Philosophical Transactions of the Royal Society of London 296B: 445–457.

Robertson, G.P. 1982b. Factors regulating nitrification in primary and secondary succession. Ecology 63: 1561–1573.

Robertson, G.P. 1984. Nitrification and nitrogen mineralization in a lowland rainforest succession in Costa Rica, central America. Oecologia 61: 99–104.

Robertson, G.P. and P.M. Vitousek. 1981. Nitrification potentials in primary and secondary succession. Ecology 62: 376–386.

Robertson, G.P. and J.M. Tiedje. 1984. Denitrification and nitrous oxide production in successional and old-growth Michigan forests. Soil Science Society of America Journal 48: 383–389.

Robertson, G.P. and J.M. Tiedje. 1987. Nitrous oxide sources in aerobic soils: Nitrification, denitrification and other biological processes. Soil Biology and Biochemistry 19: 187–193.

Robertston, G.P. and J.M. Tiedje. 1988. Deforestation alters denitrification in a lowland tropical rain forest. Nature 336: 756–759.

Robertson, G.P. and T. Rosswall. 1986. Nitrogen in west Africa: the regional cycle. Ecological Monographs 56: 43–72.

Robertson, G.P., M.A. Huston, F.C. Evans, and J.M. Tiedje. 1988. Spatial variability in a successional plant community: Patterns of nitrogen availability. Ecology 69: 1517–1524.

Robinson, D. 1986. Limits to nutrient inflow rates in roots and root systems. Physiologica Plantarum 68: 551–559.

Rodhe, H., P. Crutzen and A. Vanderpol. 1981. Formation of sulfuric and nitric acid in the atmosphere during long-range transport. Tellus 33: 132–141.

Romell, L.G. 1935. Ecological problems of the humus layers in the forest. Cornell University, Agricultural Experiment Station Memoir 170, Ithaca, New York.

Ronen, D., M. Magaritz and E. Almon. 1988. Contaminated aquifers are a forgotten component of the global N_2O budget. Nature 335: 57–59.

Rose, S.L. and C.T. Youngberg. 1981. Tripartite associations in snowbrush *(Ceanothus velutinus)*: Effect of vesicular-arbuscular mycorrhizae on growth, nodulation, and nitrogen fixation. Canadian Journal of Botany 59: 34–39.

Rosenzweig, M.L. 1968. Net primary productivity of terrestrial communities: Prediction from climatological data. American Naturalist 102: 67–74.

Roskoski, J.P. 1980. Nitrogen fixation in hardwood forests of the northeastern United States. Plant and Soil 54: 33–44.

Ross, J.E. and L.H. Aller. 1976. The chemical composition of the Sun. Science 191: 1223–1229.

Rosswall, T. 1981. The biogeochemical nitrogen cycle. pp. 25–49. *In* G.E. Likens (ed.), Some Perspectives of the Major Biogeochemical Cycles. Wiley, New York.

Rosswall, T. 1982. Microbiological regulation of the biogeochemical nitrogen cycle. Plant and Soil 67: 15–34.

Rotty, R.M. and C.D. Masters. 1985. Carbon dioxide from fossil fuel combustion: Trends, Resources, and Technological Implications. pp. 63–80. *In* J.R. Trabalka (ed.), Atmospheric Carbon Dioxide and the Global Carbon Cycle. DOE/ER-0239. U.S. Department of Energy, Washington, D.C.

Routson, R.C., R.E. Wildung and T.R. Garland. 1977. Mineral weathering in an arid watershed containing soil developed from mixed basaltic-felsic parent materials. Soil Science 124: 303–308.

Rudd, J.W.M. and C.D Taylor. 1980. Methane cycling in aquatic environments. Advances in Aquatic Microbiology 2: 77–150.

Rudd, J.W.M., C.A.Kelly and A. Furutani. 1986a. The role of sulfate reduction in long term accumulation of organic and inorganic sulfur in lake sediments. Limnology and Oceanography 3l: 1281–1291.

Rudd, J.W.M., C.A. Kelly, V. St. Louis, R.H. Hesslein, A. Furutani and M.H. Holoka. 1986b. Microbial consumption of nitric and sulfuric acids in acidified north temperate lakes. Limnology and Oceanography 31: 1267–1280.

Ruhe, R.V. 1984. Soil-climate system across the prairies in midwestern U.S.A. Geoderma 34: 201–219.

Running, S.W., R.R. Nemani, D.L. Peterson, L.E. Band, D. F. Potts, L.L. Pierce, and M.A. Spanner. 1989. Mapping regional forest evapotranspiration and photosynthesis by coupling satellite data with ecosystem simulation. Ecology 70: 1090–1101.

Russel, J.C. 1929. Organic matter problems under dry farming conditions. Journal of the American Society of Agronomy 21: 960–969.

Ryan, D.F. and F.H. Bormann. 1982. Nutrient resorption in northern hardwood forests. BioScience 32: 29–32.

Ryan, P.F., G.M. Hornberger, B.J. Cosby, J.N. Galloway, J.R. Webb and E.B. Rastetter. 1989. Changes in the chemical composition of stream water in two catchments in the Shenandoah National Park, Virginia, in response to atmospheric deposition of sulfur. Water Resources Research 25: 2091–2099.

Rychert, R., J. Skujins, D. Sorensen and D. Porcella. 1978. Nitrogen fixation by lichens and free-living microorganisms in deserts. pp. 20–30. In N.E.West and J. Skujins (eds.), Nitrogen in Desert Ecosystems. Dowden, Hutchinson and Ross, Stroudsburg, Pennsylvania.

Rycroft, D.W., D.J.A. Williams and H.A.P. Ingram. 1975. The transmission of water through peat. I. Review. Journal of Ecology 63: 535–556.

Ryther, J.H. and W.M. Dunstan. 1971. Nitrogen, phosphorus, and eutrophication in the coastal marine environment. Science 171: 1008–1013.

Sachse, G.W., R.C. Harriss, J. Fishman, G.F. Hill and D.R. Cahoon. 1988. Carbon monoxide over the Amazon Basin during the 1985 dry season. Journal of Geophysical Research 93: 1422–1430.

Sagan, C., O.B. Toon and J.B. Pollack. 1979. Anthropogenic albedo changes and the Earth's climate. Science 206: 1363–1368.

Saggar, S., J.R. Bettany and J.W.B. Stewart. 1981. Sulfur transformations in relation to carbon and nitrogen in incubated soils. Soil Biology and Biochemistry 13: 499–511.

Sala, O.E., W.J. Parton, L.A. Joyce and W.K. Lauenroth. 1988. Primary production of the central grassland region of the United States. Ecology 69: 40–45.

Salati, E. and P.B. Vose. 1984. Amazon Basin: A system in equilibrium. Science 225: 129–138.

Sanchez, P.A., D.E. Bandy, J.H. Villachica and J.J. Nicholaides. 1982a. Amazon Basin Soils: Management for continuous crop production. Science 216: 821–827.

Sanchez, P.A., M.P. Gichuru and L.B. Katz. 1982b. Organic matter in major soils of the tropical and temperate regions. International Congress of Soil Science 12: 99–114.

Sansone, F.J. and C.S. Martens. 1981. Methane production from acetate and associated methane fluxes from anoxic coastal sediments. Science 211: 707–709.

Santos, P.F., N.Z. Elkins, Y. Steinberger and W.G. Whitford. 1984. A comparison of surface and buried Larrea tridentata leaf litter decomposition in North American hot deserts. Ecology 65: 278–284.

Saunders, J.F. and W.M. Lewis. 1988. Transport of phosphorus, nitrogen, and carbon by the Apure River, Venezuela. Biogeochemistry 5: 323–342.

Savoie, D.L. and J.M. Prospero. 1989. Comparison of oceanic and continental sources of non-sea-salt sulphate over the Pacific ocean. Nature 339: 685–687.

Saxton, K.E., W.J. Rawls, J.S. Romberger and R.I. Pependick. 1986. Estimating generalized soil-water characteristics from texture. Soil Science Society of America Journal 50: 1031–1036.

Sayles, F.L. and P.C. Mangelsdorf. 1977. The equilibration of clay minerals with seawater: Exchange reactions. Geochimica et Cosmochimica Acta 41: 951–960.

Schaefer, D., Y. Steinberger and W.G. Whitford. 1985. The failure of nitrogen and lignin control of decomposition in a North American desert. Oecologia 65: 382–386.

Schaefer, D.A. and W.G. Whitford. 1981. Nutrient cycling by the subterranean termite Gnathamitermes tubiformans in a Chihuahuan desert ecosystem. Oecologia 48: 277–283.

Schaefer, M. 1990. The soil fauna of a beech forest on limestone: Trophic structure and energy budget. Oecologia 82: 128–136.

Schelske, C.L. 1985. Biogeochemical silica mass balances in Lake Michigan and Lake Superior. Biogeochemistry 1: 197–218.

Schelske, C.L. 1988. Historic trends in Lake Michigan silica concentrations. Internationale Revue der Gesamten Hydrobiologie 73: 559–591.

Schelske, C.L., E.F. Stoermer, D.J. Conley, J.A. Robbins and R.M. Glover. 1983. Early eutrophication in the lower Great Lakes: New evidence from biogenic silica in sediments. Science 222: 320–322.

Schelske, C.L., J.A. Robbins, W.S. Gardner, D.J. Conley and R.A. Bourbonniere. 1988. Sediment record of biogeochemical responses to anthropogenic perturbations of nutrient cycles in Lake Ontario. Canadian Journal of Fisheries and Aquatic Science 45: 1291–1303.

Schidlowski, M. 1980. The atmosphere. pp. 1–16. *In* O. Hutzinger (ed.), The Handbook of Environmental Chemistry, Vol. 1, Part A, The Natural Environment and the Biogeochemical Cycles. Springer-Verlag, New York.

Schidlowski, M. 1983. Evolution of photoautotrophy and early atmospheric oxygen levels. Precambrian Research 20: 319–335.

Schidlowski, M., J.M. Hayes and I.R. Kaplan. 1983. Isotopic inferences of ancient biochemistries: Carbon, Sulfur, Hydrogen, and Nitrogen. pp. 149–186. *In* J.W. Schopf (ed.), Earth's Earliest Biosphere. Princeton University Press, Princeton, New Jersey.

Schiffman, P.M. and W. C. Johnson. 1989. Phytomass and detrital carbon storage during forest regrowth in the southeastern United States piedmont. Canadian Journal of Forest Research 19: 69–78.

Schimel, D.S., W.J. Parton, F.J. Adamsen, R.G. Woodmansee, R.L. Senft and M.A. Stillwell. 1986. The role of cattle in the volatile loss of nitrogen from a shortgrass steppe. Biogeochemistry 2: 39–52.

Schimel, D., M.A. Stillwell and R.G. Woodmansee. 1985. Biogeochemistry of C, N, and P in a soil catena of the shortgrass steppe. Ecology 66: 276–282.

Schimel, J.P., M.K. Firestone and K.S. Killham. 1984. Identification of heterotrophic nitrification in a Sierran forest soil. Applied and Environmental Microbiology 48: 802–806.

Schimel, J.P. and M.K. Firestone. 1989. Nitrogen incorporation and flow through a coniferous forest soil profile. Soil Science Society of America Journal 53: 779–784.

Schindler, D.W. 1974. Eutrophication and recovery in experimental lakes: Implications for lake management. Science 184: 897–899.

Schindler, D.W. 1977. Evolution of phosphorus limitation in lakes. Science 195: 260–262.

Schindler, D.W. 1978. Factors regulating phytoplankton production and standing crop in the world's freshwaters. Limnology and Oceanography 23: 478–486.

Schindler, D.W., G.J. Brunskill, S. Emerson, W.S. Broecker and T.-H. Peng. 1972. Atmospheric carbon dioxide: Its role in maintaining phytoplankton standing crops. Science 177: 1192–1194.

Schindler, D.W., M.A. Turner, M.P. Stainton and G.A. Linsey. 1986. Natural sources of acid neutralizing capacity in low alkalinity lakes of the Precambrian shield. Science 232: 844–847.

Schindler, S.C., M.J. Mitchell, T.J. Scott, R.D. Fuller and C.T. Driscoll. 1986. Incorporation of ^{35}S-sulfate into inorganic and organic constituents of two forest soils. Soil Science Society of America Journal 50: 457–462.

Schleser, G.H. 1982. The response of CO_2 evolution from soils to global temperature changes. Zeitschrift fuer Naturforschung 37a: 287–291.

Schlesinger, M.E. and J.F.B. Mitchell. 1985. Model projections of the equilibrium climatic response to increased carbon dioxide. pp. 81–147. *In* M.C. MacCracken and F.M. Luther (eds.), Projecting the Climatic Effects of Increasing Carbon Dioxide. U.S. Department of Energy, Er-0237, Washington, D.C.

Schlesinger, W.H. 1977. Carbon balance in terrestrial detritus. Annual Review of Ecology and Systematics 8: 51–81.

Schlesinger, W.H. and P.L. Marks. 1977. Mineral cycling and the niche of Spanish moss, *Tillandsia usneoides* L. American Journal of Botany 64: 1254–1262.

Schlesinger, W.H. 1982. Carbon storage in the caliche of arid soils: A case study from Arizona. Soil Science 133: 247–255.

Schlesinger, W.H., J.T. Gray, D.S. Gill and B.E. Mahall. 1982. *Ceanothus megacarpus* chaparral: A synthesis of ecosystem processes during development and annual growth. Botanical Review 48: 71–117.

Schlesinger, W.H. 1984. Soil organic matter: A source of atmospheric CO_2. pp. 111–127. *In* G.M. Woodwell (ed.), The Role of Terrestrial Vegetation in the Global Carbon Cycle. Wiley, New York.

Schlesinger, W.H. 1985a. Decomposition of chaparral shrub foliage. Ecology 66: 1353–1359.

Schlesinger, W.H. 1985b. The formation of caliche in soils of the Mojave Desert, California. Geochimica et Cosmochimica Acta 49: 57–66.

Schlesinger, W.H. 1986. Changes in soil carbon storage and associated properties with disturbance and recovery. pp. 194–220. *In* J.R. Trabalka and D.E. Reichle (eds.), The Changing Carbon Cycle: A Global Analysis. Springer-Verlag, New York.

Schlesinger, W.H. 1989. Discussion: Ecosystem structure and function. pp. 268–274. *In* J. Roughgarden, R.M. May and S.A. Levin (eds.), Perspectives in Ecological Theory. Princeton University Press, Princeton, New Jersey.

Schlesinger, W.H. and J.M. Melack. 1981. Transport of organic carbon in the world's rivers. Tellus 33: 172–187.

Schlesinger, W.H., J.T. Gray and F.S. Gilliam. 1982. Atmospheric deposition processes and their importance as sources of nutrients in a chaparral ecosystem of southern California. Water Resources Research 18: 623–629.

Schlesinger, W.H., P.J. Fonteyn and G.M. Marion. 1987. Soil moisture content and plant transpiration in the Chihuahuan desert of New Mexico. Journal of Arid Environments 12: 119–126.

Schlesinger, W.H. and W.T. Peterjohn. 1988. Ion and sulfate-isotope ratios in arid soils subject to wind erosion in the southwestern USA. Soil Science Society of America Journal 52: 54–58.

Schlesinger, W.H., E.H. DeLucia and W.D. Billings. 1989. Nutrient-use efficiency of woody plants on contrasting soils in the western Great Basin, Nevada. Ecology 70: 105–113.

Schlesinger, W.H., J.F. Reynolds, G.L. Cunningham, L.F. Huenneke, W.M. Jarrell, R.A. Virginia and W.G. Whitford. 1990. Biological feedbacks in global desertification. Science 247: 1043–1048.

Schmidt, J., W. Seiler and R. Conrad. 1988. Emission of nitrous oxide from temperate forest soils into the atmosphere. Journal of Atmospheric Chemistry 6: 95–115.

Schoenau, J.J. and J.R. Bettany. 1987. Organic matter leaching as a component of carbon, nitrogen, phosphorus, and sulfur cycles in a forest, grassland, and gleyed soil. Soil Science Society of America Journal 51: 646–651.

Schönheit, P., J.K. Kristjansson and R.K. Thauer. 1982. Kinetic mechanism for the ability of sulfate reducers to out-compete methanogens for acetate. Archives of Microbiology 132: 285–288.

Schopf, J.W. and D. Z. Oehler. 1976. How old are the Eucaryotes? Science 193: 47–49.

Schultz, R.C., P.P. Kormanik, W.C. Bryan, and G.H. Brister. 1979. Vesicular-arbuscular mycorrhiza influence growth but not mineral concentrations in seedlings of eight sweetgum families. Canadian Journal of Forest Research 9: 218–223.

Schulze, E.-D. 1989. Air pollution and forest decline in a spruce *(Picea abies)* forest. Science 244: 776–783.

Schütz, H., R. Conrad, S. Goodwin and W. Seiler. 1988. Emission of hydrogen from deep and shallow freshwater environments. Biogeochemistry 5: 295–311.

Schütz, H., W. Seiler and R. Conrad. 1989a. Processes involved in formation and emission of methane in rice paddies. Biogeochemistry 7: 33–53.

Schütz, H., A. Holzapfel-Pschorn, R. Conrad, H. Rennenberg and W. Seiler. 1989b. A 3-year continuous record on the influence of daytime, season, and fertilizer treatment on methane emission rates from an Italian rice paddy. Journal of Geophysical Research 94: 16405–16416.

Schütz, L. 1980. Long range transport of desert dust with special emphasis on the Sahara. Annals of the New York Academy of Sciences 338: 515–532.

Schwartz, S.E. 1988. Are global cloud albedo and climate controlled by marine phytoplankton? Nature 336: 441–445.

Schwartz, S.E. 1989. Acid deposition: Unraveling a regional phenomenon. Science 243: 753–763.

Schwartzman, D.W. and T. Volk. 1989. Biotic enhancement of weathering and the habitability of Earth. Nature 340: 457–460.

Schwertmann, U. 1966. Inhibitory effect of soil organic matter on the crystallization of amorphous ferric hydroxide. Nature 212: 645–646.

Schwintzer, C.R. 1983. Nonsymbiotic and symbiotic nitrogen fixation in a weakly minerotrophic peatland. American Journal of Botany 70: 1071–1078.

Scott, R.M. 1962. Exchangeable bases of mature, well-drained soils in relation to rainfall in East Africa. Journal of Soil Science 13: 1–9.

Scriber, J.M. 1977. Limiting effects of low leaf-water content on the nitrogen utilization, energy budget, and larval growth of *Hyalophora cecropia* (Lepidoptera: Saturniidae). Oecologia 28: 269–287.

Sears, S.O. and D. Langmuir. 1982. Sorption and mineral equilibria controls on moisture chemistry in a c-horizon soil. Journal of Hydrology 56: 287–308.

Seastedt, T.R. 1985. Maximization of primary and secondary productivity by grazers. American Naturalist 126: 559–564.

Seastedt, T.R. 1988. Mass, nitrogen, and phosphorus dynamics in foliage and root detritus of tallgrass prairie. Ecology 69: 59–65.

Seastedt, T.R. and D.A. Crossley. 1980. Effects of microarthropods on the seasonal dynamics of nutrients in forest litter. Soil Biology and Biochemistry 12: 337–342.

Seastedt, T.R. and C.M. Tate. 1981. Decomposition rates and nutrient contents of arthropod remains in forest litter. Ecology 62: 13–19.

Seastedt, T.R. and D.C. Hayes. 1988. Factors influencing nitrogen concentrations in soil water in a North American tallgrass prairie. Soil Biology and Biochemistry 20: 725–729.

Sebacher, D.I., R.C. Harriss and K.B. Bartlett. 1985. Methane emissions to the atmosphere through aquatic plants. Journal of Environmental Quality 14: 40–46.

Sebacher, D.I., R.C. Harriss, K.B. Bartlett, S.M. Sebacher and S.S. Grice. 1986. Atmospheric methane sources: Alaskan tundra bogs, an alpine fen, and a subarctic boreal marsh. Tellus 38B: 1–10.

Seemann, J.R., T.D.Sharkey, J. Wang and C.B. Osmond. 1987. Environmental effects on photosynthsis, nitrogen-use efficiency, and metabolite pools in leaves of sun and shade plants. Plant Physiology 84: 796–802.

Sehmel, G.A. 1980. Particle and gas dry deposition: A review. Atmospheric Environment 14: 983–1011.

Seiler, W. and J. Fishman. 1981. The distribution of carbon monoxide and ozone in the free troposphere. Journal of Geophysical Research 86: 7255–7265.

Seiler, W. and R. Conrad. 1987. Contribution of tropical ecosystems to the global budgets of trace gases, especially CH_4, H_2, CO, and N_2O. pp. 133–162. *In* R.E. Dickinson (ed.), The Geophysiology of Amazonia. Wiley, New York.

Seiler, W. and P.J. Crutzen. 1980. Estimates of gross and net fluxes of carbon between the biosphere and the atmosphere from biomass burning. Climate Change 2: 207–247.

Seiler, W., A. Holzapfel-Pschorn, R. Conrad and D. Scharffe. 1984a. Methane emission from rice paddies. Journal of Atmospheric Chemistry 1: 241–268.

Seiler, W., R. Conrad and D. Scharffe. 1984b. Field studies of methane emission from termite nests into the atmosphere and measurements of methane uptake by tropical soils. Journal of Atmospheric Chemistry 1: 171–186.

Seitzinger, S., S. Nixon, M.E.Q. Pilson and S. Burke. 1980. Denitrification and N_2O production in near-shore marine sediments. Geochimica et Cosmochimica Acta 44: 1853–1860.

Seitzinger, S.P., S.W. Nixon and M.E.Q. Pilson. 1984. Denitrification and nitrous oxide production in a coastal marine ecosystem. Limnology and Oceanography 29: 73–83.

Seitzinger, S.P. 1988. Denitrification in freshwater and coastal marine ecosystems: Ecological and geochemical significance. Limnology and Oceanography 33: 702–724.

Servant, J. 1986. The burden of the sulphate layer of the stratosphere during volcanic 'quiescent' periods. Tellus 38B: 74–79.

Servant, J. 1989. Les sources et les puits d'oxysulfure de carbone (COS) à l'echelle mondiale. Atmospheric Research 23: 105–116.

Sexstone, A.J., N.P. Revsbech, T.B. Parkin and J.M. Tiedje. 1985a. Direct measurement of oxygen profiles and denitrification rates in soil aggregates. Soil Science Society of America Journal 49: 645–651.

Sexstone, A.J., T.B. Parkin and J.M. Tiedje. 1985b. Temporal response of soil denitrification rates to rainfall and irrigation. Soil Science Society of America Journal 49: 99–105.

Shackleton, N.J. 1977. Carbon-13 in Uvigerina: Tropical rainforest history and the equatorial Pacific carbonate dissolution cycles. pp. 201–277. In N.R. Andersen and A. Malahoff (eds.), The Fate of Fossil Fuel CO_2 in the Oceans. Plenum, New York.

Shaffer, P.W. and M.R. Church. 1989. Terrestrial and in-lake contributions to alkalinity budgets of drainage lakes: An assessment of regional differences. Canadian Journal of Fisheries and Aquatic Science 46: 509–515.

Shanks, A.L. and J.D. Trent. 1979. Marine snow: Microscale nutrient patches. Limnology and Oceanography 24: 850–854.

Shanks, A.L. and J.D. Trent. 1980. Marine snow: Sinking rates and potential role in vertical flux. Deep Sea Research 27: 137–143.

Sharkey, T.D. 1985. Photosynthesis in intact leaves of C_3 plants: Physics, physiology and rate limitations. Botanical Review 51: 53–105.

Sharkey, T.D. 1988. Estimating the rate of photorespiration in leaves. Physiologica Plantarum 73: 147–152.

Sharpley, A.N. 1985. The selective erosion of plant nutrients in runoff. Soil Science Society of America Journal 49: 1527–1534.

Sharpley, A.N., H. Tiessen and C.V. Cole. 1987. Soil phosphorus forms extracted by soil tests as a function of pedogenesis. Soil Science Society of America Journal 51: 362–365.

Shaver, G.R. and J.M. Melillo. 1984. Nutrient budgets of marsh plants: Efficiency concepts and relation to availability. Ecology 65: 1491–1510.

Shaver, G.R., F. Chapin and B.L. Gartner. 1986. Factors limiting seasonal growth and peak biomass accumulation in Eriophorum vaginatum in Alaskan tussock tundra. Journal of Ecology 74: 257–278.

Shaw, R.W. 1987. Air pollution by particles. Scientific American 255(8):96–103.

Shear, C.B., H.L. Crane and A.T. Myers. 1946. Nutrient-element balance: A fundamental concept in plant nutrition. American Society for Horticultural Science 47: 239–248.

Shearer, G., D.H. Kohl, R.A. Virginia, B.A. Bryan, J.L. Skeeters, E.T. Nilsen, M.R. Sharifi and P.W. Rundel. 1983. Estimates of N_2-fixation from variation in the natural abundance of ^{15}N in Sonoran desert ecosystems. Oecologia 56: 365–373.

Shearer, G., and D.H. Kohl. 1988. Natural ^{15}N abundance as a method of estimating the contribution of biologically fixed nitrogen to $N_2$2-fixing systems: Potential for non-legumes. Plant and Soil 110: 317–327.

Shearer, G. and D.H. Kohl. 1989. Estimates of N_2 fixation in ecosystems: The need for and basis of the ^{15}N natural abundance method. pp. 342–374. In P.W. Rundel, J.R. Ehleringer and K.A. Nagy (eds.), Stable Isotopes in Ecological Research. Springer-Verlag, New York.

Sholkovitz, E.R. 1976. Flocculation of dissolved organic and inorganic matter during mixing of river water and seawater. Geochimica et Cosmochimica Acta 40: 831–845.

Shortle, W.C. and K.T. Smith 1988. Aluminum-induced calcium deficiency syndrome in declining red spruce. Science 240: 1017–1018.

Shukla, J. and Y. Mintz. 1982. Influence of land-surface evapotranspiration on the Earth's climate. Science 215: 1498–1501.

Shultz, D.J. and J.A. Calder. 1976. Organic carbon $^{13}C/^{12}C$ variations in estuarine sediments. Geochimica et Cosmochimica Acta 40: 381–385.

Shuttleworth, W.J. 1988. Evaporation from Amazonian rainforest. Proceedings of the Royal Society of London 233B: 321–346.

Siever, R. 1974. The steady state of the Earth's crust, atmosphere and oceans. Scientific American 230(6): 72–79.

Sillén, L.G. 1966. Regulaton of O_2, N_2 and CO_2 in the atmosphere; thoughts of a laboratory chemist. Tellus 18: 198–206.

Silvester, W.B., P. Sollins, T. Verhoeven and S.P. Cline. 1982. Nitrogen fixation and acetylene reduction in decaying conifer boles: Effects of incubation time, aeration, and moisture content. Canadian Journal of Forest Research 12: 646–652.

Silvester, W.B. 1989. Molybdenum limitation of asymbiotic nitrogen fixation in forests of Pacific Northwest America. Soil Biology and Biochemistry 21: 283–289.

Simkiss, K. and K.M. Wilbur. 1989. Biomineralization: Cell Biology and Mineral Deposition. Academic Press, San Diego.

Simon, N.S. 1988. Nitrogen cycling between sediment and the shallow-water column in the transition zone of the Potomac River and estuary. I. Nitrate and ammonium fluxes. Estuarine, Coastal and Shelf Science 26: 483–497.

Simon, N.S. 1989. Nitrogen cycling between sediment and the shallow-water column in the transition zone of the Potomac River and estuary. II. The role of wind-driven resuspension and adsorbed ammonium. Estuarine, Coastal and Shelf Science 28: 531–547.

Singh, B.R. 1984. Sulfate sorption by acid forest soils: 2. Sulfate adsorption isotherms with and without organic matter and oxides of aluminum and iron. Soil Science 138: 294–296.

Singh, J.S., W.K. Lauenroth and R.K. Steinhorst. 1975. Review and assessment of various techniques for estimating net aerial primary production in grasslands from harvest data. Botanical Review 41: 181–232.

Singh, J.S. and S.R. Gupta. 1977. Plant decomposition and soil respiration in terrestrial ecosystems. Botanical Review 43: 449–528.

Singh, O.N., R. Borchers, P. Fabian, S. Lal and B.H. Subbaraya. 1988. Measurements of atmospheric BrO_x radicals in the tropical and mid-latitude atmosphere. Nature 334: 593–595.

Sirois, A. and L.A. Barrie. 1988. An estimate of the importance of dry deposition as a pathway of acidic substances from the atmosphere to the biosphere in eastern Canada. Tellus 40B: 59–80.

Sjöberg, A. 1976. Phosphate analysis of anthropic soils. Journal of Field Archaeology 3: 447–454.

Skyring, G.W. 1987. Sulfate reduction in coastal ecosystems. Geomicrobiology Journal 5: 295–374.

Slemr, F., R. Conrad and W. Seiler. 1984. Nitrous oxide emissions from fertilized and unfertilized soils in a subtropical region (Andalusia, Spain). Journal of Atmospheric Chemistry 1: 159–169.

Slemr, F. and W. Seiler. 1984. Field measurements of NO and NO_2 emissions from fertilized and unfertilized soils. Journal of Atmospheric Chemistry 2: 1–24.

Slingo, A. 1990. Sensitivity of the Earth's radiation budget to changes in low clouds. Nature 343: 49–51.

Slinn, W.G.N. 1988. A simple model for Junge's relationship between concentration fluctuations and residence times for tropospheric trace gases. Tellus 40B: 229–232.

Smeck, N.E. 1985. Phosphorus dynamics in soils and landscapes. Geoderma 36: 185–199.

Smirnoff, N., P. Todd and G.R. Stewart. 1984. The occurrence of nitrate reduction in leaves of woody plants. Annals of Botany 54: 363–374.

Smith, C.J., R.D. DeLaune and W.H. Patrick. 1983. Nitrous oxide emission from Gulf coast wetlands. Geochimica et Cosmochimica Acta 47: 1805–1814.

Smith, M.S. and J.M. Tiedje. 1979. Phases of denitrification following oxygen depletion in soil. Soil Biology and Biochemistry 11: 261–267.

Smith, R.A., R.B. Alexander and M.G. Wolman. 1987. Water-quality trends in the nation's rivers. Science 235: 1607–1615.

Smith, R.B., R.H. Waring and D.A. Perry. 1981. Interpreting foliar analyses from Douglas-fir as weight per unit of leaf area. Canadian Journal of Forest Research 11:593–598.

Smith, R.L. and M.J. Klug. 1981. Reduction of sulfur compounds in the sediments of a eutrophic lake basin. Applied and Environmental Microbiology 41: 1230–1237.

Smith, S.V. 1981. Marine macrophytes as a global carbon sink. Science 211: 838–840.

Smith, S.V. and F.T. MacKenzie. 1987. The ocean as a net heterotrophic system: Implications from the carbon biogeochemical cycle. Global Biogeochemical Cycles 1: 187–198.

Smith, S.V. and J.T. Hollibaugh. 1989. Carbon-controlled nitrogen cycling in a marine 'macrocosm': an ecosystem-scale model for managing cultural eutrophication. Marine Ecology Progress Series 52: 103–109.

Smith, V.H. 1982. The nitrogen and phosphorus dependence of algal biomass in lakes: An empirical and theoretical analysis. Limnology and Oceanography 27: 1101–1112.

Smith, V.H. 1983. Low nitrogen to phosphorus ratios favor dominance by blue-green algae in lake phytoplankton. Science 221: 669–671.

Smith, W.H. 1976. Character and significance of forest tree root exudates. Ecology 57: 324–331.

Smith, W.H. and T.G. Siccama. 1981. The Hubbard Brook Ecosystem Study: Biogeochemistry of lead in the northern hardwood forest. Journal of Environmental Quality 10: 323–333.

Snaydon, R.W. 1962. Micro-distribution of Trifolium repens L. and its relation to soil factors. Journal of Ecology 50: 133–143.

Søballe, D.M. and B.L. Kimmel. 1987. A large-scale comparison of factors influencing phytoplankton abundance in rivers, lakes, and impoundments. Ecology 68: 1943–1954.

Söderlund, R. and T. Rosswall. 1982. The nitrogen cycles. pp. 61–81. In O. Hutzinger (ed.), The Handbook of Environmental Chemistry, Vol. 1, Part B, The Natural Environment and the Biogeochemical Cycles. Springer-Verlag, New York.

Sollins, P., G. Spycher and C. Topik. 1983. Processes of soil organic-matter accretion at a mudflow chronosequence, Mt. Shasta, California. Ecology 64: 1273–1282.

Sollins, P., G. Spycher and C.A. Glassman. 1984. Net nitrogen mineralization from light- and heavy-fraction forest soil organic matter. Soil Biology and Biochemistry 16: 31–37.

Sollins, P., G.P. Robertson and G. Uehara. 1988. Nutrient mobility in variable- and permanent-charge soils. Biogeochemistry 6: 181–199.

Solomon, A.M. and M.L. Tharp. 1985. Simulation experiments with Late-Quaternary carbon storage in mid-latitude forest communities. pp. 235–250. In E.T. Sundquist and W. S. Broecker (eds.), The Carbon Cycle and Atmospheric CO_2: Natural Variations, Archean to Present. American Geophysical Union, Washington, D.C.

Solomon, D.K. and T.E. Cerling. 1987. The annual carbon dioxide cycle in a montane soil: Observations, modeling, and implications for weathering. Water Resources Research 23: 2257–2265.

Sonzogni, W.C., S.C. Chapra, D.E. Armstrong and T.J. Logan. 1982. Bioavailability of phosphorus inputs to lakes. Journal of Environmental Quality 11: 555–563.

Speidel, D.H. and A.F. Agnew. 1982. The Natural Geochemistry of our Environment. Westview Press, Boulder, Colorado.

Spencer, R.W. and J.R. Christy. 1990. Precise monitoring of global temperature trends from satellites. Science 247: 1558–1562.

Sposito, G. 1984. The Surface Chemistry of Soils. Oxford University Press, Oxford.

Spycher, G., P. Sollins and S. Rose. 1983. Carbon and nitrogen in the light fraction of a forest soil: Vertical distribution and seasonal patterns. Soil Science 135: 79–87.

Srivastava, S.C. and J.S. Singh. 1988. Carbon and phosphorus in the soil biomass of some tropical soils of India. Soil Biology and Biochemistry 20: 743–747.

Staaf, H. and B. Berg. 1982. Accumulation and release of plant nutrients in decomposing Scots pine needle litter. Long-term decomposition in a Scots pine forest II. Canadian Journal of Botany 60: 1561–1568.

Stallard, R.F. and J.M. Edmond. 1983. Geochemistry of the Amazon. 2. The influence of geology and weathering environment on the dissolved load. Journal of Geophysical Research 88: 9671–9688.

Stanley, D.W. and J.E. Hobbie. 1981. Nitrogen recycling in a North Carolina coastal river. Limnology and Oceanography 26: 30–42.

Stanley, S.R. and E.J. Ciolkosz. 1981. Classification and genesis of Spodosols in the central Appalachians. Soil Science Society of America Journal 45: 912–917.

Starkel, L. 1989. Global paleohydrology. Quaternary International 2:25–33.

Stauffer, B., G. Fischer, A. Neftel and H. Oeschger. 1985. Increase of atmospheric methane recorded in Antarctic ice core. Science 229: 1386–1388.

Stauffer, R.E. 1987. Effects of oxygen transport on the areal hypolimnetic oxygen deficit. Water Resources Research 23: 1887–1892.

Steele, L.P., P.J. Fraser, R.A. Rasmussen, M.A.K. Khalil, T.J. Conway, A.J. Crawford, R.H. Gammon, K.A. Massarie and K.W. Thoning. 1987. The global distribution of methane in the troposphere. Journal of Atmospheric Chemistry 5: 125–171.

Stetter, K.O., G. Lauerer, M. Thomm and A. Neuner. 1987. Isolation of extremely thermophilic sulfate reducers: Evidence for a novel branch of Archaebacteria. Science 236: 822–824.

Steudler, P.A. and B.J. Peterson. 1984. Contribution of gaseous sulphur from salt marshes to the global sulphur cycle. Nature 311: 455–457.

Steudler, P.A. and B.J. Peterson. 1985. Annual cycle of gaseous sulfur emissions from a New England *Spartina alterniflora* marsh. Atmospheric Environment 19: 1411–1416.

Steudler, P.A., R.D. Bowden, J.M. Melillo and J.D. Aber. 1989. Influence of nitrogen fertilization on methane uptake in temperate forest soils. Nature 341: 314–316.

Stevenson, D.J. 1983. The nature of the Earth prior to the oldest known rock record: The Hadean earth. pp. 32–40. *In* J.W. Schopf (ed.), Earth's Earliest Biosphere. Princeton University Press, Princeton, New Jersey.

Stevenson, F.J. 1982. Origin and distribution of nitrogen in soil. pp. 1–42. *In* F. J. Stevenson (ed.), Nitrogen in Agricultural Soils. American Society of Agronomy, Madison, Wisconsin.

Stevenson, F.J. 1986. Cycles of Soil. Wiley, New York.

Stock, J.B., A.M. Stock and J.M. Mottonen. 1990. Signal transduction in bacteria. Nature 344: 395–400.

Stockner, J.G. and N.J. Antia. 1986. Algal picoplankton from marine and freshwater ecosystems: A multidisciplinary perspective. Canadian Journal of Fisheries and Aquatic Science 43: 2472–2503.

Stolzy, L.H., D.D. Focht and H. Flühler. 1981. Indicators of soil aeration status. Flora 171: 236–265.

Stone, E.L. and R. Kszystyniak. 1977. Conservation of potassium in the *Pinus resinosa* ecosystem. Science 198: 192–194.

Strain, B.R. 1985. Physiological and ecological controls on carbon sequestering in terrestrial ecosystems. Biogeochemistry 1: 219–232.

Strong, A.E. 1989. Greater global warming revealed by satellite-derived sea-surface-temperature trends. Nature 338: 642–645.

Stuiver, M. 1978. Atmospheric carbon dioxide and carbon reservoir changes. Science 199: 253–258.

Stuiver, M. 1980. ^{14}C distribution in the Atlantic ocean. Journal of Geophysical Research 85: 2711–2718.

Stuiver, M., P.D. Quay and H.G. Ostlund. 1983. Abyssal water carbon-14 distribution and the age of the world oceans. Science 219: 849–851.

Subba Rao, D.V. 1981. Effect of boron on primary production of nannoplankton. Canadian Journal of Fisheries and Aquatic Science 38: 52–58.

Suberkropp, K., G.L. Godshalk and M.J. Klug. 1976. Changes in the chemical composition of leaves during processing in a woodland stream. Ecology 57: 720–727.

Sundquist, E.T., L.N. Plummer and T.M.L. Wigley. 1979. Carbon dioxide in the ocean surface: The homogeneous buffer factor. Science 204: 1203–1205.

Swank, W.T. and J.E. Douglass. 1977. Nutrient budgets for undisturbed and manipulated hardwood forest ecosystems in the mountains of North Carolina. pp. 343–362. In D.L. Correll (ed.), Watershed Research in Eastern North America, Vol. 1. Smithsonian Institution, Edgewater, Maryland.

Swank, W.T. and G. S. Henderson. 1976. Atmospheric input of some cations and anions to forest ecosystems in North Carolina and Tennessee. Water Resources Research 12: 541–546.

Swank, W.T., J.B. Waide, D.A. Crossley and R.L. Todd. 1981. Insect defoliation enhances nitrate export from forest ecosystems. Oecologia 51: 297–299.

Swank, W.T. and W.H. Caskey. 1982. Nitrate depletion in a second-order mountain stream. Journal of Environmental Quality 11: 581–584.

Swank, W.T., J.W. Fitzgerald and J.T. Ash. 1984. Microbial transformation of sulfate in forest soils. Science 223: 182–184.

Swanson, F.J., R.L. Fredriksen and F.M. McCorison. 1982. Material transfer in a western Oregon forested watershed. pp. 233–266. In R.L. Edmonds (ed.), Analysis of Coniferous Forest Ecosystems in the Western United States. Dowden, Hutchinson and Ross, Stroudsburg, Pennsylvania.

Swift, M.J., O.W. Heal and J.M. Anderson. 1979. Decomposition in Terrestrial Ecosystems. University of California Press, Berkeley.

Syers, J.K., J.A. Adams and T.W. Walker. 1970. Accumulation of organic matter in a chronosequence of soils developed on wind-blown sand in New Zealand. Journal of Soil Science 21: 146–153.

Szarek, S.R. 1979. Primary production in four North American deserts: Indices of efficiency. Journal of Arid Environments 2: 187–209.

Tabatabai, M.A. and W.A. Dick. 1979. Distribution and stability of pyrophosphatase in soils. Soil Biology and Biochemistry 11: 655–659.

Takahashi, T., D. Chipman and T. Volk. 1983. Geographical, seasonal, and secular variations of the partial pressure of CO_2 in the surface watrs of the North Atlantic Ocean. The results of the Atlantic TTO Program. pp. 125–143. Proceedings: Carbon Dioxide, Science and Consensus. CONF 820970. U.S. Department of Energy, Washington, D.C.

Tamrazyan, G.P. 1989. Global peculiarities and tendencies in river discharge and washdown of the suspended sediments—The earth as a whole. Journal of Hydrology 107: 113–131.

Tan, K.H. 1980. The release of silicon, aluminum, and potassium during decomposition of soil minerals by humic acid. Soil Science 129: 5–11.

Tan, K.H. and P.S. Troth. 1982. Silica-sesquioxide ratios as aids in characterization of some temperate region and tropical soil clays. Soil Science Society of America Journal 46: 1109–1114.

Tans, P.P., I.Y. Fung and T. Takahashi. 1990. Observational constraints on the global atmospheric CO_2 budget. Science 247: 1431–1438.

Tarafdar, J.C. and N. Claassen. 1988. Organic phosphorus compounds as a phosphorus source for higher plants through the activity of phosphatase produced by plant roots and microorganisms. Biology and Fertility of Soils 5: 308–312.

Tate, K.R., D.J. Ross and C.W. Feltham. 1988. A direct extraction method to estimate soil microbial C: Effects of experimental variables and some different calibration procedures. Soil Biology and Biochemistry 20: 327–335.

Tate, R.L. 1980. Microbial oxidation of organic matter of Histosols. Advances in Microbial Ecology 4: 169–201.

Taylor, B.R., D. Parkinson and W.F.J. Parsons. 1989. Nitrogen and lignin content as predictors of litter decay rates: A microcosm test. Ecology 70: 97–104.

Terman, G.L. 1979. Volatilization losses of nitrogen as ammonia from surface-applied fertilizers, organic amendments, and crop residues. Advances in Agronomy 31: 189–223.

Terman, G.L. 1977. Quantitative relationships among nutrients leached from soils. Soil Science Society of America Journal 41: 935–940.

Tezuka, Y. 1961. Development of vegetation in relation to soil formation in the volcanic island of Oshima, Izu, Japan. Japanese Journal of Botany 17: 371–402.

Thode-Andersen, S. and B.B. Jørgensen. 1989. Sulfate reduction and the formation of ^{35}S-labeled FeS, FeS$_2$ and S^0 in coastal marine sediments. Limnology and Oceanography 34: 793–806.

Thomas, G.E., J.J. Olivero, E.J. Jensen, W. Schroeder and O.B. Toon. 1989. Relation between increasing methane and the presence of ice clouds at the mesopause. Nature 338: 490–492.

Thompson, L.G., E. Mosley-Thompson, M.E. Davis, J.F. Bolzan, J. Dai, T. Yao, N. Gundestrup, X. Wu, L. Klein and Z. Xie. 1989. Holocene-Late Pleistocene climatic ice core records from Qinghai-Tibetan plateau. Science 246: 474–477.

Thorpe, S.A. 1985. Small-scale processes in the upper ocean boundary layer. Nature 318: 519–522.

Tiedje, J.M., A.J. Sextone, T.B. Parkin, N.P. Revsbech and D.R. Shelton. 1984. Anaerobic processes in soil. Plant and Soil 76: 197–212.

Tiedje, J.M., S. Simkins and P.M. Groffman. 1989. Perspectives on measurement of denitrification in the field including recommended protocols for acetylene based methods. Plant and Soil 115: 261–284.

Tiessen, H., J.W.B. Stewart and J.R. Bettany. 1982. Cultivation effects on the amounts and concentration of carbon, nitrogen, and phosphorus in grassland soils. Agronomy Journal 74: 831–835.

Tiessen, H., J.W.B. Stewart and C.V. Cole. 1984. Pathways of phosphorus transformations in soils of differing pedogenesis. Soil Science Society of America Journal 48: 853–858.

Tiessen H. and J.W.B. Stewart. 1983. Particle-size fractions and their use in studies of soil organic matter: II. Cultivation effects on organic matter composition in size fractions. Soil Science Society of America Journal 47: 509–514.

Tiessen, H. and J.W.B. Stewart. 1988. Light and electron microscopy of stained microaggregates: The role of organic matter and microbes in soil aggregation. Biogeochemistry 5: 312–322.

Tietze, K., M. Geyh, H. Muller, W. Stahl and H. Wehner. 1980. The genesis of the methane in Lake Kivu (Central Africa). Geologische Rundschau 69: 452–472.

Tilman, D. 1982. Resource Competition and Community Structure. Princeton University Press, Princeton, New Jersey.

Tilman, D. 1985. The resource-ratio hypothesis of plant succession. American Naturalist 125: 827–852.

Tilton, D.L. 1978. Comparative growth and foliar element concentrations of *Larix laricina* over a range of wetland types in Minnesota. Journal of Ecology 66: 499–512.

Timmer, V.R. and E.L. Stone. 1978. Comparative foliar analysis of young Balsam fir fertilized with nitrogen, phosphorus, potassium, and lime. Soil Science Society of America Journal 42: 125–130.

Ting, J.C. and M. Chang. 1985. Soil-moisture depletion under three southern pine plantations in east Texas. Forest Ecology and Mangement 12: 179–193.

Tisdall, J.M. and J.M. Oades. 1982. Organic matter and water- stable aggregates in soils. Journal of Soil Science 33: 141–163.

Titman, D. 1976. Ecological competition between algae: Experimental confirmation of resource-based competition theory. Science 192: 463–465.

Tolbert, M.A., M.J. Rossi and D.M. Golden. 1988. Antarctic ozone depletion chemistry: Reactions of N_2O_5 with H_2O and HC1 on ice surfaces. Science 240: 1018–1021.

Toon, O.B. and J. B. Pollack. 1980. Atmospheric aerosols and climate. American Scientist 68: 268–278.

Toon, O.B., J.F. Kasting, R.P. Turco and M.S. Liu. 1987. The sulfur cycle in the marine atmosphere. Journal of Geophysical Research 92: 943–963.

Topp, G.C., W.D. Zebchuk and J. Dumanski. 1980. The variation of in situ measured soil water properties within soil map units. Canadian Journal of Soil Science 60: 497–509.

Tortoso, A.C. and G.L. Hutchinson. 1990. Contributions of autotrophic and heterotrophic nitrifiers to soil NO and N_2O emissions. Applied and Environmental Microbiology 56: 1799–1805.

Travis, C.C. and E.L. Etnier. 1981. A survey of sorption relationships for reactive solutes in soil. Journal of Environmental Quality 10: 8–17.

Trefry, J.H., S. Metz, R. P. Trocine and T.A. Nelsen. 1985. A decline in lead transport by the Mississippi River. Science 230: 439–441.

Trefry, J.H. and S. Metz. 1989. Role of hydrothermal precipitates in the geochemical cycling of vanadium. Nature 342: 531–533.

Triska, F.J. and J.R. Sedell. 1976. Decomposition of four species of leaf litter in response to nitrate manipulation. Ecology 57: 783–792.

Triska, F.J., J.R. Sedell, K. Cromack, S.V. Gregory and F.M. McCorison. 1984. Nitrogen budget for a small coniferous forest stream. Ecological Monographs 54: 119–140.

Trlica, M.J. and M.E. Biondini. 1990. Soil water dynamics, transpiration, and water losses in a crested wheatgrass and native shortgrass ecosystem. Plant and Soil 126: 187–201.

Tromble, J.M. 1988. Water budget for creosotebush-infested rangeland. Journal of Arid Environments 15: 71–74.

Trudinger, P.A. 1979. The biological sulfur cycle. pp. 293–313. In P.A. Trudinger and D.J. Swaine (eds.), Biogeochemical Cycling of Mineral-Forming Elements. Elsevier Scientific, Amsterdam.

Tsutsumi, T., Y. Nishitani and Y. Kirimura. 1983. On the effects of soil fertility on the rate and the nutrient element concentrations of litterfall in a forest. Japanese Journal of Ecology 33: 313–322.

Tukey, H.B. 1970. The leaching of substances from plants. Annual Review of Plant Physiology 21: 305–324.

Turco, R.P., R.C. Whitten, O.B. Toon, J.B. Pollack and P. Hamill. 1980. OCS, stratospheric aerosols and climate. Nature 283: 283–286.

Turekian, K.K. 1977. The fate of metals in the oceans. Geochimica et Cosmochimica Acta 41: 1139–1144.

Turner, F.T. and W.H. Patrick. 1968. Chemical changes in waterlogged soils as a result of oxygen depletion. Transactions of the 9th International Congress of Soil Science 4:53–65.

Turner, J. and P.R. Olson. 1976. Nitrogen relations in a Douglas-fir plantation. Annals of Botany 40: 1185–1193.

Turner, S.M., G. Malin, P.S. Liss, D.S. Harbour and P.M. Holligan. 1988. The seasonal variation of dimethyl sulfide and dimethylsulfoniopropionate concentrations in nearshore waters. Limnology and Oceanography 33: 364–375.

Tyler, S.C., D.R. Blake and F.S. Rowland. 1987. $^{13}C/^{12}C$ ratio in methane from the flooded Amazon forest. Journal of Geophysical Research 92: 1044–1048.

Uehara, G. and G. Gillman. 1981. The Mineralogy, Chemistry and Physics of Tropical Soils with Variable Charge Clays. Westview Press, Boulder, Colorado.

Ugolini, F.C. 1968. Soil development and alder invasion in a recently deglaciated area of Glacier Bay, Alaska. pp. 115–140. In J.M. Trappe, J.F. Franklin, R.F. Tarrant and G.M. Hansen (eds.), Biology of Alder. U.S. Forest Service, Pacific Northwest Forest and Range Experiment Station, Portland Oregon.

Ugolini, F.C., R. Minden, H. Dawson and J. Zachara. 1977. An example of soil processes in the *Abies amabilis* zone of central Cascades, Washington. Soil Science 124: 291–302.

Ugolini, F.C., M.G. Stoner and D.J. Marrett. 1987. Arctic pedogenesis: I. Evidence for contemporary podzolization. Soil Science 144: 90–100.

Uhl, C. and C.F. Jordan. 1984. Succession and nutrient dynamics following forest cutting and burning in Amazonia. Ecology 65: 1476–1490.

Urban, N.R. and S.J. Eisenreich. 1988. Nitrogen cycling in a forested Minnesota bog. Canadian Journal of Botany 66: 435–449.

Urban, N.R., S.J. Eisenreich and D.F. Grigal. 1989. Sulfur cycling in a forested *Sphagnum* bog in northern Minnesota. Biogeochemistry 7: 81–109.

Valiela, I. and J.M. Teal. 1979. The nitrogen budget of a salt marsh ecosystem. Nature 280: 652–656.

Vallentyne, J.R. 1974. The Algal Bowl. Department of Environment, Miscellaneous Special Publication 22, Ottawa, Canada.

Van Breeman, N., P.A. Burrough, E.J. Velthorst, H.F. van Dobben, T. de Wit, T.B. Ridder and H.F.R. Reijnders. 1982. Soil acidification from atmospheric ammonium sulphate in forest canopy throughfall. Nature 299: 548–550.

Van Devender, T. R. and W.G. Spaulding. 1979. Development of vegetation and climate in the southwestern United States. Science 204: 701–710.

Van Dijk, H.F.G., M.H.J. de Louw, J.G.M. Roelofs and J.J. Verburgh. 1990. Impact of artificial, ammonium-enriched rainwater on soils and young coniferous trees in a greenhouse. Part II- Effects on the trees. Environmental Pollution 63: 41–59.

Vance, E.D., P.C. Brookes and D.S. Jenkinson. 1987. An extraction method for measuring soil microbial biomass C. Soil Biology and Biochemistry 19: 703–707.

Van Cleve, K. and R. White. 1980. Forest-floor nitrogen dynamics in a 60-year-old paper birch ecosystem in interior Alaska. Plant and Soil 54: 359–381.

Van Cleve, K., R. Barney and R. Schlentner. 1981. Evidence of temperature control of production and nutrient cycling in two interior Alaska black spruce ecosystems. Canadian Journal of Forest Research 11: 258–273.

Van Cleve, K., L. Oliver, R. Schlentner, L.A. Viereck and C.T. Dyrness. 1983. Production and nutrient cycling in taiga forest ecosystems. Canadian Journal of Forest Research 13: 747–766.

Vandenberg, J.J. and K.R. Knoerr. 1985. Comparison of surrogate surface techniques for estimation of sulfate dry deposition. Atmospheric Environment 19: 627–635.

Van den Driessche, R. 1974. Prediction of mineral nutrient status of trees by foliar analysis. Botanical Review 40: 347–394.

Van Sickle, J. 1981. Long-term distributions of annual sediment yields from small watersheds. Water Resources Research 17: 659–663.

Van Trump, J.E. and S.L. Miller. 1972. Prebiotic synthesis of methionine. Science 178: 859–860.

Van Veen, J.A., J.N. Ladd, J.K. Martin and M. Amato. 1987. Turnover of carbon, nitrogen and phosphorus through the microbial biomass in soils incubated with $^{14}C-$, $^{15}N-$ and ^{32}P-labelled bacterial cells. Soil Biology and Biochemistry 19: 559–565.

Venrick, E.L., J.A. McGowan, D.R. Cayan and T.L. Hayward. 1987. Climate and chlorophyll a: Long-term trends in the central North Pacific Ocean. Science 238: 70–72.

Verry, E.S. and D.R. Timmons. 1982. Waterborne nutrient flow through an upland-peatland watershed in Minnesota. Ecology 63: 1456–1467.

Viereck, L.A. 1966. Plant succession and soil development on gravel outwash of the Muldrow Glacier, Alaska. Ecological Monographs 36: 181–199.

Virginia, R.A., W.M. Jarrell, and E. Franco-Vizcaino. 1982. Direct measurement of denitrification in a *Prosopis* (mesquite) dominated Sonoran desert ecosystem. Oecologia 53: 120–122.

Virginia, R.A. and W.M. Jarrell. 1983. Soil properties in a mesquite-dominated Sonoran desert ecosystem. Soil Science Society of America Journal 47: 138–144.

Virginia, R.A. and C.C. Delwiche. 1982. Natural ^{15}N abundance of presumed N_2-fixing and non-N_2-fixing plants from selected ecosystems. Oecologia 54: 317–325.

Vitousek, P.M., J.R. Gosz, C.C. Grier, J.M. Melillo and W.A. Reiners. 1982. A comparative analysis of potential nitrification and nitrate mobility in forest ecosystems. Ecological Monographs 52: 155–177.

Vitousek, P.M., K. Van Cleve, N. Balakrishnan and D. Mueller-Dombois. 1983. Soil development and nitrogen turnover in montane rainforest soils in Hawai'i. BioTropica 15: 268–274.

Vitousek, P.M., P.R. Ehrlich, A.H. Ehrlich and P.A. Matson. 1986. Human appropriation of the products of photosynthesis. BioScience 36: 368–373.

Vitousek, P.M. 1977. The regulation of element concentrations in mountain streams in the northeastern United States. Ecological Monographs 47: 65–87.

Vitousek, P.M. and R.L. Sanford. 1986. Nutrient cycling in moist tropical forest. Annual Review of Ecology and Systematics 17: 137–167.

Vitousek, P.M., L.R. Walker, L.D. Whiteaker, D. Mueller-Dombois and P.A. Matson. 1987. Biological invasion of *Myrica faya* alters ecosystem development in Hawaii. Science 238: 802–804.

Vitousek, P.M., T. Fahey, D.W. Johnson and M.J. Swift. 1988. Element interactions in forest ecosystems: Succession, allometry and input-output budgets. Biogeochemistry 5: 7–34.

Vitousek, P.M. and P.A. Matson. 1988. Nitrogen transformations in a range of tropical forest soils. Soil Biology and Biochemistry 20: 361–367.

Vitousek, P.M. and S.W. Andariese. 1986. Microbial transformations of labelled nitrogen in a clear-cut pine plantation. Oecologia 68: 601–605.

Vitousek, P.M. and P.A. Matson. 1984. Mechanisms of nitrogen retention in forest ecosystems: A field experiment. Science 225: 51–52.

Vitousek, P.M. 1982. Nutrient cycling and nutrient-use efficiency. American Naturalist 119: 553–572.

Vitousek, P.M. 1984. Litterfall, nutrient cycling, and nutrient limitation in tropical forests. Ecology 65: 285–298.

Vitousek, P.M. and J.M. Melillo. 1979. Nitrate losses from disturbed forests: Patterns and mechanisms. Forest Science 25: 605–619.

Vitousek, P.M. and W.A. Reiners. 1975. Ecosystem succession and nutrient retention: A hypothesis. BioScience 25: 376–381.

Vogt, K.A., C.C. Grier, C.E. Meier and R.L. Edmonds. 1982. Mycorrhizal role in net primary production and nutrient cycling in *Abies amabilis* ecosystems in western Washington. Ecology 63: 370–380.

Vogt, K.A., C.C. Grier, C.E. Meier and M.R. Keyes. 1983. Organic matter and nutrient dynamics in forest floors in young and mature *Abies amabilis* stands in western Washington, as affected by fine-root input. Ecological Monographs 53: 139–157.

Vogt, K.A., C.C. Grier and D.J. Vogt. 1986. Production, turnover, and nutrient dynamics of above- and belowground detritus of world forests. Advances in Ecological Research 15: 303–377.

Vollenweider, R.A. 1968. Scientific fundamentals of the eutrophication of lakes and flowing waters, with particular reference to nitrogen and phosphorus as factors in eutrophication. Organization for Economic Cooperative Development, Technical Report DAS/CSI/68.27, Paris. 182pp.

Vollenweider, R.A., M. Munawar and P. Stadelmann. 1974. A comparative review of phytoplankton and primary production in the Laurentian Great Lakes. Journal of the Fisheries Research Board of Canada 31: 739–762.

Volz, A. and D. Kley. 1988. Evaluation of the Montsouris series of ozone measurements made in the nineteenth century. Nature 332: 240–242.

Von der Haar, T.H. and A.H. Oort. 1973. New estimate of annual poleward energy transport by northern hemisphere oceans. Journal of Physical Oceanography 3: 169–172.

Voroney, R.P. and E.A. Paul. 1984. Determination of K_c and K_n *in situ* for calibration of the chloroform fumigation-incubation method. Soil Biology and Biochemistry 16: 9–14.

Vörösmarty, C.J., B. Moore, A.L. Grace, M.P. Gildea, J.M. Melillo, B.J. Peterson, E.B. Rastetter and P.A. Steudler. 1989. Continental scale models of water balance and fluvial transport: An application to South America. Global Biogeochemical Cycles 3: 241–265.

Vossbrinck, C.R., D.C. Coleman and T.A. Woolley. 1979. Abiotic and biotic factors in litter decomposition in a semiarid grassland. Ecology 60: 265–271.

Wahlen, M., N. Tanaka, R. Henry, B. Deck, J. Zeglen, J.S. Vogel, J. Southon, A. Shemesh, R. Fairbanks and W. Broecker. 1989. Carbon-14 in methane sources and in atmospheric methane: The contribution from fossil carbon. Science 245: 286–290.

Walbridge, M.R. and P.M. Vitousek. 1987. Phosphorus mineralization potentials in acid organic soils: Processes affecting $^{32}PO_4^{3-}$ isotope dilution measurements. Soil Biology and Biochemistry 19: 709–717.

Waldman, J.M., J.W. Munger, D.J. Jacob, R.C. Flagan, J.J. Morgan and M.R. Hoffmann. 1982. Chemical composition of acid fog. Science 218: 677–680.

Waldman, J.M., J.W. Munger, D.J. Jacob and M.R. Hoffmann. 1985. Chemical characterization of stratus cloudwater and its role as a vector for pollutant deposition in a Los Angeles pine forest. Tellus 37B: 91–108.

Walker, J.C.G. 1977. Evolution of the Atmosphere. Macmillan, New York.

Walker, J.C.G. 1980. The oxygen cycle. pp. 87–104. *In* O. Hutzinger (ed.), Handbook of Environmental Chemistry, Vol. 1, Part A, The Natural Environment and the Biogeochemical Cycles. Springer-Verlag, New York.

Walker, J.C.G. 1983. Possible limits on the composition of the Archaean ocean. Nature 302: 518–520.

Walker, J.C.G. 1984. How life affects the atmosphere. BioScience 34: 486–491.

Walker, J.C.G. 1985a. Carbon dioxide on the early Earth. Origins of Life 16: 117–127.

Walker, J.C.G. 1985b. Iron and sulfur in the pre-biologic ocean. Precambrian Research 28: 205–222.

Walker, J.C.G., C. Klein, M. Schidlowski, J.W. Schopf, D.J. Stevenson, and M.R. Walter. 1983. Environmental evolution of the Archean-early Proterozoic Earth. pp. 260–290. *In* J.W. Schopf (ed.), Earth's Earliest Biosphere. Princeton University Press, Princeton, New Jersey.

Walker, T.W. and A.F.R. Adams. 1958. Studies on soil organic matter: I. Influence of phosphorus content of parent materials on accumulations of carbon, nitrogen, sulfur, and organic phosphorus in grassland soils. Soil Science 85: 307–318.

Walker, T.W. and J.K. Syers. 1976. The fate of phosphorus during pedogenesis. Geoderma 15: 1–19.

Wallerstein. G. 1988. Mixing in stars. Science 240: 1743–1750.

Walsh, J.J. 1984. The role of ocean biota in accelerated ecological cycles: A temporal view. Bioscience 34: 499–507.

Ward, B.B., K.A. Kilpatrick, P.C. Novelli and M.I. Scranton. 1987. Methane oxidation and methane fluxes in the ocean surface layer and deep anoxic waters. Nature 327: 226–229.

Ward, R.C. 1967. Principles of Hydrology. McGraw-Hill, London.

Warembourg, F.R. and E.A. Paul. 1977. Seasonal transfers of assimilated ^{14}C in grassland: Plant production and turnover, soil and plant respiration. Soil Biology and Biochemistry 9: 295–301.

Waring, R.H., J.J. Rogers and W.T. Swank. 1981. Water relations and hydrologic cycles. pp. 205–264. *In* D.E. Reichle (ed.), Dynamic Properties of Forest Ecosystems. Cambridge University Press, Cambridge.

Waring, R.H. and W.H. Schlesinger. 1985. Forest Ecosystems. Academic Press, Orlando, Florida.

Waring, R.H. 1987. Nitrate pollution: A particular danger to boreal and subalpine coniferous forests. pp. 93–105. *In* T. Fujimori and M. Kirmura (eds.), Human Impacts and Management of Mountain Forests. Forestry and Forest Products Research Institute, Ibaraki, Japan.

Warneck, P. 1988. Chemistry of the Natural Atmosphere. Academic Press, London.

Watson, A., J.E. Lovelock and L. Margulis. 1978. Methanogenesis, fires and the regulation of atmospheric oxygen. Biosystems 10: 293–298.

Watwood, M.E., J.W. Fitzgerald, W.T. Swank and E.R. Blood. 1988. Factors involved in potential sulfur accumulation in litter and soil from a coastal pine forest. Biogeochemistry 6: 3–19.

Watwood, M.E. and J.W. Fitzgerald. 1988. Sulfur transformations in forest litter and soil: Results of laboratory and field incubations. Soil Science Society of America Journal 52: 1478–1483.

Waughman, G.J. 1980. Chemical aspects of the ecology of some South German peatlands. Journal of Ecology 68: 1025–1046.

Waughman, G.J. and D.J. Bellamy. 1980. Nitrogen fixation and nitrogen balance in peatland ecosystems. Ecology 61: 1185–1198.

Weathers, K.C., G.E. Likens, F.H. Bormann, J.S. Eaton, W.B. Bowden, J.L. Anderson, D.A. Cass, J.N. Galloway, W.C. Keene, K.D. Kimball, P. Huth and D. Smiley. 1986. A regional acidic cloud/fog water event in the eastern United States. Nature 319: 657–658.

Weaver, B.L. and J. Tarney 1984. Empirical approach to estimating the composition of the continental crust. Nature 310: 575–577.

Webb, W., S. Szarek, W. Lauenroth, R. Kinerson and M. Smith. 1978. Primary productivity and water use in native forest, grassland, and desert ecosystems. Ecology 59: 1239–1247.

Webb, W.L., W.K. Lauenroth, S.R. Szarek and R.S.Kinerson. 1983. Primary production and abiotic controls in forests, grasslands, and desert ecosystems in the United States. Ecology 64: 134–151.

Weier, K.L. and J.W. Gilliam. 1986. Effect of acidity on denitrification and nitrous oxide evolution from Atlantic coastal plain soils. Soil Science Society of America Journal 50: 1202–1206.

Weir, J.S. 1972. Spatial distribution of elephants in an African National Park in relation to environmental sodium. Oikos 23: 1–13.

Wells, P.V. 1983. Paleobiogeography of montane islands in the Great Basin since the last glaciopluvial. Ecological Monographs 53: 341–382.

Wessman, C.A., J.D. Aber, D.L. Peterson and J.M. Melillo. 1988a. Remote sensing of canopy chemistry and nitrogen cycling in temperate forest ecosystems. Nature 335: 154–156.

Wessman, C.A., J.D. Aber, D.L. Peterson and J.M. Melillo. 1988b. Foliar analysis using near infrared reflectance spectroscopy. Canadian Journal of Forest Research 18: 6–11.

West, N.E. and J. Skujins. 1977. The nitrogen cycle in North American cold-winter semi-desert ecosystems. Oecologia Plantarum 12: 45–53.

Westheimer, F.H. 1987. Why nature chose phosphates. Science 235: 1173–1178.

Wetherhill, G.W. 1985. Occurrence of giant impacts during the growth of the terrestrial planets. Science 228: 877–879.

Weyl, P.K. 1978. Micropaleontology and ocean surface climate. Science 202: 475–481.

Whalen, S.C. and J.C. Cornwell. 1985. Nitrogen, phosphorus, and organic carbon cycling in an Arctic lake. Canadian Journal of Fisheries and Aquatic Science 42: 797–808.

Whalen, S.C. and W.S. Reeburgh. 1990a. A methane flux transect along the trans-Alaska pipeline haul road. Tellus 42B: 237–249.

Whalen, S.C. and W.S. Reeburgh. 1990b. Consumption of atmospheric methane by tundra soils. Nature 346: 160–162.

Whelpdale, D.M. and R.W. Shaw. 1974. Sulphur dioxide removal by turbulent transfer over grass, snow, and water surfaces. Tellus 26: 196–205.

White, C.S. 1986. Volatile and water-soluble inhibitors of nitrogen mineralization and nitrification in a ponderosa pine ecosystem. Biology and Fertility of Soils 2: 97–104.

White, C.S. 1988. Nitrification inhibition by monoterpenoids: Theoretical mode of action based on molecular structures. Ecology 69: 1631–1633.

White, T.C.R. 1984. The abundance of invertebrate herbivores in relation to the availability of nitrogen in stressed food plants. Oecologia 63: 90–105.

White, E.J. and F.Turner. 1970. A method of estimating income of nutrients in catch of airborne particles by a woodland canopy. Journal of Applied Ecology 7: 441–461.

Whitehead, D.R. 1981. Late-Pleistocene vegetation changes in northeastern North Carolina. Ecological Monographs 51: 451–471.

Whitehead, D.R., H. Rochester, S.W. Rissing, C.B. Douglass and M.C. Sheehan. 1973. Late glacial and postglacial productivity changes in a New England pond. Science 181: 744–747.

Whitfield, P.H. and H. Schreier. 1981. Hysteresis in relationships between discharge and water chemistry in the Fraser River basin, British Columbia. Limnology and Oceanography 26: 1179–1182.

Whiticar, M.J., E. Faber and M. Schoell. 1986. Biogenic methane formation in marine and freshwater environments: CO_2 reduction *vs.* acetate fermentation—Isotope evidence. Geochimica et Cosmochimica Acta 50: 693–709.

Whittaker, R.H. 1975. Communities and Ecosystems, 2nd ed. Macmillan, New York.

Whittaker, R.H. and G.E. Likens. 1973. Carbon in the biota. pp. 281–302. *In* G.M. Woodwell and E.V. Pecan (eds.), Carbon and the Biosphere. CONF 720510, National Technical Information Service, Washington, D.C.

Whittaker, R.H. and G.M. Woodwell. 1968. Dimension and production relations of trees and shrubs in the Brookhaven forest, New York. Journal of Ecology 56: 1–25.

Whittaker, R.H., F.H. Bormann, G.E. Likens and T.G. Siccama. 1974. The Hubbard Brook Ecosystem Study: Forest biomass and producation. Ecological Monographs 44: 233–254.

Whittaker, R.H. and W.A. Niering. 1975. Vegetation of the Santa Catalina Mountains, Arizona, V. Biomass, production, and diversity along the elevation gradient. Ecology 56: 771–790.

Whittaker, R.H. and P.L. Marks. 1975. Methods of assessing terrestrial productivity. pp. 55–118. *In* H. Lieth and R.H. Whittaker (eds.), Primary Productivity of the Biosphere. Springer-Verlag, New York.

Wieder, R.K. and G.E. Lang. 1988. Cycling of inorganic and organic sulfur in peat from Big Run Bog, West Virginia. Biogeochemistry 5: 221–242.

Wieder, R.K., J.B. Yavitt and G.E. Lang. 1990. Methane production and sulfate reduction in two Appalachian peatlands. Biogeochemistry 10: 81–104.

Wiegert, R.G. and F.C. Evans. 1964. Primary production and the disappearance of dead vegetation on an old field in southeastern Michigan. Ecology 45: 49–63.

Wiegert, R.G., L.R. Pomeroy and W.J. Wiebe. 1981. Ecology of salt marshes: An introduction. pp. 3–19. *In* L.R. Pomeroy and R.G. Wiegert (eds.), The Ecology of a Salt Marsh. Springer-Verlag, New York.

Wigley, T.M.L. 1989. Possible climate change due to SO_2-derived cloud condensation nuclei. Nature 339: 365–367.

Wildung, R.E., T.R. Garland and R.L. Buschbom. 1977. The interdependent effects of soil temperature and water content on soil respiration rate and plant root decomposition in arid grassland soils. Soil Biology and Biochemistry 7: 373–378.

Williams, S.N., R.E. Stoiber, N. Garcia P., A. Londono, C., J. B. Gemmell, D. R. Lowe and C. B. Connor. 1986. Eruption of the Nevado del Ruiz Volcano, Columbia, on 13 November 1985: Gas flux and fluid geochemistry. Science 233: 964–967.

Wilson, G.V., J.M. Alfonsi and P.M. Jardine. 1989. Spatial variability of saturated hydraulic conductivity of the subsoil of two forested watersheds. Soil Science Society of America Journal 53: 679–685.

Wilson, J.O., P.M. Crill, K.B. Bartlett, D.I. Sebacher, R.C. Harriss, and R.L. Sass. 1989. Seasonal variation of methane emissions from a temperate swamp. Biogeochemistry 8: 55–71.

Wilson, L.G., R.A. Bressan and P. Filner. 1978. Light-dependent emission of hydrogen sulfide from plants. Plant Physiology 61: 184–189.

Winner, W.E., C.L. Smith, G.W. Koch, H.A. Mooney, J.D. Bewley and H.R. Krouse. 1981. Rates of emission of H_2S from plants and patterns of stable sulphur isotope fractionation. Nature 289: 672–673.

Wissmar, R.C., J.E. Richey, R.F. Stallard and J.M. Edmond. 1981. Plankton metabolism and carbon processes in the Amazon River, its tributaries, and floodplain waters, Peru-Brazil, May-June 1977. Ecology 62: 1622–1633.

Wofsy, S.C., R.C. Harriss and W.A. Kaplan. 1988. Carbon dioxide in the atmosphere over the Amazon Basin. Journal of Geophysical Research 93: 1377–1387.

Wolin, M.J. and T.L. Miller. 1987. Bioconversion of organic carbon to CH_4 and CO_2. Geomicrobiology Journal 5: 239–259.

Wollast, R. 1981. Interactions between major biogeochemical cycles in marine ecosystems. pp. 125–142. In G.E. Likens (ed.), Some Perspectives of the Major Biogeochemical Cycles. Wiley, New York.

Woltemate, I., M.J. Whiticar and M. Schoell. 1984. Carbon and hydrogen isotopic composition of bacterial methane in a shallow freshwater lake. Limnology and Oceanography 29: 985–992.

Wood, T., F.H. Bormann and G.K. Voigt. 1984. Phosphorus cycling in a northern hardwood forest: Biological and chemical control. Science 223: 391–393.

Wood, W.W. and M.J. Petraitis. 1984. Origin and distribution of carbon dioxide in the unsaturated zone of the southern High Plains. Water Resources Research 20: 1193–1208.

Woodmansee, R.G. 1978. Additions and losses of nitrogen in grassland ecosystems. BioScience 28: 448–453.

Woodmansee, R.G., J.L. Dodd, R.A. Bowman, F.E. Clark and C.E. Dickinson. 1978. Nitrogen budget of a shortgrass prairie ecosystem. Oecologia 34: 363–376.

Woodmansee, R.G. and L.S. Wallach. 1981. Effects of fire regimes on biogeochemical cycles. pp. 649–669. In F.E. Clark and T. Rosswall (eds.), Terrestrial Nitrogen Cycles. Swedish Natural Science Research Council, Stockholm.

Woodwell, G.M. and W.R. Dykeman. 1966. Respiraton of a forest measured by carbon dioxide accumulation during temperature inversions. Science 154: 1031–1034.

Woodwell, G.M. 1974. Variation in the nutrient content of leaves of Quercus alba, Quercus coccinea, and Pinus rigida in the Brookhaven forest from bud-break to abscission. American Journal of Botany 61: 749–753.

Woodwell, G.M., J.E. Hobbie, R.A. Houghton, J.M. Melillo, B. Moore, B.J. Peterson, and G.R. Shaver. 1983. Global deforestation: Contribution to atmospheric carbon dioxide. Science 222: 1081–1086.

Woosley, S.E. and M. M. Phillips. 1988. Supernova 1987A!. Science 240: 750–759.

Worsley, T.R. and T.A. Davies. 1979. Sea-level fluctuations and deep-sea sedimentation rates. Science 203: 455–456.

Wright, R.F. 1976. The impact of forest fire on the nutrient influxes to small lakes in northeastern Minnesota. Ecology 57: 649–663.

Wu, J. 1981. Evidence of sea spray produced by bursting bubbles. Science 212:324–326.

Wullstein, L.H. and S.A. Pratt. 1981. Scanning electron microscopy of rhizosheaths of *Oryzopsis hymenoides*. American Journal of Botany 68: 408–419.

Yaalon, D.H. 1965. Downward movement and distribution of anions in soil profiles with limited wetting. pp. 157–164. *In* E.D. Hallsworth and D.V. Crawford (eds.), Experimental Pedology. Butterworths, London.

Yanagawa, H., Y. Ogawa, K. Kojima and M. Ito. 1988. Construction of protocellular structures under simulated primitive Earth conditions. Origins of Life 18: 179–207.

Yavitt, J.B., G.E. Lang and D.M. Downey. 1988. Potential methane production and methane oxidation rates in peatland ecosystems of the Appalachian Mountains, United States. Global Biogeochemical Cycles 2: 253–268.

Yavitt, J.B., D.M. Downey, E. Lancaster and G.E. Lang. 1990a. Methane consumption in decomposing *Sphagnum*-derived peat. Soil Biology and Biochemistry 22: 441–447.

Yavitt, J.B., D.M. Downey, G.E. Lang and A.J. Sexstone. 1990b. Methane consumption in two temperate forest soils. Biogeochemistry 9: 39–52.

Yoh, M., H. Terai and Y. Saijo. 1988. Nitrous oxide in freshwater lakes. Archiv fuer Hydrobiologie 113: 273–294.

Yoshinari, T. and R. Knowles. 1976. Acetylene inhibition of nitrous oxide reduction by denitrifying bacteria. Biochemical and Biophysical Research Communications 69: 705–710.

Young, J.R., E.C. Ellis and G.M. Hidy. 1988. Deposition of air-borne acidifiers in the western environment. Journal of Environmental Quality 17: 1–26.

Youngberg, C.T. and A.G. Wollum. 1976. Nitrogen accretion in developing *Ceanothus velutinus* stands. Soil Science Society of America Proceedings 40: 109–112.

Yung, Y.L. and M.B. McElroy. 1979. Fixation of nitrogen in the prebiotic atmosphere. Science 203: 1002–1004.

Zardini, D., R. Raynaud, S. Scharffe and W. Seiler. 1989. N_2O measurements of air extracted from Antarctic ice cores: Implication on atmospheric N_2O back to the last glacial-interglacial transition. Journal of Atmospheric Chemistry 8: 189–201.

Zehnder, A.J.B. and S.H. Zinder. 1980. The sulfur cycle. pp. 105–145. *In* O. Hutzinger (ed.), The Handbook of Environmental Chemistry, Vol. 1, Part A, The Natural Environment and the Biogeochemical Cycles. Springer-Verlag, New York.

Zimka, J.R. and A. Stachurski. 1976. Vegetation as a modifier of carbon and nitrogen transfer to soil in various types of forest ecosystems. Ekologiya Pol. 24: 493–514.

Zimmerman, P.R., R.B. Chatfield, J. Fishman, P.J. Crutzen and P.L. Hanst. 1978. Estimates on the production of CO and H_2 from the oxidation of hydrocarbon emissions from vegetation. Geophysical Research Letters 5: 679–682.

Zimmerman, P.R., J.P. Greenberg, S.O. Wandiga and P.J. Crutzen. 1982. Termites: A potentially large source of atmospheric methane, carbon dioxide, and molecular hydrogen. Science 218: 563–565.

Zimmerman, P.R., J.P. Greenberg and C.E. Westberg. 1988. Measurements of atmospheric hydrocarbons and biogenic emission fluxes in the Amazon boundary layer. Journal of Geophysical Research 93: 1407–1416.

Zinke, P.J. 1980. Influence of chronic air pollution on mineral cycling in forests. United States Forest Service, Pacific Southwest Forest and Range Experiment Station, General Technical Report 43: 88–99.

Zwally, H.J. 1989. Growth of Greenland ice sheet: Interpretation. Science 246: 1589–91.

Index